T0186608

Current Advances in Oral and Craniofacial Tissue Engineering

Editors

Vincenzo Guarino
Institute of Polymers, Composites and Biomaterials (IPCB)
National Research Council of Italy (CNR)
Naples, Italy

Marco Antonio Alvarez-Pérez
Tissue Bioengineering Laboratory
Division of Graduate Studies and Research
Faculty of Dentistry
National Autonomous University of Mexico (UNAM)
México

CRC Press
Taylor & Francis Group
Boca Raton   London   New York

CRC Press is an imprint of the
Taylor & Francis Group, an **informa** business
A SCIENCE PUBLISHERS BOOK

Cover credit:
- Top right image: Figure 11.1 from Chapter 11. Reproduced by permission of the author, Amaury Pozos-Guillén
- Top left image: Figure 8.4h from Chapter 8. Reproduced by permission of the author, Monica Sandri
- Bottom right image: Figure 4.1 from Chapter 4. Reproduced by permission of the author, David Masuoka
- Bottom left image: Figure 7.2b from Chapter 7. Reproduced by permission of the author, Roberto De Santis

CRC Press
Taylor & Francis Group
6000 Broken Sound Parkway NW, Suite 300
Boca Raton, FL 33487-2742

First issued in paperback 2022

© 2021 by Taylor & Francis Group, LLC

CRC Press is an imprint of Taylor & Francis Group, an Informa business

No claim to original U.S. Government works

Version Date: 20200217

ISBN-13: 978-0-367-62694-5 (pbk)
ISBN-13: 978-1-138-39091-1 (hbk)
ISBN-13: 978-0-429-42305-5 (hbk)

DOI: 10.1201/9780429423055

Publisher's Note
The publisher has gone to great lengths to ensure the quality of this reprint but points out that some imperfections in the original copies may be apparent.

Library of Congress Cataloging-in-Publication Data

Names: Guarino, Vincenzo, editor. | Álvarez Pérez, Marco Antonio, editor.
Title: Current advances in oral and craniofacial tissue engineering / editors, Vincenzo Guarino, Marco Antonio Alvarez-Pérez.
Description: Boca Raton : CRC Press, [2020] | Includes bibliographical references and index. | Contents: Translation of Tissue Engineering Approach from Laboratory to Clinics / Daniel Chavarría-Bolaños, José Vega-Baudrit, Bernardino Isaac Cerda-Cristerna, Amaury Pozos-Guillén and Mauricio Montero-Aguilar -- Polymer Materials for Oral and Craniofacial Tissue Engineering / Iriczalli Cruz Maya and Vincenzo Guarino -- Calcium Phosphate and Bioactive Glasses / Osmar A. Chanes-Cuevas, José L. Barrera-Bernal, Iñigo Gaitán-S. and David Masuoka -- From Conventional Approaches to Sol-gel Chemistry and Strategies for the Design of 3D Additive Manufactured Scaffolds for Craniofacial Tissue Engineering / Gloria A., Russo T., Martorelli M. and De Santis R. -- Mesenchymal Stem Cells from Dental Tissues / Febe Carolina Vázquez Vázquez, Jael Adrián Vergara-Lope Núñez, Juan José Montesinos and Patricia González-Alva -- Composite Materials for Oral and Craniofacial Repair or Regeneration / Teresa Russo, Roberto De Santis and Antonio Gloria -- Biomimetic Approaches for the Design and Development of Multifunctional Bioresorbable Layered Scaffolds for Dental Regeneration / Campodoni Elisabetta, Dozio Samuele Maria, Mulazzi Manuela, Montanari Margherita, Montesi Monica, Panseri Silvia, Sprio Simone, Tampieri Anna and Sandri Monica -- Craniofacial Regeneration : Bone / Laura Guadalupe Hernández Tapia, Lucia Pérez Sánchez, Rafael Hernández González, and Janeth Serrano-Bello -- Gingiva and Periodontal Tissue Regeneration / Avita Rath, Preena Sidhu, Priyadarshini Hesarghatta Ramamurthy, Bennete Aloysius Fernandes and Swapnil Shankargouda -- Dentin-Pulp Complex Regeneration / Amaury Pozos-Guillén and Héctor Flores -- Gene Therapy in Oral Tissue Regeneration / Fernando Suaste Olmos, Patricia González-Alva, Alejandro Luis Vega-Jiménez and Osmar Alejandro Chanes-Cuevas -- Injectable Scaffolds for Oral Tissue Regeneration / Suárez-Franco, J.L. and Cerda-Cristerna B.I. -- Clinical Progresses in Regenerative Dentistry and Dental Tissue Engineering / Rohan Shah and Manasi Shimpi.
Identifiers: LCCN 2020004487 | ISBN 9781138390911 (hardcover)
Subjects: MESH: Dental Materials | Tissue Engineering | Dentition | Mouth Mucosa | Oral Surgical Procedures | Tissue Scaffolds
Classification: LCC RK652.5 | NLM WU 190 | DDC 617.6/95--dc23
LC record available at https://lccn.loc.gov/2020004487

Visit the Taylor & Francis Web site at
http://www.taylorandfrancis.com
and the CRC Press Web site at
http://www.crcpress.com

Preface

Oral tissue engineering involves the study of current approaches for *in vitro* regeneration of soft and hard tissues located into the oral cavity. In this context, recent approaches involve the use of innovative biomaterials to replace the lost or damaged human oral tissues. Recent discoveries in material science and nanotechnology are drastically changing the traditional approach to dentistry by the design of innovative devices supporting more efficiently the natural regeneration process. The objective of this book is to highlight current progresses in tissue engineering for various dental hard/soft tissues including enamel, dentin, pulp, alveolar bone, periodontium, gum and oral mucosa, by emphasizing the role of materials and their specific applications.

The book aims to offer a large but timely overview of the current state of art in biology and clinical surgery applied to oral and craniofacial tissues.

The book includes 14 chapters, divided into three different subsections. An introductory and general section is mainly aimed at focussing upon the consolidated approaches used in tissue repair and regeneration (Chapter 1), also taking into account basic regulatory aspects (Chapter 2) and future targets in cell (Chapter 6) and gene therapy (Chapter 12) and clinical use (Chapters 13–14). The second part singularly addresses the recent discoveries on the use of biomaterials such as polymers (Chapter 3), ceramics (Chapters 4–5) and their composites (Chapters 8–9) for the design of innovative devices for repair and/or regeneration of hard and/or soft tissues in the oral or craniofacial compartments. Lastly, the third part explores the current applications of the biomaterials based devices as a function of the specific tissue target, including bone (Chapter 9), gengiva and periodontal tissue (Chapter 10), Pulp dentin complex (Chapter 11).

In each section, the peculiar role and the relevant impact of biomaterial features on the development of innovative treatments for *in vivo* surgery is strongly emphasized.

Contents

1

Introduction to Oral and Craniofacial Tissue Engineering

María Verónica Cuevas González,[1] *Eduardo Villarreal-Ramírez,*[1]
Adriana Pérez-Soria,[1] *Pedro Alberto López Reynoso,*[1]
Vincenzo Guarino[2] and *Marco Antonio Alvarez-Pérez*[1,*]

Introduction

Oral and craniofacial tissues are important in several physiological functions such as mastication, speech, facial aesthetics, and the most important in the quality of health in life. In this fundamental role, teeth are essential to these functions relying on its unique combination of hard tissues—*including enamel, dentin, root cementum and alveolar bone*—and soft tissues—*including periodontal ligament, gingiva and dental pulp* (Orsini et al. 2018a, b). The development of all organs which form from the ectodermal and endodermal sheets lining the embryo is regulated by the communication between the epithelium and underlying mesenchyme. Teeth are unique, greatly specialized organs, in which studies of classic tissue recombination in the developmental biology field established that the tooth shape, dental cells and tissues, takes place through a strict series of well-defined regulated stages that involve this kind of communication (Luukko and Kettunen 2016). This process in tooth development is characterized by complex reciprocal interactions between the epithelial and mesenchymal tissue, which occur in a stepwise process classically described, from early to late as: Lamina, Bud, Cap and Bell stages (early and late stage), in which each phase is discernible by specific histomorphological and cellular features (Yildirim et al. 2011; Mitsiadis and Harada 2015; Thesleff 2014). In recent years, with a better understanding of molecular biology and cell-tissue signaling, the process of tooth development could be broadly described as a sequential differentiation process mediated by the conserved signaling pathways, such as FGF, BMP, Hedgehog, EDA, and Wnt, where the basal cells of the dental lamina (dental epithelial tissue) undergo proliferation and form a horseshoe-shaped band that invaginates into the underlying mesenchymal tissue (this process is called epithelium invagination). The mesenchymal

[1] Tissue Bioengineering Laboratory, Division of Graduate Studies and Research, Faculty of Dentistry, National Autonomous University of Mexico (UNAM), Circuito Exterior s/n. Col. Copilco el Alto, Alcaldía de Coyoacán, C.P. 04510, CDMX, México.
[2] Institute of Polymers, Composites and Biomaterials (IPCB), National Research Council of Italy (CNR), Italy.
* Corresponding author: marcoalv@unam.mx

tissue, derived from neural crest cells, proliferated and differentiated to ameloblasts for depositing the enamel tissue, meanwhile mesenchymal tissue responds to dental epithelial signaling beginning with the process of differentiation into cementoblasts, periodontal ligament, odontoblasts and other dental pulp cells (including neurons, endothelial cells and fibroblasts) (Catón and Tucker 2009; Balic 2018; Xiao and Nasu 2014). Thus, tooth-specific tissues originating from the two principal sources gives the complete anatomy of soft and hard tissues in teeth, including dentin, dental pulp, alveolar bone and periodontal ligament.

In dentistry, it is well known that in humans these soft and hard tissues could be lost due to damage of tissue by trauma, dental caries, periodontal disease or a variety of genetic disorders that combined with age could lead to suffering a physical and mental dramatic event that compromises an individual's self-esteem and quality of life (Batista et al. 2014; Kassebaum et al. 2014). It is, therefore, necessary to develop innovative approaches for the repair/regeneration of damaged or missing alveolar bone and dental tissues (Caton et al. 2011). Recently, with the recognition of the molecular and cell biology based on the development of dental tissues structure, more and more efforts have been focused on applying the knowledge gained to design therapies to promote the dental tissues regeneration because modern dentistry is not limited to maintenance of dentition but has many subspecialties encompassing diagnosis and treatment of conditions affecting the oral and maxillofacial structures. Thus, regeneration of lost dental tissue by the rehabilitation of patients has been the ultimate dream of every clinician and dental healthcare researcher. Due to the unique, diverse role of the teeth and associated structures, there has been a sustained effort to replace the missing dentition over many centuries. This effort could be seen by several examples of proposed approaches to engineer biological dental tissues including tissue engineering scaffolds; cell-tissue recombination, gene-manipulated cell for improve tooth regeneration, dental mesenchymal stem cells co-cultures, 3D tissues culture strategies and 3D scaffold design by polymer printing, bioprinting and CAD-CAM technology (Moioli et al. 2007; Jahangirian et al. 2018; Saratti et al. 2019; Werz et al. 2018).

This chapter describes the strategies for the regeneration of oral and craniofacial tissues. First, to address the principles of the specific area related to dental tissue engineering, second to describe the components of the extracellular matrix and what strategies are used to try to mimic the natural extracellular matrix of tissue structures by using of scaffolds and, finally, what are those strategies that are being used to try to understand how to achieve the regeneration of the oral and craniofacial complex.

Dental Tissue Engineering

Oral and craniofacial tissues have a limited ability to correctly auto regenerate when the original tissue integrity has been severely damaged. Moreover, this limitation on the structural integrity of the damaged tissue is because of several common conditions requiring craniofacial and dental treatment including caries pulp, odontogenic infections, periodontal disease, tooth impactions, tissue dysfunction and malocclusion disorders are constant challenges in the dental health area (Gurtner et al. 2008; Duailibi et al. 2008). Moreover, with the high expectation of life, and with the increasing number of cases of dental pathologies related to the traumatic, inflammatory and neoplastic origin, congenital anomalies and degenerative diseases affecting the principle of oral and craniofacial structures, a new area that is getting relevant as a strategy to address these problems and developing new biological therapies to face this wide range of dental issues is dental tissue engineering.

Dental tissue engineering is a multidisciplinary approach that tries to the restore, maintain and enhance oral tissue function along with diagnostic and clinical applications. Dental tissue engineering works to combine the basic principles of physics, chemistry, tooth development, cell biology, nanotechnology, regenerative medicine and engineering to create a logical strategy approach to generate functional oral tissues at cellular and molecular levels that closely match with the physical and mechanical properties of naturally formed oral and craniofacial tissues (Rai 2015; Bossu et al. 2014). However, dental tissue engineering could be summarized in three central components of this multidisciplinarity as specific source of living cells, because one of the biggest challenges in the repair/regeneration strategies is what kind of cells must be used? Or what kind of specific phenotype of source cells must be isolated? Considering that oral and craniofacial tissue have distinct functions and coupled with this problem, also a different extracellular matrix that could be regarded as soft or hard, implies the uses of biomaterials as an artificial structure. This artificial structure also assumes the understanding of how to develop new materials that meet the physicochemical, structural and mechanical properties to instruct cellular sources to respond to deposit a specific cellular tissue, in which the bioactive signals (peptides, growth factors or genes) finally enter that have to be integrated in the artificial structure to give biological cues and regulate cell functions (Fig. 1.1).

There is no doubt that in dental tissue engineering cells play a central role in developing innovative strategies for repair/regeneration of the oral and maxillofacial region. But today not all kind of cells can be used as a source for this purpose, only specific cells that are immature, unspecialized, that have self-replication and conserve their potency and cell plasticity to differentiate into a vast variety of cells populations including dental tissue have gained attraction and are particularly important for developing tissue engineering strategies in this area. These unspecialized cells called dental mesenchymal stem cells could be identified and isolated based on their adherence to tissue-culture-treated plastic when being cultivated *in vitro*, for their rapid expansion, for showing a fibroblastic morphology, and for having a positive signal and expressing CD105, CD73 and CD90 surface markers and lacking the expression of CD45, CD34, CD14 or CD11b, CD79a or CD19 and HLA-DR surface molecules. Dental mesenchymal stem cells isolated from different tissues, for example from oral mucosa, pulp, periodontal ligament and gingiva represent an excellent source for dental cell therapy and regeneration of damaged oral and maxillofacial tissues. This option as a source is because they display multipotency when specific factors are used for inducing the differentiation showing the capacity to give rise to osteo/odontogenic cells, chondrocytes, adipocytes, neuronal cells, muscle cells,

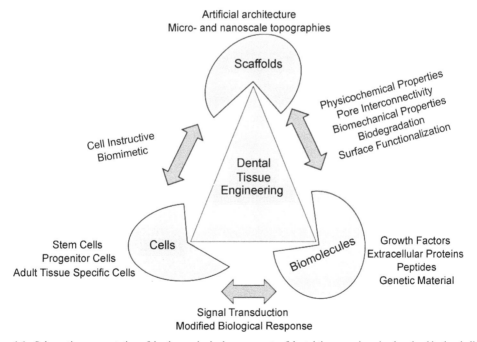

Figure 1.1. Schematic representation of the three principal components of dental tissue engineering involved in the challenge for designing new strategies for tissue repair/regeneration.

cardiomyocytes, endothelial cells, hepatocyte-like cells and islet-like cells (Bartold and Gronthos 2017; Han et al. 2014; Ge et al. 2012; Egusa et al. 2012).

A detailed chapter on this topic, will focus on mesenchymal stem cells isolated from a different source of dental tissues and also address the subject of how the mesenchymal stem cells could be used as a cell-based therapy with its potential clinical applications. Moreover, these days new cell-based approaches for enhancing the cell response on regeneration strategies that deal with replacing the defective genes with their correct analogues to produce functional proteins will be addressed in the chapter focused on gene therapy that definitely becomes a potential and promising future treatment modality of a number of diseases especially affecting the oral and craniofacial tissues.

As a second point on dental tissue engineering, important elements for tissue regeneration and for the success and efficacy of the knowledge in the cell therapy not only involve genes and dental mesenchymal stem cells but also designing appropriate scaffolds that should be biocompatible, non-toxic, under biologically safe degradation and eventually resorbed by remodeling the new tissue, allowing it to acquire similar mechanical properties and withstand mechanical forces (Zafar et al. 2015).

The importance of this kind of artificial architecture or scaffolds will be addressed in more detail in various chapters of the book that try to focus on topics such as biodegradable polymers, calcium phosphates and bioactive glasses, metal organic and hybrids, composite materials and injectable scaffolds on the applications on the regeneration strategies of oral and craniofacial tissues. Moreover, an interesting point in the synthesis of biodegradable scaffolds is that surface chemistry cues presented in the form of micro- and nanoscale topographies are significant in regulating the behavior of cells because it must be instructive and this topic will be addressed in the chapter related to biomimetic approaches for the design and development of multifunctional bioresorbable layered scaffolds for dental regeneration.

Thus, the cell instructive property implies that topography of the scaffolds gives a specific signal that allows functionality to trigger the cascade of events and also the architecture of the scaffold has to mimic the scale of the extracellular matrix creating an artificial microenvironment that improve and may direct the differentiation of stem cells towards a precise fate and function leading to regeneration (Choi et al. 2018; Darnell et al. 2018; Liu et al. 2018). Thus, nanosurface scaffolds such as nanoparticles, nanoceramic, nanofibers and nanocomposites have gained increasing interest in dental tissue engineering. This larger interest is because topography including pore size and porosity also have a direct influence on control stem cell behavior as cell attachment, cell morphology, cell proliferation, cell differentiation that allow to guide tissue repair/regeneration (Bettinger et al. 2009; Mansouri and SamiraBagheri 2016; Skoog et al. 2018).

The mechanism by which the topography exerts its effect on stem cell response is by altering focal adhesion assembly and cytoskeletal stress, resulting in adaptive gene- and protein-level changes (Kolind et al. 2014). Thus, the surface topography features of sub-micrometers to 10 μm in size act on the actin cytoskeleton, whereas features of ten to hundreds of nanometers act on integrin receptors (Miyoshi and Adachi 2014). However, the cell-surface interaction regarding the link between topographical cues and changes at proteomic, genomic and epigenomic levels are not completely understood.

Furthermore, topography of the scaffold could also be relevant for hydrophilicity and for specific protein adsorption, as shown by the selective take up of proteins relevant for cell attachment, such as fibronectin and vitronectin, on fibrous meshes with nanoscale fiber diameters (Bai et al. 2018).

Indeed, since most types of mesenchymal stem cells are dependent on anchoring and could die if there is not a favorable cell adhesion by the surface instruction of the scaffold; the search for functionalization of the scaffold for enhancing a favorable cell–material interaction are paramount for inducing a specific signaling of cellular respond at the cell–biomaterial interface. This could be achieved by integrating bioactive signals as cell adhesion peptides or growth factors or by specific adsorption of endogenous extracellular matrix proteins, as functionalization of the surface scaffold by the different methods to synthesize the scaffolds as solvent evaporation, salt-leaching, molecular self-assembly, gas foaming, phase separation, emulsification/lyophilization, 3D rapid prototyping and electrospinning, for help to regulate cell function.

Finally, the third point in dental tissue engineering, signaling biomolecules as growth factors are the choice for functionalization because these proteins activate the cellular communications network and influence functions, such as cell adhesion, cell proliferation, matrix deposition and differentiation of tissues. Also, growth factors have been shown to play a key role on the induction of fracture and wound healing, formation and repair of all craniofacial tissues and there are at least six growth factors that could be involved in oral and craniofacial tissue engineering that have been used *in vitro* and *in vivo* as Platelet-Derived Growth Factor (PDGF), basic Fibroblast Growth Factor (bFGF), Insulin-like Growth Factor (IGF), Transforming Growth Factor beta (TGFb), Vascular Endothelial Growth Factor (VEGF), and Bone Morphogenetic Proteins (BMPs) (Mandla et al. 2018). Moreover, structural fibrous protein of the extracellular matrix provides adhesive ligands such as collagen, elastin, keratin, laminins, fibronectin and vitronectin that direct cell function and could allow a more precise regulation of cell function and tissue formation (Cruz-Maya et al. 2018).

Extracellular Matrix Mimicry Dental Tissue Engineering Scaffolds

The loss of tissues due to trauma, disease or congenital anomalies is a health care issue. Therefore, the development of materials that could be implanted or incorporated into tissues to regain function is of utmost importance. The global market for biomaterials was valued at US $ 66.2 billion in 2015 and is likely to expand to 14% by 2020 (Nagrath et al. 2018). A large part of this market is centralized on mineralized tissues such as orthopedic and craniofacial/dental implants. In dentistry, tooth loss is not seen as an aesthetic or psychological problem; this is definitively a problem with serious physiological consequences. Tooth loss should be considered as an imbalance in the masticatory system. Edentulism can lead directly to impairment, functional limitation, gum damage and alveolar bone resorption (Polzer et al. 2010). For many years, autologous implants were considered the gold standard. However, this requires a more significant discomfort for the patient. Here we try to describe the recent advances in dental tissue engineering which will likely produce alternatives to autogenous bone grafts that may well exceed existing clinical outcomes and replace traditional indications for its use (Misch 2010).

The alternative to the gold standard toward the fundamental understanding of structure-function relationships in normal and pathological oral and craniofacial tissues could be the development of biological substitutes to restore, maintain or improve tissue function (O'Brien 2011). This biological substitute named biomaterial scaffold could be defined as either naturally occurring materials in living tissues or materials designed, fabricated, tested, applied and synthetic to replace, and repair oral and craniofacial tissues, which can perform, enhance or replace a specific biological function (Stupp and Braun 1997; Ratner et al. 2013). Moreover, these biomaterial scaffolds should be able to interact with cells to perform their function with an appropriate host response in a specific clinical application (Williams 2014).

In dental tissue engineering, the biomaterial scaffold basically acts as a template for cell growth and recolonization of injured tissue areas as a conductor tissue growth. Scaffolds not only provide a mechanical support, but they also may present biochemical signals like growth factors to encourage cell attachment, migration onto or within the scaffold, cell proliferation, cell differentiation and modulate cell behavior (Hutmacher 2001). Some scaffolds are frequently seeded with stem cells to accelerate cell recruitment and homing (Hussey et al. 2018). Scaffolds are made from several materials like polymers, metals, ceramics or composites. Therefore, before making the selection of the material for the scaffold design, it is necessary to understand the mechanical properties—structure relationships of the tissue to replace. The scaffolds are designed to replace a wide variety of soft tissues such as skin, tendons, blood vessels, among others, and they are generally composed of synthetic or natural polymers. Replacement of mineralized tissues such as craniofacial tissues, bone and tooth are usually designed scaffolds built up from metallic, ceramics, polymers and composites materials and must be designed to be permanently implanted. Nevertheless, long-term complications such as stress shielding, wear debris, loosening and mechanical or chemical breakdown of the material itself, encouraged scientists to develop new materials. Scaffolds, in addition, must provide high porosity, high surface area, structural strength, three-dimensional *micro- and nano*-structure, and biodegradability if needed (Vats et al. 2003). However, an important key to design a cellular scaffold and it should be pointed out as essential for the biological response is if the scaffold could replicate or mimic the extracellular matrix properties. To understand the challenge of replicating the extracellular matrix, of oral and craniofacial tissues we will briefly explain the extracellular matrix functions and mention their most representative members.

Extracellular Matrix

The extracellular matrix is a non-cellular structure present in all vertebrate tissues. The origin of multicellularity in metazoans is intimately related to the development of the extracellular matrix. The evolutionary transition from unicellular to multicellular organisms was a fundamental change in the history of living beings on Earth. At the beginning of cellular cooperation, the cells were able to work in a cohesive way to perform more complex and sophisticated tasks; this was due to the emergence of the proteins secreted by the cells and their particular structural arrangement (Czaker 2000).

Extracellular matrix dictates several cellular functions as migration, differentiation, proliferation, morphogenesis, growth, survival and maintains tissue homeostasis (Frantz et al. 2010; Bonnans et al. 2014). Extracellular matrix maintains a highly organized three-dimensional heterogenous fibrillar network structure with tissue-specific composition and topology, that provides an indispensable physical scaffolding for the cellular constituents and is under constant and highly coordinated remodeling throughout the entire lifetime, in accordance with physiological or pathological conditions (Theocharis et al. 2016).

The extracellular matrix is composed of water and macromolecules like fibrous proteins and proteoglycans (Frantz et al. 2010). The most abundant proteins in the extracellular matrix are collagens, elastins, fibronectin, tenascin, laminin and fibrillins. Besides, extracellular matrix contains significant contents of polysaccharides as glycosaminoglycan (GAG) chains covalently attached to a specific protein core (PG-proteoglycans), except for the hyaluronic acid (Rozario and DeSimone 2010). The GAG chains are generally long linear and negatively charged with disaccharide repeats. PG make up a hydrogel due to their extended conformations and remarkably hydrophilic character. This cross-linked biopolymer gel is needed to hold up high compressive forces (Kwansa et al. 2014). The PG more abundant in the extracellular matrix are perlecan, syndecan, decorin, aggrecan, glypican and lumican. The three principal families are Small Leucine-Rich Proteoglycans (SLRPs), modular proteoglycans and cell-surface proteoglycans (Schaefer and Schaefer 2010; Kresse and Schönherr 2001).

Collagens are the most abundant family of proteins in the animal kingdom with at least 45 different collagen genes that code for collagen polypeptides, and account for up to 30% of total protein mass present in the interstitial of the extracellular matrix (Rozario and DeSimone 2010). Moreover, collagen fibrillar and non-fibrillar proteins that can be assembled into supramolecular structures, to give organization and rigidity to the extracellular matrix. The mechanical properties of the extracellular matrix can be determined by the organization, distribution and density of the collagen fibers in the tissues (Birk and Brückner 2011). Mutations in the genes of collagen can be fatal in embryonic development, even embryos lacking collagen I that reach late stages of development suffer from abrupt rupture of the aorta due to lack of tissue compliance and it is striking that one collagen type can predominate overwhelmingly in tissues that are extremely diverse (Löhler et al. 1984).

All collagen members share common features. The collagen proteins have at least one triple alpha helix collagenous domain (COL), which is a rod-like structure to provide stiffness and the content or number of repetitions of this COL domain depends on the specific type of the collagen. The alpha helix chains are supercoiled to form a triple helix. The COL domain frequently has a triplet sequence Gly-X-Y, where Gly corresponds to the glycine residues and generally by steric constraints occupies the central positions in the triple helix. Meanwhile, X and Y can be any amino acid but frequently are found in these positions to proline and hydroxyproline, respectively. Hydroxyproline residues are essential to triple helix stability (Birk and Brückner 2011).

In the extracellular matrix, there are mainly two types of elastic fibers, chemically and morphologically different: elastins and microfibrils. Elastin fibers comprise approximately 90% of all of them. Elastin fibers provide the properties of resilience and elasticity to the tissues preventing deformation when they are under repeated stretching (Mithieux and Weiss 2005). However, the elastin fibers stretch is limited due to its close structural association with collagen fibers. Poles apart to collagen family that is encoded by several genes, elastin is encoded by a single gene in mammals and secreted to the extracellular medium as a monomer with a molecular weight of 60–70 kDa, called tropoelastin (Frantz et al. 2010). In the extracellular space, the secreted tropoelastin molecules are processed to elastin fibers; this process is catalyzed by a member of the lysine oxidase family, > 80% of the lysine residues form covalent cross-links between and within the elastin molecules. The arrangement and size of elastin fibers vary according to the tissue, but elastin fibers are associated with different proteins such as fibulins and fibrillins, which are essential to preserving the integrity of elastin fibers (Muiznieks and Keeley 2013).

Another essential fibrous protein in the matrix for playing a crucial role in the organization of the interstitial of the extracellular matrix and for interactions with cells is fibronectin. Fibronectin is directly related to central cellular events related to the matrix, such as cell adhesion, migration, growth and differentiation. Fibronectin is a protein made up of a dimer that is held together by a pair of disulfide linkages in the carboxyl terminus, and each monomer has a molecular weight around of ~ 250 kDa. Similar to elastin, fibronectin is a protein encoded by a single gene. However, fibronectin may occur in different versions due to alternative splicing, generating up to 20 known isoforms. Fibronectin can be stretched several times over its resting length by a neighboring cellular traction force. Owing to the tensile strength over fibronectin, it undergoes to conformational changes to expose integrin-binding sites recognized by integrins on the cellular surface. Consequently, the adhesion of integrins to fibronectin is allowed and promotes fibronectin-fibril assembly, which implies that fibronectin is also a mechano-regulator of the extracellular matrix (Frantz et al. 2010; Pankov and Yamada 2002; Xu and Mosher 2011).

In contrast to the predominantly fibrillar structure of collagen and elastin, proteoglycans adopt highly extended conformations that are essential for hydrogel constitution (Mouw et al. 2014). The hydrogel molecules support the high compression forces in the extracellular matrix, around the cells and interstitial matrix. As stated above, the biological function of proteoglycans derives from the biochemical and hydrodynamic properties of the GAG macromolecules, which combine chemically with water to provide hydration and compression resistance. GAG has been classified into subtypes according to the function and structure of these carbohydrate chains, as well as the distribution, density and length of these chains concerning the core protein (Iozzo and Schaefer 2015). The most common GAG chains found in the ECM are heparin sulfate, chondroitin sulfate, dermatan sulfate, hyaluronan and keratin sulfate. The proteoglycans are encoded by a little less than 50 genes, besides several of them can be subject to alternative splicing, which shows the enormous variety of members of proteoglycans present in the matrix. They also have a vast range of functions such as cell adhesion, migration, proliferation, signaling, communication, morphogenesis and growth. In turn, proteoglycans also participate in angiogenesis, the inflammatory response to pathogens and injuries (Kresse and Schönherr 2001; Iozzo and Schaefer 2015). After having explained the unique features, functions and essential members of the extracellular matrix some techniques to replicate or mimic the extracellular matrix need to be shown.

Biomimetic Extracellular Matrix

The scaffolds designed to replicate the properties of the extracellular matrix must achieve several premises described above as morphology and mechanical properties. To set up a simple classification, they could be divided into two broads categories: organic scaffold materials and inorganic scaffold materials.

The inorganic scaffold materials can also be classified into different classes according to: (a) type and whole number of polymers constituent, (b) ability to be reabsorbed and replaced by the host tissue, (c) capability of chemical modification and bio-functionalization on their surface, (d) capability of covalent bond forming with some other materials (organic or inorganic), and finally, (e) type of manufacturing. The biodegradable polymers are preferred because they reduce the immune reaction to the foreign body, and also allow the recolonization of the damaged area allowing a more efficient regeneration. Polymers more frequently used for preparing inorganic scaffolds are poly(L-lactic acid) (PLLA), poly(glycolic acid) (PGA), poly(caprolactone) (PCL) and poly(lactic-co-glycolic acid) (PLGA) (Hussey et al. 2018; Stratton et al. 2016).

The Organic Scaffold Materials (OSM), on the other hand, are those formed of macromolecules from a natural source from a living being, sometimes purified or treated as an extract. OSM also can include macromolecules as proteins over-expressed and purified in a cell host. Organic scaffold materials have several advantages, producing a low foreign body reaction, which is why, in general, they have a moderate inflammatory response. Besides, organic materials are biodegradable by enzymes that are naturally found in the body. Moreover, organic materials can easily be mixed with inorganic materials to form composite materials with new physical and chemical properties. Among them, biopolymers more frequently used for preparing natural scaffolds are collagen, fibronectin, glycosaminoglycans, fibrin, silk proteins, alginate, chitosan, hyaluronic acid and cellulose (Mano et al. 2007; Guarino et al. 2012).

Nanofibers

One of the goals of dental tissue engineering is to mimic the extracellular matrix; to achieve this goal, different technologies have been developed including Air-Jet Spinning (AJS) and electrospinning (ES). Several studies have demonstrated that AJS and ES can be successfully used to fabricate biomimetic analogs able to respond to the fibrillar architecture of the extracellular matrix (Braghirolli et al. 2014; Stojanovska et al. 2016; Guarino et al. 2016).

Air-jet spinning (AJS) is a technique to produce spun microfibers and nanofibers of polymeric materials and composite materials. This technique is based on the use of air regulated pressure to apply as a jet or spray the polymer solution as thin fibers onto the surface and may also be passed or conducted under controlled pulsations directly through the nozzle. As the polymer solution leaves the AJS nozzle, the solvent evaporates during the formation and deposition of the fibers onto the surface. The system has many advantages such as: easy to use, not requiring sophisticated equipment, scalability, short period for manufacturing and low cost (Suarez-Franco et al. 2018; Lasprilla et al. 2012).

Electrospinning is a technique which uses the basic principles of AJS. However, the driving force to generate the fibers from a polymer solution consists of the high voltage electric field applied between a metallic nozzle and the grounded collector where the fibers will be deposited. Electrospinning shows an essential advantage related to the capability to reduce fiber size down to several nanometers in diameter, despite this some limitations still remain, such as low production rates and high dependency on polymer properties. However, this technique, i.e., AJS, is easy to use and with high adaptability to several conditions and different materials (Guarino et al. 2018; Sill and von Recum 2008). For these reasons, AJS and electrospinning have gained considerable attention in the past several years as technology for mimicking the fibrous structure and sizes of the extracellular matrix with several applications on oral and craniofacial tissue engineering. Due to the versatility, adaptability and easy handling of the fibers several studies were performed to increase the bioactivity and biofunctionality of fibers, by incorporating different dopants. Scaffold doping may be divided into two categories depending on whether this is non-covalent or covalent bonding between the fibers and the dopant (Ji et al. 2011). Non-covalent binding to the fibers could be guided by distinct forces as electronegativity, van der Waals forces, hydrogen bonds or the dopants could be adsorbed or wrapped by the polymer (Park and Ji 2011). The covalent bonding to the fibers is accomplished by chemical modification to produce a reactive functional group to immobilize a bioactive dopant to the polymer (Yoo et al. 2009). These processes can change the behavior of the fiber dramatically and improve the cellular response for healing.

Some non-covalent composite scaffolds are polylactic acid (PLA)-hydroxyapatite (HA), copolymer PLGA [PLA + polyglycolide (PGA)]-titania, poly(methyl methacrylate) (PMMA)-carbon nanotubes (CNTs) (Gloria et al. 2010; Luong et al. 2008). These composite scaffolds have different physical and chemical properties in comparison with the pure polymer scaffolds. PLA-HA shows a high surface free energy and low water contact angle, because this degradation rate was slow as compared with PLA scaffold. Also, fibers were doped with different drugs, and they even have added organic compounds such as proteins, peptides and growth factors to develop other types of non-covalent composite scaffolds. In a recent study using a syringe pump coupled to two gas-brush devices, demonstrated depositing PLGA fibers with proteins physisorbed and/or adsorbed as collagen and fibrin. This scaffold with deposited proteins shows capability of fibroblast migrations towards the fibers, arrangements of cells in similar conformations to the connective tissues and high expression of ECM proteins (Kaufman et al. 2018). However, several studies have demonstrated that surface structure, hydrophobicity and charge affect the protein structure when they are adsorbed to a solid-liquid interface. The most crucial step in protein absorption on the biomaterial surface is the release of water. In this way, the interface protein biomaterial surface is separated by two interfaces: on the one hand surface and water interface and the other hand protein and water interface. The process of protein adsorption over the surface could cause changes in the protein structure, which could induce the denaturalization of proteins (Ostuni et al. 2001). These structural changes on protein adsorption are due to the native folding of the protein that corresponds to the free energy minimum in solution, and does not compare to the free energy minimum once the protein is in contact with the surface (Rabe et al. 2011).

Hydrogels

Hydrogels are cross-linked polymeric structures, with three-dimensional morphology composed by covalent bonds produced by the reaction of one or more comonomers, physical cross-links are due to chain entanglements. Furthermore, there are chain interactions due to weak bonds such as hydrogen bonds and van der Waals forces. Hydrogels are also hydrophilic materials with high capacity to absorb a large amount of water or biological fluids up to more than 90% (Peppas et al. 2000). Classification of hydrogels can be made according to the preparation method, the overall charge, and the mechanical and structural characteristics (Van Vlierberghe et al. 2011). Gelation involves the formation of a three-dimensional network. Several stimuli can promote gelation of the hydrogel, such as ion concentration, UV exposure or temperature (Carballo-Molina et al. 2016). Hydrogels are frequently synthesized based on inorganic scaffold materials, such as Poly(Propylene Fumarate-co-Ethylene Glycol) [P(PF-co-EG)], Poly(Vinyl Alcohol) (PVA), Poly(Acrylic Acid) (PAA) and polyethylene oxide (PEO). However, methodologies have been developed to use biopolymers derived from the extracellular matrix to form a more biomimetic hydrogel. Among the most frequent biopolymers used for the formation of hydrogels are collagen, gelatin, elastin, silk fibroin and polysaccharides (Re et al. 2019). Hydrogels designed from biopolymers, unlike synthetic polymers, have a high affinity and ability to adsorb proteins from biological fluids or plasma. Also, hierarchical multiphase porous can mimic the critical extracellular matrix properties, to provide mechanical support for tissues and to regulate cellular behaviors, such as adhesion, proliferation and differentiation. Moreover, micro/nanoscale-topography of the hydrogels can be degraded by enzymes such as metalloproteases, allowing cell migration, colonization and growth (Zhu et al. 2019).

Proteins and Peptides

Proteins and peptides are used as primary constituents to construct biofunctional scaffolds with properties to mimic the extracellular matrix (Tsomaia 2015). Protein-based biomaterials have been modeled from the structures found within the extracellular matrix to replicate the three-dimensional microenvironment of the native tissue. Advances in molecular biology and synthetic biology now allow for precise control and manipulation of amino acid sequences and the modular architecture of proteins (Tang and Lampe 2018). Besides, these macromolecules have self-assembly properties that can be exploited to build supramolecular structures with *micro-* and *nano-*structured dimensions and used to design three-dimensional nanofibrous scaffold. The modular nature of proteins could be used to create chimeras with modified features to have a better performance or even create new functions by swapping modules from their different parental proteins. Theoretically, proteinaceous biomaterials could be understood as a protein LEGO set, which would allow controlling the supramolecular properties from the macroscopic to nanoscopic properties (Sandhya et al. 2016). Generally, these proteinaceous biomaterials present a structure like biopolymer-based hydrogels, which promotes cytocompatibility, cell interactions and minimizes foreign body reaction. Furthermore, current knowledge of the interactions between cells and extracellular matrix allows knowing which are the signal transduction pathways of different cellular events such as cell adhesion, proliferation and differentiation through an integrin-binding Arg-Gly-Asp (RGD) motif (De Santis and Ryadnov 2016).

Molecular Dynamics Simulations as a Tool for Designing Scaffolds

Most of the research on development scaffolds carried out around the world has the following approach to develop new data and materials: (A) These researches are begun with a solid base of current knowledge. (B) Proficient skills on the techniques for material synthesis. (C) A robust material characterization in physical and chemical properties. (D) Extensive observation of the biological response to the material *in vitro* and *in vivo*. Nevertheless, to improve our understanding of the features of these materials, it is necessary to decipher the behavior at the atomic level and the physicochemical interactions that occur in the biomaterial-water-ECM interface (Zhou 2015). Proteins and biomacromolecules, biopolymers, hydrogels and other small solutes, are often interfacially active and adsorb to a variety of interfaces.

Unwanted adsorption is a primary complication for phenomena as wettability, bioactivity and biocompatibility on implants (Firkowska-Boden et al. 2018). Ideally, the surface of biomaterials should be precisely controlled and designed to obtain good coordination with water molecules (Liping et al. 2008), controlling protein adsorption and avoiding the integrity of secondary structure elements is compromised upon absorption (Ge et al. 2011). Molecular dynamics simulations are an alternative method to directly explore the structural and dynamic properties of scaffolds at an atomic level and to analyze their interactions with macromolecules and water. Therefore, MDS is a versatile tool to obtain *in silico* data from platforms, and this serves to limit the experimental conditions in the laboratory. By using molecular dynamics simulations it is possible to study different sizes, shapes and surface chemistries of the scaffolds, these properties can vary the toxicity, gene regulation and clearance due to the interatomic interactions between materials and biological molecules (Qun et al. 2015). The spatial and temporal resolutions that computational techniques currently allow enable the investigation of the specific interactions and dynamics that scaffolds induce in biological molecules (Zhou 2014).

Figure 1.2. Molecular dynamics to understand the interaction between biomolecules with polymer surface. (A) Collagen fiber deposited on polylactic acid (PLA) surface. (B) Fibrinogen protein deposited on a PLA surface.

Therefore, the use of computational methods is becoming more frequent to elucidate the fundamental aspects of intermolecular interactions within the biomaterial-water-ECM system (Fig. 1.2).

Strategies for the Regeneration of the Oral and Craniofacial Complex

The regeneration strategies for the complex craniofacial structures is a highly promising field and in recent years focuses mainly on the regeneration of oral tissues such as oral mucosa, periodontal, bone and dental pulp tissues.

Oral Mucosa

The mucosa is the most abundant tissue that lines the oral cavity, its damage or loss is mainly caused by different diseases, inflammatory, traumatic, neoplastic or congenital origin, that result in defects of the buccal mucosa, among which are included gingival recessions, vestibuloplasties, cleft palate, traumatisms and the removal of tumors. Although various surgical procedures have been proposed for the treatment of these defects, there remains a need to find functional, anatomical and aesthetically similar substitutes for the tissue to be replaced, as well as solutions that reduce the morbidity associated with obtaining tissue from donor areas, which represents a clinical challenge for periodontists and oral and maxillofacial surgeons. Most surgical techniques and even the replacement of affected tissue include the use of partial or full-thickness skin grafts or flaps and heterologous transplants or reconstruction techniques. However, these techniques are not an option when the patient requires more than one surgical procedure due to the extension of the injury. For this reason, dental tissue engineering has tried to venture into developing new oral mucosal substitutes that are aesthetic and functional at the same time. In this case numerous studies have allowed the development of cell culture strategies or development of scaffolds to regenerate the function of the oral mucosa (Yoshizawa et al. 2012; Moharamzadeh et al. 2012; Scheller et al. 2009).

One of the first approaches was the use of monolayers of keratinocytes whose clinical applications have been used in palate wounds and mucogingival defects. However, among its drawbacks is the reduced resistance to mechanical stress, its fragility due to the absence of the connective tissue (Rheinwald and Green 1975). Another approach to cell culture strategies was tissue-engineered 3D culture systems of the oral mucosa, which provide an organizational complexity and several advantages that include a high degree of differentiation, the potential for histological assessment of the process under study and the potential for monitoring tissue growth or damage (Dongari-Bagtzoglu and Kashleva 2006). Therefore, a bilayer substitution has been proposed in which an extracellular matrix analog that serves as a scaffold is sought. Among the extracellular matrix analogues, one finds the use of fibrin, elastin, collagen, chitosan, agarose, gelatin in combination with several synthetic polymers as poly(ethylene terephthalate), polycarbonate membrane, electrospun membranes of poly(L-lactic) acid, polycaprolactone, polystyrene and polylactic glycolic acid which seeded and cultured with gingival fibroblast, epithelial or keratinocytes generate an artificial equivalent of partial and total thickness that allow to obtain the biocomplexity of the tissue-engineered oral mucosa (Moharamzadeh et al. 2008; Peña et al. 2010; San Martin et al. 2013).

Periodontal and Bone Tissues

The periodontium involves all the tissues that are surrounding and supporting the tooth, and can be divided into union tissues composed by the Root Cementum (RC), Alveolar Bone (AB) and periodontal ligament (PDL), and dentogingival tissues (Posnick and Posnick 2014). The periodontium is composed of dynamic tissues because it is a mixed combination of mineralized and soft tissue that makes the periodontium a complex tissue in the field of dental tissue engineering and a challenge for the strategies in its regeneration for the mixed-function development.

Root Cementum (RC) is a mineralized tissue that surrounds the superficial root of the tooth; their function is to support the tooth through the PDL and alveolar bone (Yamamoto et al. 2016). Alveolar Bone (AB) is another mineralized tissue and is associated with the formation of membranous bone of both mandibular and maxillary tissues during the development of the first dentition, two components form this kind of bone, the first belong to the alveolar process, which in turn is composed by the cortical and cancellous bone tissue, the last one stores Haversian systems required for maintenance and remodeling of the bone; the second component is the alveolar bone itself which corresponds to the bone portion that covers the dental surface and serves as a union site to the Sharpey fibers from PDL (Chu et al. 2014). Periodontal ligament (PDL) is formed by collagen fibers which could be classified according to their localization of the fibers onto the alveolar crest, oblique, transseptal, horizontal, inter-radicular or apical (Maheaswari et al. 2015). The union of these fibers to the soft tissue provides a natural coupling of the roots of the tooth in the alveolus: the union of the PDL to the RC or the AB facilitates the transfer of loads of the teeth towards the bone, because the bone-cement/PDL-binding sites contain areas between 10–15 μm rich in biochemical gradients, which are known as enthesis sites that facilitate cell-cell interactions and communications (Lee et al. 2015).

The periodontal tissue can be affected by multiple factors destroying the connective tissue and alveolar bone triggering the loss of the dental organ. The periodontitis has been described in the last 4000 years by Egyptians and Chinese, who mentioned that the periodontal disease is an inflammatory condition (Loe 2000). At present, periodontal disease is a chronic inflammation, clinically characterized, in the early stages by gum inflammation, loss insertion to probing, periodontal pockets, bleeding to probing and loss of bone level which vary in size according to the degree of affectation (Chapple et al. 2018; Papapanou et al. 2018). Moreover, a series of risk factors have been described for the development of periodontal disease, these factors are divided into modifiable factors such as tobacco, poor oral hygiene, hormonal changes, diabetes mellitus, medication and stress, on the other hand, non-modifiable factors include age and genetics; however their pathogenesis has also been attributed to the presence of multiple diseases such as cardiovascular, metabolic diseases, rheumatoid arthritis, respiratory diseases, cancer among others (Kinane et al. 2017; Nazir 2017; Shewale et al. 2016).

Due to the above, periodontal disease has been considered a public health problem; therefore, the prevention, treatment and regeneration strategies have played a crucial role.

In the treatment of tissue defects caused by periodontal disease it is necessary to promote tissue regeneration with the main objective of restoring the lost or injured tissues structure and function of the periodontium. Within the methods used for periodontal regeneration, specific biomaterials such as bone grafts from diverse origins (allograft, xenograft, alloplastic) and cell-occlusive barrier membranes, are used in guided tissue regeneration (Tassi et al. 2017).

One of the first approaches for wound healing to target the restoration of tooth-supporting bone, periodontal ligament and root cementum has been the use of growth factors that have shown significant growth in the field of periodontal regenerative medicine. Advances in molecular cloning have yielded an unlimited availability of recombinant growth factors for applications in tissue engineering. The growth factors are a group of small polypeptides involved in the stimulation of different cellular signaling pathways through its association with specific membrane receptors and promoting its phosphorylation in tyrosine, threonine or serine aminoacidic residues which in turn activates a complex system of transcriptional regulation inside the cell. Growth factors associated with soft and hard tissue regeneration have been used at a preclinical and clinical level, based on scientific evidence: Platelet-Derived Growth Factor (PDGF), Fibroblast Growth Factor (FGF) and Bone Morphogenetic Proteins (BMPs).

Platelet-Derived Growth Factor (PDGF) is a soluble protein has reported four additional forms (PDGFa, PDGFb, PDGFc, PDGFd) a cystine knit motif characterizes this family of proteins and its function depends on its self-association in homodimers (AA, BB, CC, DD) and heterodimers (AB). The signaling mechanism of the five isoforms of PDGF depends on the interaction with some of the two receptors PDGFRα and PDGFRβ. Platelets are composed of multiple storage granules, i.e., lysosomes and alpha granules which are delivered after its activation (coagulation), through this process of degranulation PDGF is released and performs its function locally in an autocrine fashion or in other tissues such as a paracrine mode. The cellular mechanism described for the PDGF associated with periodontium regeneration involves binding to the cell membrane tyrosine kinase receptors and subsequently exerts the activation of its effects on chemotaxis, cell proliferation, migration, extracellular matrix synthesis and anti-apoptosis via the Rac GTPase which modulates the actin cytoskeleton and the lamellopodia formation (Trofin et al. 2013; McGuire et al. 2006; Mellonig et al. 2009).

In a systematic review, Li noted that the use of 0.3 mg/ml of rhPDGF-BB (recombinant human platelet-derived growth factor-BB) has a positive impact on the bone fill of periodontal defects, linear bone growth, clinical attachment level gain and probing depth reduction. In patients treated with this growth factor, bone fill was 22.71% higher than the control patients. Regarding the clinical attachment level, there was a gain of 0.76% compared to the control groups (Li et al. 2017).

For the transportation matrix, PDGF-BB has been used in combination with an allograft, such as FDBA (freeze-dried bone allograft) and β-TCP (β-tricalcium phosphate). The latter showed positive outcomes and a mean gain of 4.1 mm in the 50 treated periodontal defects (Rosen et al. 2011).

An optimal dose is warranted so that growth factors can exert the appropriate clinical effect. The 0.3 mg/ml dose of rhPDGF-BB results in a mean bone gain of 2.6 mm, compared to 1.5 mm with the 1.0 mg/ml dose, according to a multicenter study. The results suggest that PDGF high doses might reduce the tissue healing outcomes due to feedback inhibition of the local delivery high doses (Nevins et al. 2005).

According to Khoshkram's systematic review and metanalysis, topical delivery of PDGF resulted in a statistically significant high linear bone fill in periodontal defects (0.95 mm or 20.17%) than control groups (Khoshkram et al. 2015).

Bone Morphogenetic Proteins (BMP) are a group of proteins involved in multiple development processes which include skeletal formation, embryogenesis, hematopoiesis and neurogenesis. These proteins belong to the Transforming Growth Factor Beta superfamily, and over 20 members have been characterized. Four groups are formed to classify these proteins based on its amino acid sequence similarity: BMP2/4, BMP5/6/7/8a/8b, BMP9/10, and BMP12/13/14. These proteins are synthetized as a large precursor from 400–500 aa, which has three main domains, N-terminal secretion signal, a prodomain and a C-terminal region that constitutes the mature protein. Most of BMPs have seven cysteine residues in the C-terminal region, which are involved in its self-assembly and is known as a cysteine knot. BMPs functioning depends on the structural arrangement as homo- or heterodimers, which in turn are associated with specific membrane serine/threonine receptors denoted as type I and type II to trigger two main signal pathways: Smad (mothers against decapentaplegic) dependent pathway and Mitogen-Activated Protein Kinase (MAPK) pathway. This BMPs-MAPK signal pathway has shown its potential as an inductor of mesenchymal stem cells differentiation into osteoblasts, this extracellular signal is transduced inside the nucleus via the activation of ERK1/2, p38, JNK 1/2/3 cascades which activate specific transcriptional factors (RUNX2, DLX5, and Osterix) related with the osteoblastic commitment and initiate the production of bone matrix proteins, leading to bone morphogenesis (Ripamonti 2019; Anusuya et al. 2016).

For craniofacial indications, the bone development induction exerted by rhBMP-2 has been assessed in rat calvaria critical defects, segmental mandibular canine defects, peri-implant defects and sinus augmentation in non-human primates. Several studies have indicated BMP-2's effectivity to correct infraosseous, supra-alveolar, furcation and fenestration defects in canine models, resulting in bone and root cementum formation, as well as the formation of fibrovascular tissue. In humans, it has shown significant roles in periodontal regeneration, bone healing, acceleration in osseointegration, oral surgery applied to orthodontics, repair as a sequel of bone pathology and distraction osteogenesis. The long-term evaluation of rhBMP-2 has displayed an increment in bone density after functional loads (Batool et al. 2018; Carreita et al. 2014). From a clinical standpoint, the new formation of autologous bone upon the rhBMP-2 application has been accomplished in sinus augmentation procedures, from a 1.6 mm baseline height to a final bone height of 10–12 mm (combining the remaining bone and new bone), in relation to the ridge baseline measurements for the implant placement protocol. In this study, it was also demonstrated that the dose-dependent effect of the protein of 1.5 mg/mL shows better results than 0.75 mg/mL. Thus, better outcomes have been noted in sinus bone augmentation procedures (De Frietas et al. 2015; van Hout et al. 2011; Wikesjö et al. 2009).

Fibroblast Growth Factor (FGF) is a family of proteins consisting of 23 members with related structural characteristics. These proteins are characterized by its capacity to bind to the Heparin-like glycosaminoglycans, FGFs trigger different cellular functions which include cellular proliferation, adhesion, migration, and differentiation of different target tissues. Specifically, members involved with FGFs play a role in PDL regeneration by enhancing cell proliferation, inhibition of alkaline phosphatase and angiogenesis through the modulation of the FGF-1 and FGF-2. Moreover; FGF-2 is the most extensively studied in regenerative medicine and periodontal tissue regeneration. This protein has been used in the treatment of ulcers and bone fractures due to its potential to facilitate revascularization.

Additionally, *in vivo* studies have shown that FGF-2 promotes osteoblast proliferation and bone formation acceleration.

Moreover, animal studies have confirmed that the local application of FGF-2 significantly enhances periodontal regeneration compared to control sites. Among its effects, FGF-2 promotes endothelial cell proliferation and possesses a potent angiogenic and mitogenic activity on mesenchymal cells within the periodontal ligament (Li et al. 2017; Suarez-López del Amo et al. 2015).

FGF-2 has not been thoroughly studied yet in the field of periodontology. However, several animal studies have shown that this factor is useful in terms of periodontal regeneration improvement in class II furcation defects. Other pre-clinical studies have shown that FGF-2 induces the formation of new root cementum in periodontal regeneration, with Sharpey's fibers functionally oriented towards the alveolar bone. In human periodontal regeneration studies, it has been observed that FGF-2 has improved the bone fill percentage compared to control groups. It has been noted that with different FGF-2 concentrations, the bone fill percentage is higher in FGF-2 treated sites at 36 weeks than controls without growth factors. Specifically, the 0.3% dose of rhFGF-2 shows a significant positive impact on the bone fill percentage and in the linear bone growth. As it can be noted with preliminary data of this concentration, a clinical attachment level gain is also noteworthy (Murakami 2011; Khoshkam et al. 2015).

A second approach for tissue regeneration has been subsequently introduced to the periodontal community as active biomaterial for natural and synthetic scaffolds. For example; collagen membranes, fibrin nanofibers or synthetic materials (i.e., poly(lactic acid), polycaprolactone, polyglycolide, polylactide polymers and copolymers). These materials may also be designed with *nano-* or *micro* structure that can mimic the stem cell niche to alter the regulation of stem cells or to release molecules to induce and accelerate the periodontal regeneration cascade events (Skoog et al. 2018; Sheikh et al. 2017).

The reconstruction of critical size bony defects remains a challenge in oral and maxillofacial surgery, and the strategies have been focused on the use of xenografts, allografts and autografts (Johnson et al. 2011). The approach is called guided bone, and periodontal tissue regeneration and its concept advocate the reconstruction of defective periodontium tissues using occlusive membranes that exclude undesirable cell types (fibroblasts or epithelial cells). The membrane acts as a physical barrier to prevent rapidly growing tissues (fibrous or epithelial) from invading the defect space, maintaining space and guiding the success of the regeneration of defective tissues. The structural integrity of the membrane must be maintained during the maturation of the newly formed tissue and varies according to the application. Moreover, the membranes used as a barrier could be non-resorbable and resorbable where the last could be of natural or synthetic materials (Elgali et al. 2016).

Within the non-absorbable membranes are polytetrafluoroethylene (PTFE) and titanium mesh. These membranes offer an effective barrier function in terms of biocompatibility, can maintain the space for a sufficient period, have a more predictable clinical behavior, have a lower risk and are easy to manipulate. However, one drawback in the use of this type of membrane is the need for its removal with a second stage of a surgical procedure (Naung et al. 2019; Rakhmatia et al. 2013).

Polytetrafluoroethylene is considered a stable polymer, chemically and biologically inert, and able to resist enzymatic and microbiological attack based on its structure; there are two different types of membranes named as high-dense polytetrafluoroethylene (n-PTFE) and expanded-polytetrafluoroethylene (e-PTFE). The e-PTFE membranes have a higher porosity range (5–30 μm) which is believed to enhance regeneration by improving wound stability, support and isolation of soft tissues, creation of a space occupied by the clot, exclusion of non-osteogenic cells and improving bone formation. In studies carried out in animals and humans, histological analysis indicates that no inflammatory, epithelial cells or foreign body reactions have been found, unlike the presence of highly calcified osteoid matrix, invasion of fibroblasts cells, thin collagen fibers and small capillaries. However, the drawbacks of e-PTFE are that an early bacterial infection can occur and require soft tissue coverage or primary closure to prevent soft tissue ingrowth. Meanwhile, the n-PTFE have low porosity ranges (0.2–0.3 μm), non-expanded, non-permeable and is not necessarily the primary closure of the flaps, there is no need of additional surgical intervention because the membrane could be extracted with a clamping from the site, impairs bacterial penetration and also prevents cell adhesion (Soldatos et al. 2017; Greenstein and Carpentieri 2015; Carbonell et al. 2014).

Nowadays, cell-based approaches in combination with growth factors and lyophilized bone, membranes or metallic prostheses are probably the most abundantly researched and demonstrate the formation of healthy lamellar bone, without complications up to 6 months after treatment.

Finally, as a third approach and the most recent advance in regeneration strategies is to explore the manipulation by gene therapy that could be used in combination with polymeric scaffolds or fuse biomaterials for improving the gene expression and allowing to regenerate healthy tissue. Gene therapy has generated an alternative parallel to cell therapy, since its main component is the delivery of genes, which encode mainly for growth and transcription factors, allowing the activation of the differentiation process of undifferentiated mesenchymal cells present at the site of injury to the corresponding cell lineage inducing the deposition of collagen, hydroxyapatite, osteogenic peptides according to clinical needs.

The primary limitations that this methodology faces are associated with the type of vectors used for the release of the gene or genes of interest. The genes used in gene therapy are released through the use of viral vectors such as adenovirus, adenoassociated viruses, retroviruses and lentiviruses. These vectors can integrate stably and randomly to the host genome (lentivirus and retrovirus), generating mutations that can result in severe genetic disorders. On the other hand, in the adenoassociated virus and adenoviruses that have a very low integration efficiency to the host genome, they generate an immune response when producing viral proteins (Jooss and Chirnule 2003).

For decades the role of the RNA molecule had been considered merely as a transition element between DNA and proteins, however, the discovery that double-stranded RNA molecules (dsRNA) could lead to the silencing of genes post-transcriptionally in the Caenorhabditis elegans model, led to the development of a molecular tool known as interference RNA (RNAi), which later proved its usefulness in the manipulation of mammalian cells (Fire et al. 1998; Caplen et al. 2001). In general, this post-transcriptional control system works by means of the initial synthesis of a primary micro RNA called pri-miRNA in the nucleus, which acquires a secondary structure of stem-loop type, later this pri-miRNA is processed by the DROSHA-DGCR8 protein complex, generating a short double-stranded RNA (shRNA) called pre-miRNA from ~ 20–30 bp, which binds to the protein Exportin 5, which in turn translocates it to the cytoplasm of the cell where it is released and associated with the proteins DICER/TAR RNA-binding protein (TRBP) which remove the terminal loop of the pre-miRNA, allowing the assembly of the RISC silencing complex (RNA-induced silencing complex) that include the protein argonaute. The RISC complex finally selects one of the strands of pre-miRNA, which will act as a guide for the silencing of the target transcript.

The RNAi system has been considered a therapeutic promise in multiple fields of medicine, however, it is a system composed of various components controlling their functioning at the molecular level, its use in humans has been limited to animal models. Recently in 2018 the FDA approved the use of patisiran (Onpattro; Alnylam Pharmaceuticals), a siRNA that acts on the liver for the treatment of hATTR (hereditary transthyretin amyloidosis with polyneuropathy) (Setten et al. 2019).

The therapeutic use of RNAi involves a series of strict controls regarding its design: Avoiding toxicity of RNAi-based drugs, immunogenic reactions to dsRNA, toxicity of excipients, unintended RNAi activity, and on-target RNAi activity in non-target tissues (Setten et al. 2019). In the tissue regeneration field, particularly in the dental tissue engineering area, the use of RNAi has been proposed as a means of controlling cellular processes that affect the natural process of tissue repair. In particular, its use has been intended against genetic targets involved in the process of inflammation, cell proliferation and apoptosis (Intini 2010). At the molecular level, RNAi silencing studies on BMP inhibitors such as nogging and chording have been extensively studied in bone repair and regeneration. As an additional example silencing by RNAi of the gremlin glycoprotein (antagonist of the BMP protein), had shown the activation of the signaling cascade mediated by BMP-2 in osteoblastic cells, which led to the expression of bone differentiation markers such as Runx2 and osteocalcin (Gazzerro et al. 2007). However, even though this has proved the efficiency of using RNAi to induce cell differentiation of dental stem cells to bone tissue, one of the main challenges of this methodology is the release of siRNAs from the extracellular space into the cytosol of the cell and the ensuing formation of the RISC complex. These methods include the use of naked siRNA, coupled ligands to siRNA which allow it to be internalized into the cell (GalNAc), lipid nanoparticles, nanohydrogels, aptamer-siRNA conjugates (Ghadakzadeh et al. 2016). Moreover, a separate chapter of this book focuses on periodontal and bone tissue engineering and gives a more indepth approach to these topics.

Dentin-Pulp Complex

In cases of severe pulpitis, the capability of self-regeneration or repair is limited because the odontoblasts that produce reparative dentin are destroyed, therefore root canal treatment is the treatment of choice. However, regeneration of pulp tissue instead of its total elimination, is the current priority, giving rise to the birth of the 'biological endodontics' that try to replace the necrotic pulp with scaffolds, healing promoting factors and cell therapies with the aim of regenerating new pulp and dentine within the root canal system (Rosa et al. 2013; Gotlieb et al. 2008). This strategy promises excellent versatility in the design of constructs, depending on the needs of each patient (Chandrahasa et al. 2011; Demarco et al. 2010). Several authors argue that the next therapy will be 'regenerative endodontics', which works in the neoformation of vascularized tissue very similar to pulp tissue. They suggest that such a treatment is the most fundamental approach for clinical translation (Hecksher et al. 2018; Galler et al. 2011). Both strategies are promising, need to refine some aspects such as the construct insertion level, the irrigant type that better promotes cellular maintenance, the acceptable pulp tissue remotion amount and antibiotic therapy. However; a separate chapter will give an indepth approach to this exciting topic of regeneration of the dentine-pulp complex with revitalization/revascularization therapy.

Brief information of some strategies for the regeneration of the oral and craniofacial complex are described. Two specific chapters in the book will focus on the clinical progress in regenerative dentistry and how the results from different points of view address the regeneration of dental tissue engineering strategies that could be used for translation of the tissue engineering approach from laboratory to clinical applications.

Acknowledgments

The authors wish to thank the financial support by the DGAPA-UNAM program PAPIIT IT203618 project, and the CONACYT with the particular program of Fondo Sectorial de Investigación para la Educación A1-S-9178 project and MVCG want to thank CONACYT scholarship support (No. 326127 with CVU: 537149) for her PhD studies in the Programa de Doctorado en Ciencias Odontológicas of the PMDCMOS, UNAM.

References

Anusuya, G.S., Kandasamy, M., Jacob Raja, S.A., Sabarinathan, S., Ravishankar, P. and Kandhasamy, B. 2016. Bone morphogenetic proteins: Signaling periodontal bone regeneration and repair. J. Pharm. Bioallied. Sci. 8(Suppl. 1): S39–S41.

Bai, M., Xie, J., Liu, X., Chen, X., Liu, W., Wu, F. et al. 2018. Microenvironmental stiffness regulates dental papilla cell differentiation: implications for the importance of fibronectin-paxillin-β-catenin axis. ACS Appl. Mater. Interfaces 10(32): 26917–26927.

Balic, A. 2018. Biology explaining tooth repair and regeneration: a mini-review. Gerontology 64: 382–388.

Bartold, P.M. and Gronthos, S. 2017. Standardization of criteria defining periodontal ligament stem cells. J. Dent. Res. 96(5): 487–490.

Batista, M.J., Lawrence, H.P. and de Sousa, M.R. 2014. Impact of tooth loss related to number and position on oral health quality of life among adults. Health Qual. Life Out. 12: 165.

Batool, F., Strub, M., Petit, C., Bugueno, I.M., Bornert, F., Clauss, F. et al. 2018. Periodontal tissues, maxillary jaw bone, and tooth regeneration approaches: from animal models analyses to clinical applications. Nanomaterials (Basel) 8(5): pii: E337.

Bettinger, C.J., Langer, R. and Borenstein, J.T. 2009. Engineering substrate topography at the micro- and nanoscale to control cell function. Angew. Chem. Int. Ed. Engl. 48(30): 5406–5415.

Birk, D.E. and Brückner, P. 2011. Collagens, suprastructures, and collagen fibril assembly. *In*: Mecham, R. (ed.). The Extracellular Matrix: An Overview. Biology of Extracellular Matrix. Springer, Berlin, Heidelberg.

Bonnans, C., Chou, J. and Werb, Z. 2014. Remodelling the extracellular matrix in development and disease. Nat. Rev. Mol. Cell Bio. 15: 786.

Bossu, M., Maurizio, B., Andrea, P., Daniele, C., Gianluca, T., Gaetano, I. et al. 2014. Today prospects for tissue engineering therapeutic approach in dentistry. Sci. World J. 14: 151–252.

Braghirolli, D.I., Steffens, D. and Pranke, P. 2014. Electrospinning for regenerative medicine: a review of the main topics. Drug Discov. Today 19: 743–753.

Caplen, N.J., Parrish, S., Imani, F., Fire, A. and Morgan, R.A. 2001. Specific inhibition of gene expression by small double-stranded RNAs in invertebrate and vertebrate systems. Proc. Natl. Acad. Sci. USA 98: 9742–9747.

Carballo-Molina, A., Sánchez-Navarro, A., López-Ornelas, A., Lara-Rodarte, R., Salazar, P., Campos-Romo, A. et al. 2016. Semaphorin 3C released from a biocompatible hydrogel guides and promotes axonal growth of rodent and human dopaminergic neurons. Tissue Eng. PT. A 22: 850–861.

Carbonell, J.M., Sanz Martín, I., Santos, A., Pujol, A., Sanz-Moliner, J.D. and Nart, J. 2014. High-density polytetrafluoroethylene membranes in guided bone and tissue regeneration procedures: a literature review. Int. J. Oral Maxillofac. Surg. 43: 75–84.

Carreita, A.C., Lojudice, F.H., Halsick, E., Navarro, R.D., SOayar, M.C. and Granjeiro, J.M. 2014. Bone morphogentic proteins: facts, challenges, and future perspectives. J. Dent. Rest. 1–11.

Catón, J. and Tucker, A.S. 2009. Current knowledge of tooth development: patterning and mineralization of the murine dentition. J. Anat. 214: 502–515.

Caton, J., Bostanci, N., Remboutsika, E., De Bari, C. and Mitsiadis, T.A. 2011. Future dentistry: cell therapy meets tooth and periodontal repair and regeneration. J. Cell Mol. Med. 15(5): 1054–1065.

Chandrahasa, S., Murray, P.E. and Namerow, K.N. 2011. Proliferation of mature *ex vivo* human dental pulp using tissue engineering scaffolds. J. Endod. 37(9): 1236–1239.

Chapple, I.L.C., Mealey, B.L., Van Dyke, T.M., Bartold, P.M., Dommisch, H., Eickholz, P. et al. 2018. Periodontal health and gingival diseases and conditions on an intact and a reduced periodontium: consensus report of workgroup 1 of the 2017 world workshop on the classification of periodontal and peri-implant diseases and conditions. J. Periodontol. 89: S74–84.

Choi, B., Kim, D., Han, I. and Lee, S.H. 2018. Microenvironental regulation of stem cell behavior through biochemical and biophysical stimulation. Adv. Exp. Med. Biol. 1064: 147–160.

Chu, T.M.G. and Sean, S.Y.L. Babler. 2014. Craniofacial biology, orthodontics, and implants. Basic Appl. Bone Biol. 225–42.

Cruz-Maya, I., Guarino, V. and Alvarez-Perez, M.A. 2018. Protein based devices for oral tissue repair and regeneration. AIMS Mater. Sci. 5(2): 156–170.

Czaker, R. 2000. Extracellular matrix (ECM) components in a very primitive multicellular animal, the dicyemid mesozoan Kantharella antarctica. Anat. Rec. 259: 52–59.

Darnell, M., O'Neil, A., Mao, A., Gu, L., Rubin, L.L. and Mooney, D.J. 2018. Material microenvironmental properties couple to induce distinct transcriptional programs in mammalian stem cells. Proc. Natl. Acad. Sci. 115(36): E8368–E8377.

De Frietas, R., Spin-Nieto, R., Junior, E.M., Violin, L.A., Wikesjo, U.M. and Susin, C. 2015. Alveolar ridge and maxillary sinus augmentation using rhbmp-2: a systematic review. Clin. Imp. Dent. and Rel. Res. 17(Supp. 1): E192–e201.

De Santis, E. and Ryadnov, M.G. 2016. Self-assembling peptide motifs for nanostructure design and applications. pp. 199–238. *In*: Amino Acids, Peptides and Proteins: Volume 40: RSC.

Demarco, F.F., Casagrande, L., Zhang, Z., Dong, Z., Tarquinio, S.B., Zeitlin, B.D. et al. 2010. Effects of morphogen and scaffold porogen on the differentiation of dental pulp stem cells. J. Endod. 36(11): 1805–1811.

Dongari-Bagtzoglu, A. and Kashleva, H. 2006. Development of a highly reproducible three-dimensional organotypic model of oral mucosa. Nat. Protoc. 1(4): 2012–2018.

Duailibi, S.E., Duailibi, M.T., Zhang, W., Asrican, R., Vacanti, J.P. and Yelick, P.C. 2008. Bioengineered dental tissues grown in the rat jaw. J. Dent. Res. 87(8): 745–750.

Egusa, H., Sonoyama, W., Nishimura, M., Atsuta, I. and Akiyama, K. 2012. Stem cells in dentistry-part I: stem cell sources. J. Prosthodont. Res. 56: 151–165.

Elgali, I., Turri, A., Xia, W., Norlindh, B., Johansson, A., Dahlin, C. et al. 2016. Guided bone regeneration using resorbable membrane and different bone substitutes: Early histological and molecular events. Acta Biomater. 29: 409–423.

Fire, A., Xu, S., Montgomery, M.K., Kostas, S.A., Driver, S.E. and Mello, C.C. 1998. Potent and specific genetic interference by double-stranded RNA in Caenorhabditis elegans. Nature 391(6669): 806–11.

Firkowska-Boden, I., Zhang, X. and Jandt, K.D. 2018. Controlling protein adsorption through nanostructured polymeric surfaces. Adv. Healthc. Mater. 7: 1700995.

Frantz, C., Stewart, K.M. and Weaver, V.M. 2010. The extracellular matrix at a glance. J. Cell Sci. 123: 4195–4200.

Galler, K.M., D'Souza, R.N., Federlin, M., Cavender, A.C., Hartgerink, J.D., Hecker, S. et al. 2011. Dentin conditioning codetermines cell fate in regenerative endodontics. J. Endod. 37: 1536–1541.

Gazzerro, E., Smerdel-Ramoya, A., Zanotti, S., Stadmeyer, L., Durant, D., Economides, A.N. et al. 2007. Conditional deletion of gremlin causes a transient increase in bone formation and bone mass. J. Biol. Chem. 282(43): 31549–31557.

Ge, C., Du, J., Zhao, L., Wang, L., Liu, Y., Li, D. et al. 2011. Binding of blood proteins to carbon nanotubes reduces cytotoxicity. Proc. Natl. Acad. Sci. USA 108: 16968–73.

Ge, S., Mrozik, K.M., Menicanin, D., Gronthos, S. and Bartold, P.M. 2012. Isolation and characterization of mesenchymal stem cell-like cells from healthy and inflamed gingival tissue: potential use for clinical therapy. Regen. Med. 7(6): 819–32.

Ghadakzadeh, S., Mekhail, M., Aoude, A., Hamdy, R. and Tabrizian, M. 2016. Small players ruling the hard game: siRNA in bone regeneration. J. Bone Miner. Res. 31(3): 475–87.

Gloria, A., De Santis, R. and Ambrosio, L. 2010. Polymer-based composite scaffolds for tissue engineering. J. Appl. Biomater. Biom. 8: 57–67.

Gotlieb, E.L., Murray, P.E., Namerow, K.N., Kuttler, S. and Garcia-Godoy, F. 2008. An ultrastructural investigation of tissue-engineered pulp constructs implanted within endodontically treated teeth. J. Am. Dent. Assoc. 139(4): 457–465.

Greenstein, G. and Carpentieri, J.R. 2015. Utilization of d-PTFE barriers for post-extraction bone regeneration in preparation for dental implants. Compend. Contin. Educ. Dent. 36(7): 465–73.

Guarino, V., Gloria, A., Raucci, M.G. and Ambrosio, L. 2012. Hydrogel-based platforms for biomedical applications. Polymers 4(3): 1590–1612.

Guarino, V., Cirillo, V. and Ambrosio, L. 2016. Bicomponent electrospun scaffolds to design ECM tissue analogues. Exp. Rev. Med. Dev. 13(1): 83–102.

Guarino, V., Cruz-Maya, I., Altobelli, R., Abdul Khodir, W.K., Ambrosio, L., Alvarez Perez, M.A. et al. 2018. Electrospun polycaprolactone nanofibres decorated by drug loaded chitosan nano-reservoirs for antibacterial treatments. Nanotechnol. 28(50): 50510.

Gurtner, G.C., Werner, S., Barrandon, Y. and Longaker, M.T. 2008. Wound repair and regeneration. Nature 453: 314–21.

Han, J., Menicanin, D., Gronthos, S. and Bartold, P.M. 2014. Stem cells, tissue engineering and periodontal regeneration. Aust. Dent. J. 59(Suppl. 1): 117–30.

Hecksher, F., Vidigal, B., Coelho, P., Otoni, D., Alvarenga, C. and Nunes, E. 2018. Endodontic treatment in artificial deciduous teeth by manual and mechanical instrumentation: a pilot study. Int. J. Clin. Pediatr. Dent. 11(6): 510–512.

Hussey, G.S., Dziki, J.L. and Badylak, S.F. 2018. Extracellular matrix-based materials for regenerative medicine. Nat. Rev. Mater. 3: 159–173.

Hutmacher, D.W. 2001. Scaffold design and fabrication technologies for engineering tissues—state of the art and future perspectives. J. Biomater. Sci. Polym. Ed. 12: 107–124.

Intini, G. 2010. Future approaches in periodontal regeneration: gene therapy, stem cells, and RNA interference. Dent. Clin. North Am. 54(1): 141–155.

Iozzo, R.V. and Schaefer, L. 2015. Proteoglycan form and function: A comprehensive nomenclature of proteoglycans. Matrix Biol. 42: 11–55.

Jahangirian, H., Lemraski, E.G., Rafiee-Moghaddam, R. and Webster, T.J. 2018. A review of using green chemistry methods for biomaterials in tissue engineering. J. Nanomed. 13: 5953–5969.

Ji, W., Sun, Y., Yang, F., van den Beucken, J., Fan, M.W., Chen, Z. et al. 2011. Bioactive electrospun scaffolds delivering growth factors and genes for tissue engineering applications. Pharm. Res. 28: 1259–1272.

Johnson, E.O., Troupis, T. and Soucacos, P.N. 2011. Tissue-engineered vascularized bone grafts: basic science and clinical relevance to trauma and reconstructive microsurgery. Microsurgery 31(3): 176–82.

Jooss, K. and Chirmule, N. 2003. Immunity to adenovirus and adeno-associated viral vectors: implications for gene therapy. Gene Ther. 10(11): 955–63.

Kassebaum, N.J., Bernabé, E., Dahiya, M., Bhandari, B., Murray, C.J.L. and Marcenes, W. 2014. Global burden of severe tooth loss a systematic review and meta-analysis. J. Dent. Res. 93(7 Suppl.): 20S–28S.

Kaufman, G., Whitescarver, R.A., Nunes, L., Palmer, X.L., Skrtic, D. and Tutak, W. 2018. Effects of protein-coated nanofibers on conformation of gingival fibroblast spheroids: potential utility for connective tissue regeneration. Biomed. Mater. 13: 025006.

Khoshkam, V., Chan, H.L., Lin, G.H., Mailoa, J., Giannobile, W.V., Wang, H.L. et al. 2015. Outcomes of regenerative treatment with rhPDGF-BB and rhFGF-2 for periodontal intra-bony defects: a systematic review and meta-analysis. J. Clin. Periodontol. 42(3): 272–80.

Kinane, D.F., Stathopoulou, P.G. and Papapanou, P.N. 2017. Periodontal diseases. Nat. Rev. Dis. Primers 3: 17038.

Kolind, K., Kraft, D., Bøggild, T., Duch, M., Lovmand, J., Pedersen, F.S. et al. 2014. Control of proliferation and osteogenic differentiation of human dental-pulp-derived stem cells by distinct surface structures. Acta Biomater. 10(2): 641–650.

Kresse, H. and Schönherr, E. 2001. Proteoglycans of the extracellular matrix and growth control. J. Cell. Physiol. 189: 266–274.

Kwansa, A.L., De Vita, R. and Freeman, J.W. 2014. Mechanical recruitment of N- and C-crosslinks in collagen type I. Matrix Biol. 34: 161–169.

Lasprilla, A.J.R., Martinez, G.A.R., Lunelli, B.H., Jardini, A.L. and Filho, R.M. 2012. Poly-lactic acid synthesis for application in biomedical devices—A review. Biotechnol. Adv. 30: 321–328.

Lee, J.H., Pryce, B.A., Schweitzer, R., Ryder, M.I. and Ho, S.P. 2015. Differentiating zones at periodontal ligament-bone and periodontal ligament-cementum entheses. J. Periodontal Res. 50(6): 870–880.

Li, F., Yu, F., Xu, X., Li, C., Huang, D., Zhou, X. et al. 2017. Evaluation of recombinant human FGF-2 and PDGF-BB in periodontal regeneration: A systematic review and meta-analysis. Sci. Rep. 7(65): 1–10.

Liping, T., Paul, T. and Wenjing, H. 2008. Surface chemistry influences implant biocompatibility. Curr. Top. Med. Chem. 8: 270–280.

Liu, N., Zhou, M., Zhang, Q., Yong, L., Zhang, T., Tian, T. et al. 2018. Effect of substrate stiffness on proliferation and differentiation of periodontal ligament stem cells. Cell Prolif. 51(5): e12478.

Loe, H. 2000. Periodontal diseases: a brief historical. Periodontol. 2: 7–12.

Löhler, J., Timpl, R. and Jaenisch, R. 1984. Embryonic lethal mutation in mouse collagen I gene causes rupture of blood vessels and is associated with erythropoietic and mesenchymal cell death. Cell 38: 597–607.

Luong, N.D., Moon, I.S., Lee, D.S., Lee, Y.K. and Nam, J.D. 2008. Surface modification of poly(l-lactide) electrospun fibers with nanocrystal hydroxyapatite for engineered scaffold applications. Mater. Sci. Eng. C 28: 1242–1249.

Luukko, K. and Kettunen, P. 2016. Integration of tooth morphogenesis and innervation by local tissue interactions, signaling networks, and semaphorin 3A. Cell Adh. Migr. 10(6): 618–626.

Maheaswari, R., Usha, R. and Selvam, A. 2015. Transseptal fibers-crosslinking convolutes: a review. Int. J. Contemp. Dent. Med. Rev. 031015: 1–5.

Mandla, S., Davenport Huyer, L. and Radisic, M. 2018. Review: Multimodal bioactive material approaches for wound healing. APL Bioeng. 2(2): 021503.

Mano, J.F., Silva, G.A., Azevedo, H.S., Malafaya, P.B., Sousa, R.A., Silva, S.S. et al. 2007. Natural origin biodegradable systems in tissue engineering and regenerative medicine: present status and some moving trends. J. R. Soc. Interface 4: 999–1030.

Mansouri, N. and Samira Bagheri. 2016. The influence of topography on tissue engineering perspective. Mater. Sci. Eng. C Mater. Biol. Appl. 61: 906–921.

McGuire, M.K., Kao, R.T., Nevins, M. and Lynch, S.E. 2006. rh PDGF-BB promotes healing of periodontol defects: 24-month clinical and radiografic observations. Int. J. Periodonics and Rest an Rest Dent. 26: 223–231.

Mellonig, J.T., Valderrama Mdel, P. and Cochran, D.L. 2009. Histological and clinical evaluation of recombinathuman platelet-drrived growth factor combined with beta tricalcium phosphate for the treatment of humans Class III furcations defects. Int. J. Periodontics and Rest Dent. 29: 169–177.

Misch, C.M. 2010. Autogenous bone: Is it still the gold standard? Implant. Dent. 19: 361.

Mithieux, S.M. and Weiss, A.S. 2005. Elastin. pp. 437–461. *In*: Advances in Protein Chemistry: Academic Press.

Mitsiadis, T.A. and Harada, H. 2015. Regenerated teeth: the future of tooth replacement. An update. Reg. Med. 10: 5–8.

Miyoshi, H. and Adachi, T. 2014. Topography design concept of a tissue engineering scaffold for controlling cell function and fate through actin cytoskeletal modulation. Tissue Eng. Part B Rev. 20(6): 609–27.

Moharamzadeh, K., Brook, I.M., Van Noort, R., Scutt, M., Smith, K.G. and Thornhill, M.H. 2008. Development, optimization and characterization of a fullthickness tissue engineered human oral mucosal model for biological assessment of dental biomaterials. J. Mater. Sci. Mater. Med. 19(4): 1793–1801.

Moharamzadeh, K., Colley, H., Murdoch, C., Hearnden, V., Chai, W.L., Brook, I.M. 2012. Tissue-engineered oral mucosa. J. Dent. Res. 91(7): 642–50.

Moioli, E.K., Clark, P.A., Xin, X., Lal, S. and Mao, J.J. 2007. Matrices and scaffolds for drug delivery in dental, oral and craniofacial tissue engineering. Adv. Drug Deliv. Rev. 59(4-5): 308–324.

Mouw, J.K., Ou, G. and Weaver, V.M. 2014. Extracellular matrix assembly: a multiscale deconstruction. Nat. Rev. Mol. Cell Biol. 15: 771.

Muiznieks, L.D. and Keeley, F.W. 2013. Molecular assembly and mechanical properties of the extracellular matrix: A fibrous protein perspective. Biochim. Biophys. Acta—Mol. Basis Dis. 1832: 866–875.

Murakami, S. 2011. Periodontal tissue regeneration by signaling molecule(s): what role does basic fibroblast growth factor (FGF-2) have in periodontal therapy? Periodontol. 2000, 56(1): 188–208.

Nagrath, M., Sikora, A., Graca, J., Chinnici, J.L., Rahman, S.U., Reddy, S.G. et al. 2018. Functionalized prosthetic interfaces using 3D printing: Generating infection-neutralizing prosthesis in dentistry. Mater. Today Commun. 15: 114–119.

Naung, N.Y., Shehata, E. and Van Sickels, J.E. 2019. Resorbable versus nonresorbable membranes: When and why? Dent. Clin. North Am. 63(3): 419–431.

Nazir, M.A. 2017. Prevalence of periodontal disease, its association with systemic diseases and prevention. Int. J. Health Sci. 11(2): 72–80.

Nevins, M., Giannobile, W.V., McGuire, M.K., Kao, R.T., Mellonig, J.T., Hinrichs, J.E. et al. 2005. Platelet-derived growth factor stimulates bone fill an rate of attachment level gain: results of a large multicenter randomized controlled trial. J. Periodontol. 76: 2205–2215.

O'Brien, F.J. 2011. Biomaterials & scaffolds for tissue engineering. Mater. Today 14: 88–95.

Orsini, G., Pagella, P. and Mitsiadis, T.A. 2018. Modern trends in dental medicine: an update for internists. The Am. J. Med. 131: 1425–1430.

Orsini, G., Pagella, P., Putignano, A. and Mitsiadis, T.A. 2018. Novel biological and technological platforms for dental clinical use. Front. Physiol. 9: 1102.

Ostuni, E., Chapman, R.G., Holmlin, R.E., Takayama, S. and Whitesides, G.M. 2001. A survey of structure–property relationships of surfaces that resist the adsorption of protein. Langmuir 17: 5605–5620.

Pankov, R. and Yamada, K.M. 2002. Fibronectin at a glance. J. Cell Sci. 115: 3861–3863.

Papapanou, P.N., Sanz, M., Buduneli, N., Dietrich, T., Feres, M., Fine, D.H. et al. 2018. Periodontitis: consensus report of workgroup 2 of the 2017 world workshop on the classification of periodontal and peri-implant diseases and conditions. J. Clin. Periodontol. 45: S162–70.

Park, S.J. and Jin, F.L. 2011. A review of the preparation and properties of carbon nanotubes-reinforced polymer composites. Carbon Lett. 12: 57–69.

Peña, I., Junquera, L.M., Meana, Á., García, E. and García, V. 2010. *In vitro* engineering of complete autologous oral mucosa equivalents: Characterization of a novel scaffold. J. Periodontal. Res. 45: 375–380.

Peppas, N.A., Bures, P., Leobandung, W. and Ichikawa, H. 2000. Hydrogels in pharmaceutical formulations. Eur. J. Pharm. Biopharm. 50: 27–46.

Polzer, I., Schimmel, M., Müller, F. and Biffar, R. 2010. Edentulism as part of the general health problems of elderly adults. Int. Dent. J. 60: 143–155.

Posnick, J.C. and Posnick, J.C. 2014. Periodontal considerations in the evaluation and treatment of dentofacial deformities. Orthognath. Surg. 171–208.

Qun, W., Meng-hao, W., Ke-feng, W., Yaling, L., Hong-ping, Z., Xiong, L. et al. 2015. Computer simulation of biomolecule-biomaterial interactions at surfaces and interfaces. Biomed. Mater. 10: 032001.

Rabe, M., Verdes, D. and Seeger, S. 2011. Understanding protein adsorption phenomena at solid surfaces. Adv. Colloid Interface Sci. 162: 87–106.

Rai, R. 2015. Tissue engineering: step ahead in maxillofacial reconstruction. J. Int. Oral Health 9(7): 138–142.

Rakhmatia, Y.D., Ayukawa, Y., Furuhashi, A. and Koyano, K. 2013. Current barrier membranes: titanium mesh and other membranes for guided bone regeneration in dental applications. J. Prosthodont. Res. 57(1): 3–14.

Ratner, B.D., Hoffman, A.S., Schoen, F.J. and Lemons, J.E. 2013. Introduction—biomaterials science: an evolving, multidisciplinary endeavor. pp. xxv–xxxix. *In*: Ratner, B.D., Hoffman, A.S., Schoen, F.J. and Lemons, J.E. (eds.). Biomaterials Science (Third Edition): Academic Press.

Re, F., Sartore, L., Moulisova, V., Cantini, M., Almici, C., Bianchetti, A. et al. 2019. 3D gelatin-chitosan hybrid hydrogels combined with human platelet lysate highly support human mesenchymal stem cell proliferation and osteogenic differentiation. J. Tissue Eng. 10: 2041731419845852.

Rheinwald, J.G. and Green, H. 1975. Serial cultivation of strains of human epidermal keratinocytes: The formation of keratinizing colonies from single cells. Cell 6: 331–344.

Ripamonti, U. 2019. Developmental pathways of periodontal tissue regeneration: Developmental diversities of tooth morphogenesis do also map capacity of periodontal tissue regeneration? J. Periodontal Res. 54(1): 10–26.

Rosa, V. and Zhang, Z. 2013. Grande RH, Nör JE. Dental pulp tissue engineering in full-length human root canals. J. Dent. Res. 92(11): 970–975.

Rosen, P.S., Toscano, N., Hozclaw, D. and Reynolds, M.A. 2011. A retrospective consecutive case series using mineralized allografth combined with recombinant human platelet-derived growth factor BB to treat moderate to severe osseos lesions. Int. J. Perio. and Res. Dent. 31: 335–342.

Rozario, T. and DeSimone, D.W. 2010. The extracellular matrix in development and morphogenesis: A dynamic view. Dev. Biol. 341: 126–140.

San Martin, S., Alaminos, M., Zorn, T.M., Sanchez-Quevedo, M.C., Garzon, I., Rodriguez, I.A. et al. 2013. The effects of fibrin and fibrin agarose on the extracellular matrix profile of bioengineering oral mucosa. J. Tissue Eng. Regen. Med. 7(1): 10–19.

Sandhya, S., Mudgal, R., Kumar, G., Sowdhamini, R. and Srinivasan, N. 2016. Protein sequence design and its applications. Curr. Opin. Struct. Biol. 37: 71–80.

Saratti, C.M., Rocca, G.T. and Krejci, I. 2019. The potential of three-dimensional printing technologies to unlock the development of new 'bio-inspired' dental materials: an overview and research roadmap. J. Prosthodont. Res. 63(2): 131–139.

Schaefer, L. and Schaefer, R.M. 2010. Proteoglycans: from structural compounds to signaling molecules. Cell Tissue Res. 339: 237.

Scheller, E.L., Krebsbach, P.H. and Kohn, D.H. 2009. Tissue engineering: state of the art in oral rehabilitation. J. Oral Rehabil. 36(5): 368–89.

Setten, R.L., Rossi, J.J. and Han, S.P. 2019. The current state and future directions of RNAi-based therapeutics. Nat. Rev. Drug. Discov. 18(6): 421–446.

Sheikh, Z., Hamdan, N., Ikeda, Y., Grynpas, M., Ganss, B. and Glogauer, M. 2017. Natural graft tissues and synthetic biomaterials for periodontal and alveolar bone reconstructive applications: a review. Biomater. Res. 21: 9.

Shewale, A.H., Gattani, D.R., Bhatia, N., Mahajan, R. and Saravanan, S.P. 2016. Prevalence of periodontal disease in the general population of india—A systematic review. J. Clin. Diagn. Res. 10(6): ZE04–9.

Sill, T.J. and von Recum, H.A. 2008. Electro spinning: Applications in drug delivery and tissue engineering. Biomaterials 29: 1989–2006.

Skoog, S.A., Kumar, G., Narayan, R.J. and Goering, P.L. 2018. Biological responses to immobilized microscale and nanoscale surface topographies. Pharmacol. Ther. 182: 33–55.

Soldatos, N.K., Stylianou, P., Koidou, V.P., Angelov, N., Yukna, R. and Romanos, G.E. 2017. Limitations and options using resorbable versus nonresorbable membranes for successful guided bone regeneration. Quintessence Int. 48(2): 131–147.

Stojanovska, E., Canbay, E., Pampal, E.S., Calisir, M.D., Agma, O., Polat, Y. et al. 2016. A review on non-electro nanofibre spinning techniques. RSC Adv. 6: 83783–83801.

Stratton, S., Shelke, N.B., Hoshino, K., Rudraiah, S. and Kumbar, S.G. 2016. Bioactive polymeric scaffolds for tissue engineering. Bioact. Mater. 1: 93–108.

Stupp, S.I. and Braun, P.V. 1997. Molecular manipulation of microstructures: Biomaterials, ceramics, and semiconductors. Science 277: 1242–1248.

Suarez-Franco, J.L., Vázquez-Vázquez, F.C., Pozos-Guillen, A., Montesinos, J.J., Alvarez-Fregoso, O. and Alvarez-Perez, M.A. 2018. Influence of diameter of fiber membrane scaffolds on the biocompatibility of hPDL mesenchymal stromal cells. Dent. Mater. J. 37(3): 465–473.

Suarez-López deL Amo, F., Monje Alberto, Padial-Molina, M., Tang, Z. and Wng, H.L. 2015. Biologic agents for periodontal regeneration and implant site development. Biomed. Res. Int. 9575518: 1–10.

Tang, J.D. and Lampe, K.J. 2018. From *de novo* peptides to native proteins: advancements in biomaterial scaffolds for acute ischemic stroke repair. Biomed. Mater. 13: 034103.

Tassi, S.A., Sergio, N.Z., Misawa, M.Y.O. and Villar, C.C. 2017. Efficacy of stem cells on periodontal regeneration: systematic review of pre-clinical studies. J. Periodontal Res. 52(5): 793–812.

Theocharis, A.D., Skandalis, S.S., Gialeli, C. and Karamanos, N.K. 2016. Extracellular matrix structure. Adv. Drug Deliv. Rev. 97: 4–27.

Thesleff, I. 2014. Current understanding of the process of tooth formation: transfer from the laboratory to the clinic. Aust. Dent. J. 59(Suppl. 1): 48–54.

Trofin, E.A., Monsarrat, P. and Philippe Kemoun. 2013. Cell therapy of periodontum: from animal to human? Front. Physiol. 4: 1–11.

Tsomaia, N. 2015. Peptide therapeutics: Targeting the undruggable space. Eur. J. Med. Chem. 94: 459–470.

van Hout, W.M., Mink van der Molen, A.B., Breugem, C.C., Koole, R. and Van Cann, E.M. 2011. Reconstruction of the alveolar cleft: can growth factor-aided tissue engineering replace autologous bone grafting? A literature review and systematic review of results obtained with bone morphogenetic protein-2. Clin. Oral Investig. 15(3): 297–303.

Van Vlierberghe, S., Dubruel, P. and Schacht, E. 2011. Biopolymer-based hydrogels as scaffolds for tissue engineering applications: A review. Biomacromolecules 12: 1387–1408.

Vats, A., Tolley, N.S., Polak, J.M. and Gough, J.E. 2003. Scaffolds and biomaterials for tissue engineering: a review of clinical applications. Clin. Otolaryngol. Allied Sci. 28: 165–172.

Werz, S.M., Zeichner, S.J., Berg, B.I., Zeilhofer, H.F. and Thieringer, F. 2018. 3D printed surgical simulation models as educational tool by maxillofacial surgeons. Eur. J. Dent. Educ. 22(3): e500–e505.

Wikesjö, U.M., Qahash, M., Huang, Y.H., Xiropaidis, A., Polimeni, G. and Susin, C. 2009. Bone morphogenetic proteins for periodontal and alveolar indications; biological observations—clinical implications. Orthod. Craniofac. Res. 12(3): 263–70.

Williams, D.F. 2014. There is no such thing as a biocompatible material. Biomaterials 35: 10009–10014.

Xiao, L. and Nasu, M. 2014. From regenerative dentistry to regenerative medicine: progress, challenges, and potential applications of oral stem cells. Stem Cells Cloning Adv. Appl. 7: 89–99.

Xu, J. and Mosher, D. 2011. Fibronectin and other adhesive glycoproteins. *In*: Mecham, R. (ed.). The Extracellular Matrix: An Overview. Biology of Extracellular Matrix. Springer, Berlin, Heidelberg.

Yamamoto, T., Hasegawa, T., Yamamoto, T., Hongo, H. and Amizuka, N. 2016. Histology of human cementum: its structure, function, and development. Jpn. Dent. Sci. Rev. 52(3): 63–74.

Yildirim, S., Fu, S.Y., Kim, K., Zhou, H., Lee, Ch. L., Li, A. et al. 2011. Tooth regeneration: a revolution in stomatology and evolution in regenerative medicine. Int. J. Oral Sci. 3: 107–116.

Yoo, H.S., Kim, T.G. and Park, T.G. 2009. Surface-functionalized electrospun nanofibers for tissue engineering and drug delivery. Adv. Drug Deliv. Rev. 61: 1033–1042.

Yoshizawa, M., Koyama, T., Kojima, T., Kato, H., Ono, Y. and Saito, C. 2012. Keratinocytes of tissue-engineered human oral mucosa promote re-epithelialization after intraoral grafting in athymic mice. J. Oral Maxillofac. Surg. 70(5): 1199–214.

Zafar, M.S., Khurshid, Z. and Almas, K. 2015. Oral tissue engineering progress and challenges. Tissue Eng. Regen. Med. 12(6): 387–397.

Zhou, R. 2014. Molecular Modeling at the Atomic Scale: Methods and Applications in Quantitative Biology. Taylor & Francis, ISBN 9781466562950.

Zhou, R. 2015. Modeling of Nanotoxicity: Molecular Interactions of Nanomaterials with Bionanomachines. Springer International Publishing ISBN 978-3-319-15382-7.

Zhu, Y., Zhang, Q., Shi, X. and Han, D. 2019. Hierarchical hydrogel composite interfaces with robust mechanical properties for biomedical applications. Adv. Mater. e1804950.

2

Translation of Tissue Engineering Approach from Laboratory to Clinics

Daniel Chavarría-Bolaños,[1,]* José Vega-Baudrit,[2] Bernardino Isaac Cerda-Cristerna,[3] Amaury Pozos-Guillén[4] and Mauricio Montero-Aguilar[1]

Introduction

Biotechnological advances in the dental and maxillofacial field have shown exponential growth in the last decade in terms of research funding and product development, resulting in novel more effective clinical applications. Now, more than ever, the industry is working with academia to produce new biomaterials allowing the possibility for techniques to either repair or regenerate biological tissues. The regeneration process is described as a slow replacement of a lost or damaged structure with the identical original tissue, a process observed naturally only during the first stages of life. On the other hand, repairing tissue is a much faster process involving the inflammatory cell cascade, following the deposition of a matrix and remodeling of the tissues in the damaged site.

Tissue engineering makes it possible for a clinician to repair or regenerate an affected tissue using cells fixed to biological or synthetic matrices or scaffolds which will guide the growth of the new tissue (Oakes 2004). In the near future, clinicians will be able to replace or regenerate completely lost or damaged oral structures with some kind of engineered product coming from a laboratory. Currently, scientific research is focusing on expanding our knowledge regarding the biological processes involved in tissue regeneration and repairment. This new knowledge has the potential to be translated clinically in more biological approaches for the final benefit of patients (Simon et al. 2011). Unfortunately, the commercially available products represent a small proportion of what has been tested and developed in the laboratory. The elevated costs and the complex regulatory pathway are responsible for the disparity between the amount of research investment and the number of products available in the market (Mishra et al. 2016).

[1] Faculty of Dentistry, Universidad de Costa Rica, Costa Rica. Ciudad Universitaria Rodrigo Facio, San José Costa Rica. Email: maumontero@gmail.com
[2] National Laboratory of Nanotechnology, Costa Rica.
[3] Faculty of Dentistry, Universidad Veracruzana, México.
[4] Faculty of Dentistry, Universidad Autónoma de San Luis Potosi, México.
* Corresponding author: danielchava2@gmail.com

New inventions and advances in tissue engineering commonly raise the same question among clinicians… *WHEN?* When will it be available in the market? And when can it be used in our clinical practice? It seems that other questions such as, how does it work? Who could benefit from it? And what adverse effects can we expect, play a secondary role. For researchers, it is impossible to predict when a new material or method will be available in the market, since the road from the laboratory to the final consumer is regularly a long way, full of complex challenges which not always depend on the research team's efforts or capacity. Zafar et al., discuss some of the challenges in the process of translating tissue engineering research from the laboratory to the clinic. These challenges could be classified into two main groups: scientific and non-scientific challenges. Among the scientific challenges, they highlight the need for a multidisciplinary approach as one of the major concerns. Also, the complex nature of the oral tissues, the lack of an ideal scaffold material, and sterilization protocols are some of the challenges scientists and research groups must deal with in order to advance in bringing new therapeutic options to the clinic. On the other hand, regulatory and ethical issues regarding the use of some of the basic elements needed in tissue engineering like stem cells, funding opportunities, cost-effectiveness analysis, and the ideal packing and storage of these tissues are listed as the non-scientific challenges (Zafar et al. 2015). Most of the answers for overcoming these challenges are still to be discovered, others need a change in the perception of scientific progress from authorities and regulatory agencies. Now more than ever, the industry and academia need to work together, and scarce resources need to be used effectively.

Clinical Setting

Let us consider a likely real-life clinical situation. After cancer surgery, a patient loses a segment of the mandible and now needs the replacement of the lost bone structure. The treating clinician could opt for one of the available current options which include a standard titanium prosthesis or an autologous bone graft from a different donor site, for example. Also, he may figure out a new therapeutic option, not yet properly tested or developed, which may bring more clinical benefits and fewer costs. After reading a press article, he thinks of a novel idea, a hypothetical 3D printable scaffold made of shrimp crust residues, which is also functionalized with an analgesic molecule to theoretically improve the postoperative application. As a clinician, the fundamental question still remains, *how much time and effort would it take to turn this concept into an available commercial product ready to use?* Figures 2.1 and 2.2 will lead us in exploring the pathway for how to turn this concept into an available product.

The Rise of New Ideas—The Key to the Laboratory

Tissue engineering has been defined as "an interdisciplinary field which applies the principles of engineering and life sciences toward the development of biological substitutes that restore, maintain or improve tissue function" (Langer and Vacanti 1993). This definition is valid to describe the concept of oral and craniofacial tissue engineering with a focus on dental and maxillofacial structures. As an interdisciplinary research field, this involves many scientific areas such as biology, mathematics, physics, chemistry, mechanics and informatics, among others. Hence, the final product is the result of the combined knowledge of these areas. In an attempt to simplify it, tissue engineering is a system mainly composed by a scaffold, cells and active molecules; consequently, every compound and also the combined compounds should be evaluated with *in vitro* and *in vivo* tests. The translation from the bench to the clinic consequently implies a hard-scientific work as well as an ethical challenge.

Translational research refers to all the steps needed for bringing the concept of an idea to advanced preclinical and clinical testing and, ultimately, to the development of new therapies for patients (Chen et al. 2012; Ungerleider and Christman 2014). As it can be exemplified (Fig. 2.1a), every new invention rises from a clinical need, and depending on the context, this clinical need can have different

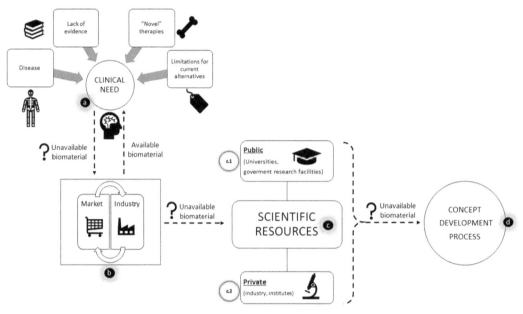

Figure 2.1. Schematic representation of the initial steps for the concept development process.

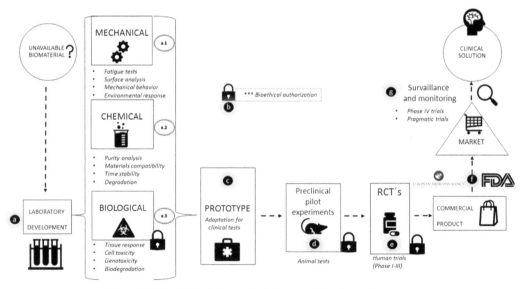

Figure 2.2. Translational flowgraph of the concept development process.

origins. In most cases, the main source of a clinical need is the patient itself who presents a certain disease or condition. Usually if there is an effective treatment available, the clinician will treat the patient with the existing approach; but, if this treatment is not ideal or could be enhanced, then the clinician should look for an alternative treatment. Some of the available 'novel' therapies may not fit every single case or may lack enough evidence to assure a safe and effective outcome. In such cases, the clinician may finally consider, once again, the traditional strategy. Of course, this will drag the limitations or side-effects already known, and the clinical need may still be unsatisfactory.

After recognizing the need for a new material, the clinician will play a key role in the translational process, which is to search in the market for available options; subsequently, they will turn to industry when they realize the material is not available (Fig. 2.1b). Let us consider again our example, the new 3D printable scaffold, which from now on we will call 3D-SCAFF! The treating clinician will search the literature and look in the market for a scaffold with analgesic purposes that also acts better than the available treatment options, just to find out that none exists yet. The market offers several kinds of bone grafts, but these are not ideal for this case. Then the market will ask the companies and industry if this product exists, with no positive response.

At this point, we can recognize a defining moment where the translational process could be interrupted. When the idea is presented to a pharmaceutical or medical business incubator or a startup company, the amount of resources the company invests in research and development (R&D) of new medical products and its experience in the regulatory process, is crucial (Bergek and Norrma 2008). For instance, a local or artisanal company might consider the idea interesting, but no further steps are taken since they do not have the capacity to develop the product. Nonetheless, if the company has an advanced program for R&D and usually invests money in testing new ideas, it could identify an emerging market and the possibility to create a new product that will strengthen its position in the market is conceivable. These kind of companies will make a deep analysis of the development opportunity, and if the project is considered profitable, then the translational process will continue, looking for either private or public sponsors to materialize the new idea. This is finally the key for the laboratory.

Activation of Scientific Resources—Where the Industry and the Academy Meet

The unavailable material has now spun from a problem to a proposal. The company in charge of the development process will shape the concept into a formal business-oriented document. At this point, hermeticism and confidentiality are mandatory, and conflict of interests should be analyzed. The patient's welfare should always be prioritized and kept in mind as the driving force during this translational process. Other interests, such as financial benefits and professional prestige should be considered as secondary endpoints when taking a new product from the laboratory to the clinic (Bekelman et al. 2003). Now, the idea can be a mere intention of gaining the intellectual property of the invention, or in turn, it can be worth millions of dollars if the product is finally developed. Secrecy and strict bioethical conducts between the different stakeholders are crucial when searching for the required scientific resources (Fig. 2.1c).

Biomaterial researchers need to provide meaningful endpoints to start the translation process (Tracey 2014), thus developing the idea should be scientifically and conceptually sound enough to gain the trust of potential investors. The prospect to create a new product will depend on two key issues: financial and academic support. Even when a big company may cover both factors, usually the amount of human and economic resources needed can make the project unfeasible. Hence, the creation of interdisciplinary collaborations is decisive and should involve private and public contributors.

Public resources include the participation of the universities or academic public groups (such as government research facilities), which may be in charge of the experiments to develop the prototype of the desirable product (Fig. 2.1c). These institutions usually require the option of research grants to cover the expenses of the research, and most of these arrangements may require confidentiality agreements between the parties. In some cases, the academic groups themselves may play a leading role, creating an inner industry that will benefit the universities' budget. This kind of participation is usually known as *"spin-off companies"*. Spin-off refers to how a company is created from another pre existing entity. In this case, entrepreneurial universities may play a key role in the economic and social development of a country (Miranda et al. 2018).

The private sector may contribute to the process as active developers financing the first experimental stages or only as mere suppliers of scientific resources. When time is vital and an unavailable advanced byproduct is needed, skilled suppliers are essential to minimize costs and time.

By reaching out for these suppliers with specific technology and advanced manufacturing processes, the developers could access the required piece of the puzzle, avoiding side-developing investments (Martinelli-Lopes et al. 2015).

The 3D SCAFF is a good example to show this process. Once the industry identifies the need for this new scaffold, they may look at different scientific databases for a competent academic or research group, who will satisfy this demand. This group could offer not only the equipment and technology but also the needed experience to guarantee an optimum beginning of the research process. If the developer recognizes that the project is too expensive to be covered by the company, the option of a public/private grant, to share the intellectual property with a sponsor or to activate a university spin-off company could be a good idea. Also, if the analgesic drug needed to load the 3D-SCAFF or a source of purified shrimp crust is needed, an outsourcer should be contacted to speed up the process. Once all the pieces are placed in order, the development process may begin (Fig. 2.1d).

Getting into the Laboratory

The concept development is ready to activate the laboratory protocols (Fig. 2.2a). This step, far from being a recipe to follow, is a complex multidisciplinary collaboration that will change depending on the desired product. For instance, some biomaterials for tissue engineering will require more mechanical than chemical evaluation, while others will need a deeper analysis of their biological behavior. However, one step cannot suppress the other, and a complete understanding of every factor is needed; from a full physicochemical characterization (Fig. 2.2a.1), to a dynamic mechanical evaluation (Fig. 2.2a.2) and a biological guarantee of safety for further *in vivo* experiments (Fig. 2.2a.3). It is advisable for the research team to consult different international standards (such as ISO protocols) to assure that every step is sound and accepted by others.

Since all the experiments needed vary from one product to another, let us discuss the possible steps to develop 3D-SCAFF. First, all the single materials to be used must be characterized individually, determining their chemical composition, purity and behavior under different conditions (such as temperature changes, variable pH, pressure modifications or light exposure). Understanding each material will predict how they will behave once combined and may help to select and adapt the fabrication method. 3d printing—as an example—sometimes requires modifying the polymer temperature or the printing pattern; especially when an active molecule is added to the polymer. Once a possible combination is obtained, several mechanical tests will determine how the biomaterial will behave in the body and will predict the maximum load of force or deformation that may suffer. Different chemical analysis will also analyze the stability of the combinations, and the integrity of the added active molecules, such as drugs. Some vulnerable molecules may become inactive after preparation processes, so controlled release assays and bioactivity research must be conducted to determine if the loaded drug is still present.

Once some samples are obtained, the best laboratory options will be selected for biological tests to determine its biocompatibility (Fig. 2.2a.3). Biocompatibility states that the material must be non-toxic, non-allergenic, non-carcinogenic and non-mutagenic; and that it does not influence the fertility of the patient (Rogero et al. 2003). Specific experiments will determine the cellular response over the new product, in order to predict further responses from a living organism. In this context, the degradation and the by-products generated should be determined and understood as well. Although the first attempts to evaluate cytotoxicity depends on animal models, current advances in cell culture techniques have allowed the use of faster and standardized analyses to determine specific cell responses such as proliferation, adhesion, metabolism changes and differentiation, among others (Hanks et al. 1996). The obtained biological information; along with the mechanical and physicochemical data will lead to obtaining an accepted prototype: a theoretical ideal candidate (Fig. 2.2c).

Even when laboratory research is evolving, *in vitro* testing is essential to safely investigate the biological performance of newly developed devices when implanted in a living system. The animal model chosen should achieve the expected answers and should try to mimic the human response (Costa-Pinto et al. 2016). Unfortunately, a single animal model will only be used to evaluate a limited number of variables, and different tests and animal sources should be included to fully understand the biological response (Fig. 2.2d). In order to minimize the number of animals and tests required, the planning of each experiment must be meticulous, trying to obtain every possible data from each intervention. Life is valuable, and this does not exclude laboratory animals. The design of the study should consider the aim of each phase (i.e., evaluation of bone response, determination of inflammatory reaction and/or related pain and expression of biochemical mediators, among others) as well as specific biomaterial features (size, shape, degradation time and byproducts) animal factors (species, age, sex, genetic background) and technical aspects (housing conditions surgical procedure, time of evaluation, sacrifice method) in order to guarantee a good experiment (Costa-Pinto et al. 2016; Pearce et al. 2007).

Contextualizing these steps to our 3D-SCAFF, specific experiments will include the characterization of both the natural polymer (including purity, stability, manipulation, physicochemical properties, etc.), and the analgesic drug. Compatibility between the materials and stability of the obtained product must be guaranteed before biological assays. Then, the cellular models will determine which of the prepared batches shows better behavior, and all data will then be translated into an animal model. In this case, an experiment to determine the bone reaction and replacement; and a second study to determine degradation and biocompatibility may be needed. However, as for *in vitro* biological assays, every experiment should have the corresponding bioethical authorization (Fig. 2.2b). It will be described that bioethical permission is not a license that allows a 'free behavior' in research. Single authorizations must be needed for every single step that involves living organisms; from samples to animals and of course human beings.

Bioethical Issues—The Research Team Obligations

All the different parties involved during the process of bringing an experimental product from the laboratory to the clinic, including academia, sponsors and pharmaceutical/medical industry, as well clinicians and the entire research group, all have great fundamental responsibilities and ethical obligations with the scientific community and with the potential final beneficiaries of the developed product. Good clinical practice principles must be considered during every step of the process and an independent ethics committee must review and approve each study or experiment conducted (Corley et al. 2016). In a simple definition, ethics is "the study of the nature and meaning of human activity", and it is not an exclusive field for philosophers or ethicists. The ethical process of translating research, specifically in tissue engineering, is a daily challenge because humans and human/animal-derived products will be inherently used in these processes (Michman 1990). Bioethical concerns are present in basic or preclinical research studies, during human clinical trials, in the adoption of best practices in the community, and in refinement of best practices in the community (Shapiro and Layde 2008).

Ethical behavior should start from the conception of the idea (Fig. 2.2b). No matter how the experimental product is conceptualized, all the ideas and perceptions around it should be understood from an ethical point of view. Available resources should be maximized, and human safety measures should always be prioritized along the process. An efficient translation process should be well planned and well designed to reduce failed processes in the early stages. To impact and contribute to the clinical stage, good translation planning should be considered from the laboratory (Kimmelman and London 2015). For instance, the early stages of tissue engineering product development include the use of cells from human extracted teeth, stem cells from the apical papilla, human dental pulp cells and human umbilical vein endothelial cells, among others. These harvested cells from donated human tissues should be considered by scientists as an ethical issue, and the research protocol, even on basic research, has to be revised by an ethics committee. Usually, universities and research centers have an ethics committee within their organizations and independent/private research teams should submit their protocols to an independent ethics committee. Especially in the clinical stage, it will be common that protocols include international research teams located in different locations or countries. These are called multi-center or multi-site trials, and an ethics committee should review and approve the protocol locally for every participating center. During the course of a multi-center trial, it is of utmost importance that data is constantly reviewed and any rising issue should be discussed by the research organizing team. Regularly, the research protocol may require many amendments during the study and each center should keep its local ethics committee with an updated protocol (Cerda-Cristerna et al. 2014).

The use of animal models in research is sometimes controversial and discussions exploring the rights of animals and the responsibilities of scientists to animals are imperative to understand and to achieve a rational decision for the use of these models (Baumans 2004). For some biomaterials, 3D cell models have been developed to substitute testing with animals; nonetheless, when experimenting in the preclinical stages of tissue engineering, *in vivo* models are mandatory since the regenerative process of tissue comprises intricate biological interactions that are difficult or sometimes impossible to reproduce *in vitro*. Hence, rats, mice, dogs and other *in vivo* animal models are common in oral and craniofacial tissue engineering studies (Zang et al. 2014; Shamma et al. 2017; Chien et al. 2018). To date, it seems impossible to move forward on translational research by only applying *in vitro* models or *in silico* models, but strict policies and ethical regulations limit unethical procedures on research of animals and great efforts are made to limit the unnecessary use of these models (Combrisson 2017). Since 1956, the concept of the 3R on animal research has impacted positively on animal welfare and also on the efficiency of translational research moving knowledge faster from the bench to clinics, giving more reproducible results and improving cost-effective results. These 3R conventional definitions have been recently discussed and analyzed, and contemporary approaches have been proposed: (i) replacement should be understood as the need to accelerate the development and use of tools relevant to the target species (usually humans) based on the latest technologies, (ii) reduction describes how protocols should aim to use appropriately designed and considerate animal experiments that are robust and reproducible, (iii) refinement finally stands for employing new *in vivo* technologies that can benefit both animal welfare and science, including methods to minimize pain and distress, as well as to deliver enhancements in animal care, housing, training and use (MacArthur 2018). The 3Rs have helped to create an international consensus on animal welfare, and have guided discussions to create international laws of the use of animals in experimental models. Although the legal requirements and practices for animal models in preclinical research usually vary between countries, currently a common fundamental rule is that scientists carrying out research involving the participation of animals must undergo special training in order to plan and conduct experiments on animal models. Thus, translational research in tissue engineering requires scientists to develop skills in management, development and data analysis of experiments with animals. Scientists trained in animal welfare should apply the 3R concepts and understand the ethical implications to their laboratories (Franco et al. 2018).

During the entire translational process of developing an experimental product in oral and craniofacial tissue engineering, the stage where most ethical concerns arise is definitely during the clinical trial's I–IV phases (Fig. 2.2e), where human testing of the product is conducted. The ethical regulation for clinical research must be understood as a dynamic evolving regulation, since this a more recent concept as compared to drug/device developing, for example. The regulatory agencies responsible for supervision and evaluation of new drugs and medical/dental devices, the US Food and Drug Administration (FDA) and the European Medicines Agency (EMA), have created guidelines on ethical conduct, in an effort to allow a faster arrival of novel tissue engineering products into the market. The ethical issues for clinical studies are many because a tissue engineering product is a combination of human, animal and/or synthetic subproducts (scaffold, cells and active molecules), which will be used as part of clinical treatment in human beings. Before the experimental product could be tested clinically, its safety and efficacy must be demonstrated satisfactorily on *in vitro* and *in vivo* studies (Lu et al. 2015). According to Li et al., independent of the socio-cultural context where the study is taking place when writing and/or reviewing the ethics section of a clinical research protocol, 11 essential elements should be addressed and discussed to consider all possible aspects involved during the study. These elements are (1) addressing a relevant question, (2) choice of control and standard of care, (3) choice of study design, (4) choice of subject population, (5) potential benefits and harms, (6) informed consent, (7) community engagement, (8) return of research results and incidental findings, (9) post-trial access, (10) payment for participation, and (11) study-related injury (Li et al. 2016). If these elements are adequately discussed in the research protocol, the approval of the document could be achieved faster and more efficiently.

Researchers and drug developers should be aware and consider the complexity of an experimental product and which regulations oversee its clinical evaluation, before starting the clinical stages. Each country has its own regulations and laws for approving the use of specific drugs, active molecules and autologous or xenogenous cells employed in the research of medical/dental biomaterials. Moreover, the legislation for treating patients with tissue engineering products is not the same between countries and cultures (Chen et al. 2013). It is widely recommended that all the parties involved in the development of experimental products in tissue engineering be locally guided in the corresponding ethical and legal regulation for conducting research of such products.

From the First Prototype to Pre-Clinical Data

Going back to the 3D-SCAFF example, to evaluate the preclinical behavior of this experimental product, two bioethical concerns must be considered. First, obtaining cell lines and biological tissues (such as human blood for coagulation assays) where the biocompatibility and cytotoxicity will be tested. All human tissue samples (even when they are as tiny as a cluster of cells from disposable tissue like human pulp from extracted teeth) must be acquired following informed consent from the donor, and the final manipulation of these samples after each assay must assure no biological hazard to the researchers and the environment. The second concern is related to the evaluation itself. The acquiring, manipulation and sacrifice of animals involved in biological tests should be considered. Some of the tests can be expected innocuous or less harmful for the animals, while others are invasive and may include experiments that affect the integrity of the specimen. The researcher must count with the approval from a bioethical committee, to assure the best and minimum amount of experiments to assure high-quality data. If the 3D-SCAFF succeeds in being considered as biocompatible, analgesic and non-cytotoxic, then a human clinical trial can be designed.

The "Jump" to Clinical Trials—First Contact with Human Beings

Once all the needed preclinical data has been collected and all the required initial quality and biocompatibility tests have been cleared, *IT'S TIME TO GO CLINICAL!* (Fig. 2.2e). But, what exactly does this mean? What steps must the research group take in order to initiate human clinical trials?

As previously discussed, new information on disease mechanisms has been gained through animal testing and experimental cellular designs, and new methods and materials for treating the targeted condition have been developed in the laboratory. But, usually, the regulatory requirements to initiate human testing could be overwhelming, especially if not properly guided. Tracy points out five key elements that are desired to shorten the gap of taking a new medical drug/device to a clinical trial phase (i) the researchers and clinicians involved in the product development must be adequately educated in the translation process; (ii) it is desirable that this translation process be standardized across different institutions; (iii) network connections between interdisciplinary teams must be promoted; (iv) optimized infrastructure for the translation process; and (v) proper funding for project management, bioethical regulatory process management, intellectual property management, informatics support, and for enabling industry-academy liaisons. If such elements are achieved, the process of taking a laboratory product for human testing should be much faster and the final step to becoming a commercially available product is much closer (Fig. 2.2f) (Tracy 2014).

Randomized Clinical Trials (RCTs) remain the gold standard study design to compare the effectiveness and safety of new biomaterials, medical/dental devices and clinical procedures, to prevent, diagnose or treat oral health conditions. Although, designing and executing RCTs could be more difficult and expensive than it may seem. For RCTs to provide significant input and really contribute to the translational process, these must be implemented with robust scientific rigor and all legal and bioethical matters must be considered. As stipulated by the FDA and EMA, the road to developing a new drug/device must consider safety and efficacy, in that order of priority. With all the required preclinical data, a sponsor or drug developer, usually from industry or academia, will submit an Investigational New Drug (IND) application and then clinical research could begin (Kashyap et al. 2013).

RCTs are classified according to their purpose and should follow a specific order. Phase I studies are designed to test the safety and maximum tolerated dose (when testing a drug). Since this will be the first contact of the experimental product with humans, this study design usually involves a small number of healthy test subjects (20–100) and the product is open-labeled. Volunteers are very closely monitored for signs of toxicity and, bioethical issues must be especially addressed during informed consent, in order to avoid the misconception that participants will receive a therapeutic intervention. The FDA estimates that approximately 70% of these studies move to the next phase. Phase II trials also include a small number of volunteers (100–300), but unlike Phase I, participants have the condition or disease of interest. This design is used to understand more of the pharmacokinetics and the pharmacodynamics of the tested product including optimal doses, frequency of intake, administration routes and endpoints. These trials could be designed to provide valuable information for a much larger Phase III trial. Phase II trials will use some exploratory methods to understand the therapeutic efficacy of the product, but since only a small number of participants are recruited, they lack statistical power to infer any effect. FDA reports that 33% of Phase II studies will move to the next phase, and usually sponsors, researchers and the regulatory agency will meet at this point to discuss the preliminary data, IND, methodology of Phase III trials and any safety concerns (Umscheid et al. 2011).

Phase III RCTs are the next step in the road to an IND approval and are ethically justified only after enough data has been collected to satisfy the rigorous standards for product safety and potential efficacy. Phase III trials will recruit a number of participants (sample size) large enough to have statistical power to confirm therapeutic efficiency (300–3000). These trials will not only attempt to demonstrate and confirm efficacy in a significantly much larger sample of variedly ill volunteers, but also will identify common adverse reactions and how often these occur. Some characteristic RCT elements appear in this phase, including randomization, stratification of participants and double-blinding. A Phase III trial could take between 1–4 years to be completed, and approximately 25% of these studies will go to Phase IV. The 'placebo-controlled trial' is the most common design of Phase III RCTs. The experimental product's efficacy and safety are compared to standard therapy or a sometimes controversially used placebo group. Another design for Phase III studies is the 'equivalency trial' and these will establish if the experimental product has equal efficacy than the available therapy. This equal efficacy is defined by researchers usually in a more clinical than statistically way. Many other designs exist depending on the purpose of the study, such as cross-over, factorial design and split-mouth (Lesaffre 2008; Pozos-Guillén et al. 2017; Garrocho-Rangel et al. 2019). Although Phase III RCTs are the gold standard for the drug/device approval process, the regulatory agency will require more than one Phase III trial to establish drug safety and efficacy, and in the pathway of clinical testing, this phase is where most of the resources (time and money) need to be invested.

For the researchers in charge of the clinical evaluation of the innovative 3D-SCAFF example, the first will be addressed to design a Phase I trial that will evaluate the behavior of the scaffold in a small group of healthy patients, obtaining important information to design further experiments. All trials should be evaluated by the pertinent bioethical committees and should be registered as well. Then, the following steps may include larger trials, as well as comparative experiments to evaluate the benefits between this new material and earlier similar options. A comparison will determine if the new biomaterial is ready to be introduced into the market. However, the complexity of this stage will need long processes and observation periods, so patience of the researches and the expecting witnesses is advised.

Surveillance and Monitoring—When the Ball is in the Court

Based on all the data generated from preclinical studies and Phases I, II and III clinical trials (premarketing studies), the regulatory agency approves the new drug/device (Fig. 2.2f). The most common adverse events have already been identified, and efficacy has been statistically demonstrated. However, the sponsor will usually be asked to run a Phase IV study (Fig. 2.2g), also referred to as 'post-marketing' or 'pharmacovigilance' studies. These are observational studies specifically designed to monitor the newly approved experimental product, concentrating on safety and effectiveness on a much larger scale. Just as Phase I is the first contact of the experimental product with humans, Phase IV studies are the first contact of the marketed product with the real world and, usually, several thousand volunteers who have the condition/disease participate in these long-lasting studies (Suvarna 2010). Although much of the required information about the experimental product is already known, most of the in-depth understanding of the product will be attained during Phase IV studies. These studies will identify less common adverse reactions, evaluate cost and drug effectiveness across a larger range of methodological factors than those investigated previously. Different study designs could be implemented in the post-marketing phase, according to the targeted goal. Either the industry or the regulatory agency could be interested in looking into specific data regarding drug-drug interactions, formulation advancement, special safety, special populations (elderly, pediatrics, etc.), superiority vs. equivalence testing, pharmacovigilance studies, drug utilization studies and large sample trial (also called Phase V), among other existing designs. Moreover, these studies could contribute directly to the implementation of the product through labeling changes, pricing negotiations and marketing (Glasser et al. 2007).

An alternative post-marketing design that can be utilized by sponsor companies or drug developers is the pragmatic trial. Unlike the RCT designs discussed so far, which aim to test whether an intervention works under optimal situations, pragmatic trials are designed to evaluate the effectiveness of interventions in real-life routine practice conditions. The results of these trials can be generalized and applied in daily practice settings and the strong internal validity (control for most potential biases) of RCT designs in exchange for a solid external validity (ability to generalize the results) of the pragmatic design (Patsopoulos 2011). Table 2.1 compares the basic characteristics of traditional RCTs and pragmatic trails.

Although the pragmatic trial concept has been used since 1967, when it was first introduced (Schwartz and Lellouch 1967), it was only in recent years when the scientific community started to be conscious of its pros and cons. It is important to highlight that implementing this design in more surveillance and monitoring studies will contribute to expanding the understanding of the behavior of newly approved medical products in real-life settings, encircling the full spectrum of the target population. They should not replace the exploratory premarketing trials, but rather be a continuum of the pathway necessary to fully comprehend the safety and efficacy of a new product. These studies could be especially relevant for tissue engineering product development since the clinical application of these biomaterials could vastly vary among patients. The input pragmatic studies could open the door to new clinical necessities and products.

If the 3D SCAFF obtained the corresponding permissions, it could probably be available in the market in the short term. Now, every clinician will determine if this new material offers the ideal properties that were offered by the developers. The same clinicians and patients that ask for this new

Table 2.1. Comparison of RCTs vs pragmatic trial characteristics.

Randomized clinical trials	Pragmatic trials
High internal validity	High external validity
Smaller sample size	Large sample size
Sophisticated design	Simple design
Controlled environment	Diverse settings
Premarketing studies	Post-marketing studies

option will play a new role as observers and 'anonymous evaluators'. They will report possible adverse effects or will even contribute by publishing clinical cases sharing their experience with 3D-SCAFF. Some users and researchers may also get involved in a pragmatic trial to fully understand the day-by-day behavior of the biomaterial. Curiously, with the new materials new clinical needs will also emerge, and maybe the cycle and the development of new materials will start up once again.

Conclusions

Tissue engineering provides a new era for therapeutic medicine; it is progressing very rapidly and extends to include all tissues in our physique. Three decades ago, tissue engineering was an idea and today it has become a potential treatment for numerous conditions. However, the road from the conception of new ideas to a fully developed biomaterial is complex, and all the right steps to achieve a good product cannot be avoided. Not all the ideas will become a tangible option, many will be part of the laboratory 'experience-stock', but definitely every attempt will contribute to the process.

References

Abou Neel, E.A., Chrzanowski, W., Salih, V.M., Kim, H.W. and Knowles, J.C. 2014. Tissue engineering in dentistry. J. Dent. 42: 915–928.

Baumans, V. 2004. Use of animals in experimental research: an ethical dilemma? Gene Ther. 11(Suppl. 1): S64–S66.

Bekelman, J.E., Li, Y. and Gross, C.P. 2003. Scope and impact of financial conflicts of interest in biomedical research. JAMA 289: 454–465.

Bergek, A. and Norrma, C. 2008. Incubator best practice: A framework. Technovation 28: 20–28.

Cerda-Cristerna, B.I., Garrocho-Rangel, A. and Pozos-Guillén, A.J. 2014. Research ethics committees: The conscience of the researcher to guarantee the protection of participants in clinical research in dentistry and oral medicine. ADM 71: 256–260.

Chen, F.A., Zhao, Y.M., Jin, Y. and Shi, S. 2012. Prospects for translational regenerative medicine. Biotechnol. Adv. 30: 658–672.

Chen, L., Wang, C. and Xi, T. 2013. Regulation challenge of tissue engineering and regenerative medicine in China. Burns Trauma 1: 56–62.

Chien, K.H., Chang, Y.L., Wang, M.L., Chuang, J.H., Yang, Y.C., Tai, M.C. et al. 2018. Promoting induced pluripotent stem cell-driven biomineralization and periodontal regeneration in rats with maxillary-molar defects using injectable BMP-6 hydrogel. Sci. Rep. 8: 114.

Combrisson, H. 2017. Animal experiment, can we replace? Transfus Clin. Biol. 24: 93–95.

Corley, E.A., Kim, Y. and Scheufele, D.A. 2016. Scientists' ethical obligations and social responsibility for nanotechnology research. Sci. Eng. Ethics 22: 111–132.

Costa-Pinto, A., Santos, T.C., Neves, N.M. and Reis, R.L. 2016. Testing natural biomaterials in animal models. pp. 562–579. *In*: Neves, N.M. and Reis, R.L. (eds.). Biomaterials from Nature for Advanced Devices and Therapies. John Wiley & Sons, Inc.

Franco, N.H., Sandoe, P. and Olsson, I.A.S. 2018. Researchers' attitudes to the 3Rs—An upturned hierarchy? PLoS One 13: e0200895.

Garrocho-Rangel, A., Ruiz-Rodríguez, S., Gaitán-Fonseca, C. and Pozos-Guillén, A. 2019. Randomized clinical trials in pediatric dentistry: application of evidence-based dentistry through the CONSORT statement. J. Clin. Pediatr. Dent. In Press.

Glasser, S.P., Salas, M. and Delzell, E. 2007. Importance and challenges of studying marketed drugs: what is a Phase IV study? Common clinical research designs, registries, and self-reporting systems. J. Clin. Pharmacol. 47: 1074–1086.

Hanks, C.T., Wataha, J.C. and Sun, Z. 1996. *In vitro* models of biocompatibility: a review. Dent. Mater. 12: 186–193.

Kashyap, U.N., Gupta, V. and Raghunandan, H.V. 2013. Comparison of drug approval process in United States & Europe. J. Pharm. Sci. Res. 5: 131–136.

Kimmelman, J. and London, A.J. 2015. The structure of clinical translation: efficiency, information, and ethics. Hastings Cent. Rep. 45: 27–39.

Langer, R. and Vacanti, J.P. 1993. Tissue engineering. Science 260: 920–926.

Lesaffre, E. 2008. Superiority, equivalence, and non-inferiority trials. Bull. NYU Hosp. Jt. Dis. 66: 150–154.

Li, R.H., Wacholtz, M.C., Barnes, M., Boggs, L., Callery-D'Amico, S., Davis, A. et al. 2016. Incorporating ethical principles into clinical research protocols: a tool for protocol writers and ethics committees. J. Med. Ethics 42: 229–234.

Lu, L., Arbit, H.M., Herrick, J.L., Segovis, S.G., Maran, A. and Yaszemski, M.J. 2015. Tissue engineered constructs: perspectives on clinical translation. Ann. Biomed. Eng. 43: 796–804.

MacArthur Clark, J. 2018. The 3Rs in research: a contemporary approach to replacement, reduction and refinement. Br. J. Nutr. 120(Suppl. 1): S1–S7.

Martinelli-Lopes, K., Amato-Neto, J. and de Senzi-Zancul, E. 2015. Suppliers participation in new product development: A case study in a financial organization. Conference: 22nd European Operations Management Association (EurOMA) Conference At: Neuchatel, Switzerland.

Michman, C. 1990. Ethics in bioengineering. J. Bus. Ethics 9: 227–231.

Miranda, F.J., Chamorro, A. and Rubio, S. 2018. Re-thinking university spin-off: a critical literature review and a research agenda. J. Technol. Transfer 43: 1007–1038.

Mishra, R., Bishop, T., Valerio, I.L., Fisher, J.P. and Dean, D. 2016. The potential impact of bone tissue engineering in the clinic. Regen. Med. 11: 571–587.

Oakes, B.W. 2004. Orthopaedic tissue engineering: from laboratory to the clinic. Med. J. Australia 180: S35–S38.

Patsopoulos, N.A. 2011. A pragmatic view on pragmatic trials. Dialogues Clin. Neurosci. 13: 217–224.

Pearce, A.I., Richards, R.G., Milz, S., Schneider, E. and Pearce, S.G. 2007. Animal models for implant biomaterial research in bone: a review. Eur. Cell Mater. 13: 1–10.

Pozos-Guillén, A., Chavarría-Bolaños, D. and Garrocho-Rangel, A. 2017. Split-mouth design in paediatric dentistry clinical trials. Eur. J. Paediatr. Dent. 18: 61–65.

Rogero, S.O., Malmonge, S.M., Lugão, A.B., Ikeda, T.I., Miyamaru, L. and Cruz, A.S. 2003. Biocompatibility study of polymeric biomaterials. Artif. Organs 27: 424–427.

Schwartz, D. and Lellouch, J. 1967. Explanatory and pragmatic attitudes in therapeutical trials. J. Chronic. Dis. 20: 637–648.

Shamma, R.N., Elkasabgy, N.A., Mahmoud, A.A., Gawdat, S.I., Kataia, M.M. and Abdel Hamid, M.A. 2017. Design of novel injectable *in-situ* forming scaffolds for non-surgical treatment of periapical lesions: *In-vitro* and *in-vivo* evaluation. Int. J. Pharm. 521: 306–117.

Shapiro, R.S. and Layde, P.M. 2008. Integrating bioethics into clinical and translational science research: a roadmap. Clin. Transl. Sci. 167–170.

Simon, S.R.J., Berdal, A., Cooper, P.R., Lumley, P.J., Tomson, P.L. and Smith, A.J. 2011. Dentin-pulp complex regeneration: from lab to clinic. Adv. Dent. Res. 23: 340–345.

Suvarna, V. 2010. Phase IV of drug development. Perspect. Clin. Res. 1: 57–60.

Tracy, S.L. 2014. From bench-top to chair-side: How scientific evidence is incorporated into clinical practice. Dent. Mater. 30: 1–15.

Umscheid, C.A., Margolis, D.J. and Grossman, C.E. 2011. Key concepts of clinical trials: a narrative review. Postgrad. Med. 123: 194–204.

Ungerleider, J.L. and Christman, K.L. 2014. Concise review: Injectable biomaterials for the treatment of myocardial infarction and peripheral artery disease: Translational challenges and progress. Stem Cells Transl. Med. 3: 1090–1099.

Zafar, M.S., Khurshid, Z. and Almas, K. 2015. Oral tissue engineering progress and challenges. Tissue Eng. Regen. Med. 12: 387–397.

Zang, S., Dong, G., Peng, B., Xu, J., Ma, Z., Wang, X. et al. 2014. A comparison of physicochemical properties of sterilized chitosan hydrogel and its applicability in a canine model of periodontal regeneration. Carbohydr. Polym. 113: 240–248.

3

Polymer Materials for Oral and Craniofacial Tissue Engineering

Iriczalli Cruz Maya and *Vincenzo Guarino**

Introduction

The oral and maxillofacial regions are complex areas since they are composed of different tissues. These regions can be affected by a broad range of pathologies, congenital defects, oncologic resection, trauma and infections (Susarla et al. 2011). The current strategies involve the use of allogenic, xenogeneic and autogenic grafts (Wang et al. 2005). However, there are drawbacks regarding the graft rejection, transmission of diseases and infection causing regeneration failure. Moreover, the complexity of cranio-maxillofacial tissues is a challenge due to the interactions between different types of tissues, including epithelium, mineralized and non-mineralized connective tissues (Bartold et al. 2000; Aurrekoetxea et al. 2015). Researchers focused on the study of new strategies based on the basic principle of tissue engineering—which means the use of cells, scaffolds and bioactive molecules to regenerate damaged tissues.

In this scheme, the use of stem cells derived from different sources alone or in combination with growth factors or gene therapy plays an essential role in tissue engineering because stem cells are self-renewable and can differentiate into different cell lines (Horst et al. 2012). Meanwhile, the scaffolds should provide a porous and stable structure to mimic native ECM and allow cells to synthesize their own matrix (Kim et al. 2016; Chanes-Cuevas et al. 2018). The proper biological, physical and chemical characteristics of scaffolds have led researchers to develop bio-inspired devices with tunable morphology and bioactivity using different technologies and materials.

For this purpose, biodegradable materials have been studied as a promising approach to design biomaterials for tissue engineering. Polymers have been extensively proposed for their use in biomedical applications due to their biodegradability and biocompatibility (Ozdil and Aydin 2014). Polymers can be classified into two major groups depending on their source: natural and synthetic polymers respectively. Polymers have been extensively used as biomaterials for different applications, i.e., wound healing, bone, nerve, tendon, muscle, cartilage and cardiac tissue engineering. A wide

Institute of Polymers, Composites and Biomaterials (IPCB), National Research Council of Italy, Mostra d'Oltremare Pad. 20, V.le J.F. Kennedy 54, I-80125 Naples, Italy.
* Corresponding author: vguarino@unina.it, vincenzo.guarino@cnr.it

variety of biomaterials have been used for craniofacial and alveolar bone, gingival tissue, periodontal ligament, dental pulp, cementum, dentin and drug delivery systems (Gupte and Ma 2012; Kim et al. 2014).

Synthetic Polymers

Biocompatible synthetic polymers are promising materials with tunable properties in terms of degradation rate, biocompatibility and mechanical properties. The most commonly used synthetic degradable polymers for tissue engineering and drug delivery are mainly aliphatic polyesters including Poly(Lactic Acid) (PLA), Poly(Glycolic Acid) (PGA), Poly(Caprolactone) (PCL) and their copolymers. These polymers are approved by the U.S. Food and Drug Administration (FDA) for clinical applications (Table 3.1).

PLA is a biodegradable, bioabsorbable and thermoplastic polyester, with excellent mechanical properties, obtained from renewable resources (Gupta et al. 2007). The monomer and acid lactic, have two optical isomers, L- and D-lactic acid, and PLA has three stereoisomers, poly(L-lactic acid) (PLLA), poly(D-lactic acid) (PDLLA) and poly(DL-lactic acid). The isomer L-lactic acid is found in living organisms and is the main fraction of PLA obtained from renewable sources (Fukushima and Kimura 2008). PLLA has gained attention for its use in tissue engineering due to its excellent biocompatibility and mechanical properties. PLA can be degraded by non-enzymatic hydrolysis into lactic acid, a bio-product which can be metabolized by normal cells. Due to the crystalline nature, PLA is a long-time degradable material, that can form crystalline fragments during the degradation process, causing an inflammatory response. That is why the combination of L- and D-isomer could be used to control the degradation time, crystallinity and melting point (Carrasco et al. 2010; Lopes et al. 2014; Santoro et al. 2016a).

PGA is characterized by low solubility in most organic solvents, as a function of the molecular weight. However, as molecular weight is too low, the solubility increases but the mechanical properties can be lost (Ma and Langer 1995). PGA was first used for absorbable sutures, due to its relative hydrophilicity nature which allows its rapid degradation and loss of its mechanical integrity in around 2–3 weeks (Ma and Langer 1995). PLA and PGA are biocompatible materials, with controllable degradation rate depending on chemical structure, molecular weight and crystallinity of polymers (Pamuła et al. 2001). In order to extend the applicative use of PGA, it has been used to form co-polymer with PLA, a more hydrophobic polymer, suitable to regulate the degradation of PGA. Poly(lactic acid-co-glycolic acid) (PLGA) may combine the advantages of both polymers, by modulating the co-polymer composition. For instance, high fractions of glycolic acid make the PLGA degradation faster, higher portions of PLA make it stiffer (Makadia and Siegel 2011). PLGA has been used for tissue engineering scaffolds and particularly to fabricate nanoparticles for drug delivery systems, due to the broad range of the degradation time (Moioli et al. 2007; Makadia and Siegel 2011; Danhier et al. 2012).

PCL is a semi-crystalline polymer that can be easily processed and dissolved with organic solvents. PCL is a biocompatible polymer with a slow degradation rate driven by hydrolysis of ester linkages, that allows addressing its use to a long regeneration process as in the case of the bone (Lam

Table 3.1. Principal synthetic polymers for oral and craniofacial tissue regeneration.

Synthetic polymers	Application	Reference
Poly(ε-caprolactone) (PCL)	Alveolar bone, GBR, periodontal ligament, drug delivery	(Cirillo et al. 2014; Guarino et al. 2017b; Cruz-Maya et al. 2019)
Poly(lactic acid) (PLA)	Bone, periodontal ligament	(Suarez-Franco et al. 2018; Vazquez-Vazquez et al. 2019)
Poly(glycolic acid) (PGA)	Periodontal ligament, pulp regeneration	(Mooney et al. 1996)

et al. 2009; Xu et al. 2015). PCL has also been frequently proposed in the form of a drug carrier, being able to uniformly distribute bioactive molecules and drugs into the matrix, and releasing them by a controlled degradation mechanism for sustained delivery (Luong-Van et al. 2006). In order to overcome some limitations of PCL, due to its hydrophobic properties, PCL was frequently blended with hydrophilic polymer phases, such as collagen, gelatin, keratin, elastin, silk among others suitable to enhance cell-material interactions (Guarino et al. 2011; Zhang et al. 2015; Aguirre-Chagala et al. 2017; Cruz-Maya et al. 2019).

Natural Polymers

The increased interest to reproduce the extracellular microenvironment to promote tissue regeneration has led to developing materials based on natural polymers (Cruz-Maya et al. 2018). Among them, proteins—including collagen, gelatin, fibrin, elastin, keratin, silk, zein—and polysaccharides such as chitosan, hyaluronic acid, alginate; and polynucleotides have been differently used for the design of materials in oral tissue engineering (Table 3.2) (Smith et al. 2016).

Collagen is the most abundant protein in the human body and the main compound of ECM. Collagen is a triple helix protein, characterized by a repetitive sequence of amino acids, glycine, proline, and hydroxyproline (Sherman et al. 2015). Currently, it can be identified by more than 20 types of collagen types. Collagen type I is the most abundant in nature followed by collagen type III (Parenteau-Bareil et al. 2010). Collagen and its denaturated form, gelatin, have been intensely studied for their use in tissue engineering because of their excellent biocompatibility and the presence of RGD sequences to promote integrin-mediated cell adhesion (Parenteau-Bareil et al. 2010). To improve the properties, collagen has been blended with other polymers for their use in periodontal tissue regeneration. Gelatin is a protein derived from the partial hydrolysis of collagen and depending on the extraction and manufacturing method there are type A, acidic treatment, and type B, alkaline treatment (Aldana and Abraham 2016). Gelatin has several advantages for tissue regeneration, as RGD-like sequences to promote cell adhesion, biocompatibility, and because it is a denatured form of collagen, gelatin is less antigenic (Su and Wang 2015).

Elastin is the second most common protein in ECM, being responsible for the elasticity and resilience of tissues. Elastin is characterized by its hydrophobic nature due to several amino acids in its composition. The precursor of elastin is tropoelastin, characterized by hydrophobic domains, composed basically by alanine, proline, valine, leucine, isoleucine and glycine conferring the elasticity of protein (Rodríguez-Cabello et al. 2018). The elastin lacks RGD motifs, however, it is recognized

Table 3.2. Principal natural polymers for oral and craniofacial tissue regeneration.

Natural polymers	Application	Reference
Collagen and gelatin	Alveolar and bone, GBR, periodontal ligament	(Tal et al. 1996; Stoecklin-Wasmer et al. 2013)
Elastin	Bone, periodontal ligament, salivary glands	(Foraida et al. 2017)
Silk fibroin and sericin	Periodontal ligament, pulp regeneration, maxillofacial bone defects, implant osseointegration	(Mooney et al. 1996; Sangkert et al. 2017)
Keratin	Periodontal ligament, coating of transmucosal implants, drug release	(Ferraris et al. 2017; 2018; Cruz-Maya et al. 2019)
Zein	Bone, periodontal regeneration, drug delivery	(Zhou et al. 2014; Yang et al. 2017; Bonadies et al. 2019)
Polysaccarides (Chitosan, hyaluronic acid, alginate, cellulose)	Gingiva augmentation, dental pulp regeneration, bone regeneration, drug delivery, antibacterial activity, periodontal regeneration	(Becker et al. 2010; Inuyama et al. 2010; Moshaverinia et al. 2012; Zhou et al. 2014; 2017; Chamieh et al. 2016; Ni et al. 2019)

by integrins, as integrin $\alpha_v\beta_5$ allowing integrin-mediated cell adhesion (Lee et al. 2014). Elastin has been widely studied to create biomaterials for applications where elasticity is required, as wound dressing, arterial or vascular applications, lung and cartilage (Grover et al. 2012; Milleret et al. 2012; Minardi et al. 2016).

Silk fibroin is a natural fiber protein that has gained attention for biomedical applications requiring an improvement of mechanical properties (i.e., flexibility and high tensile strength). Silk fibers extracted from domesticated silkworm *Bombix mori* (*B. mori*) are the best characterized. The amino acid composition of silk consists of glycine, alanine, serine (Vepari and Kaplan 2007; Ma et al. 2018). Silk is composed of a filament core coated with sericin, a hydrophilic protein. Sericin is degummed during the silk purification process leaving the core fibers corresponding to silk. For tissue engineering applications, silk fibroin has shown to have better mechanical properties than other natural polymers, excellent biocompatibility and its degradation products are non-toxic (Bai et al. 2015).

On the other hand, sericin was considered to promote hypersensitivity reactions, however subsequent studies have shown that sericin as silk fibrin were immunologically inert in the culture of murine macrophage cells (Panilaitis et al. 2003). Sericin can be used as a biomaterial since some studies have demonstrated that there is no cytotoxicity for several cell lines when sericin is added to the culture media (Kunz et al. 2016).

Keratin is a fibrous protein, found in hair, wool, feathers, nails and horns of mammals, reptiles and birds. Keratin proteins can be classified in intermediate filament proteins and the matrix proteins. The characteristic secondary structure of intermediate filaments is α-helix, also known as α-keratins and are low in sulfur content. The matrix proteins are globular, have high sulfur content and are surrounding the intermediate filament proteins interacting through disulfide bonds (Magin et al. 2007). Keratin is characterized by the presence of sequences as RGD (Arg-Gly-Asp) and LDV (Leu-Asp-Val) found in several ECM proteins for cell adhesion. Thus, keratin has been proposed as an alternative to collagen for developing biomaterials for tissue regeneration (Srinivasan et al. 2010). Besides, several studies have shown that the addition of keratin and adjusting its concentration, improved the mechanical properties of biomaterials (Zhang et al. 2014; Wang et al. 2015).

Zein is a vegetable protein found in the endosperm of corn that has been explored for tissue engineering and drug delivery application due to its excellent biocompatibility (Dong et al. 2004; Zhang et al. 2016). The amino acid sequence is characterized by hydrophobic and neutral amino acids, and sole polar amino acids. Due to its composition, zein is a hydrophobic protein, which may contribute to controlling the material degradation for tissue engineering, and allowing longer and sustained release of drugs as carrier (Ali et al. 2014; Zhang et al. 2016).

Polysaccharides have been studied as promising biomaterials to design medical gels, scaffolds for tissue regeneration and controlled drug delivery. Polysaccharides are natural polymers that can be extracted from plant, alga, animal and microbial sources. This group of polymers is composed of monosaccharides linked by *O*-glycosidic linkages. They include chitosan, hyaluronic acid, alginate, cellulose and their derivates. All these materials show good biocompatibility, biodegradability and low toxicity, high hydrophilicity, excellent mucoadhesive properties and tailored chemistry (Guarino et al. 2017a; Ahmad et al. 2018).

Chitosan is derived from the *N*-deacetylation of chitin, found in the exoskeleton of crustaceans. Chitosan has been widely used for oral treatments due to its properties as bioadhesivity, biodegradability, biocompatibility and antimicrobial activity (Şenel 2010). Chitosan properties have led to developing different biomaterials as films, hydrogels and fibers for tissue engineering (Lan Levengood and Zhang 2015). Chitosan is a cationic polysaccharide because its primary amino groups are responsible for the sustained release of molecules, thus have been used widely for local drug delivery systems, and also for periodontal applications (Felt et al. 1998; Guarino et al. 2015).

Hyaluronic acid is a non-sulfated glycosaminoglycan, consisting of disaccharide units of D-glucuronic acid and n-acetyl-D-glucosamine. It is found in extracellular matrix of connective tissue, synovial fluid, vitreous humor, embryonic mesenchyme skin and other tissues in human body (Tiwari and Bahadur 2019). The high hydrophilicity of hyaluronic acid confers a particular rheological behavior and inherent pharmacological properties (Liu et al. 2017).

Lastly, alginate is a natural polysaccharide extracted from brown sea algae, composed of D-manuronic acid and L-guloronic acid. Alginate is a biocompatible, non-toxic, biodegradable and low-cost material (Kolambkar et al. 2011; Westhrin et al. 2015). Alginate-based materials have been used in the form of hydrogels, microspheres, microcapsules and fibers for tissue engineering and drug delivery systems. Alginate hydrogels or scaffolds can be prepared in the presence of cations such as Ca^{2+} at low concentrations via the ionic interaction between the cation and the carboxyl functional group of alginate (Guarino et al. 2015; Barron and He 2017). To improve the biological and mechanical properties of alginate materials, it has been used in combination with other synthetic or natural polymers and growth factors (Bonino et al. 2011; Lan et al. 2018). For encapsulation of cells, alginate is also a promising material for its ability to form microgels that can serve as a 3D reservoir (Yao et al. 2012).

Applications for Oral and Cranio-Facial Tissue Engineering

Both natural and synthetic polymers have been studied individually or combined with inorganic materials, principally for oral tissue engineering (Berahim et al. 2011) (Fig. 3.1).

PCL has been widely used in bone tissue engineering, due to their peculiar mechanical properties and their *in vivo* stability, that provides a guided regeneration process after implantation. Moreover, it has been combined with natural proteins as keratin (Cruz-Maya et al. 2019). The advantage of blended polymers is the improvement of the biological response with good stability that allows cell proliferation, thus further investigations should be one for periodontal regeneration applications. The use of PCL and gelatin for the fabrication of scaffolds has shown to improve the biological properties. Moreover, electrospun fibers of PCL and gelatin can be aligned to provide a morphological cue to cells and guide their morphology and synthesis of collagen fibers for periodontal or nerve regeneration (Cirillo et al. 2014).

PLA has been widely considered as a biomaterial for tissue engineering applications, the use of PLA as fibers is preferred due to its mimicking the extracellular environment of cells. *In vitro* studies have demonstrated the good biocompatibility nanometric fibers of PLA at different concentrations. Meanwhile on *in vivo* experiments it was possible to observe the synthesis of collagen after 100 days of implantation in rats (Granados-Hernández et al. 2018). For periodontal tissue regeneration, PLA

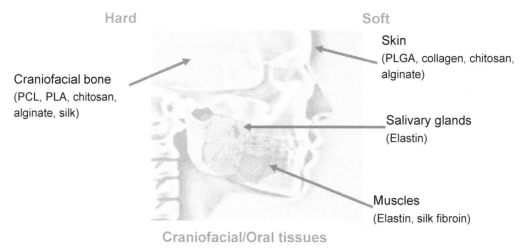

Figure 3.1. Polymers commonly used in craniofacial applications.

fibers were evaluated *in vitro* by using periodontal ligament mesenchymal stem cells for periodontal applications (Suarez-Franco et al. 2018). Results have shown that cells adhered and proliferated in PLA fibers, and most importantly, that morphological properties of PLA fibers in nanometric ranges allow the viability of mimicking the ECM of periodontal environment of periodontal ligament stem cells. Moreover, 3D scaffolds of PLA coated with fibers were non-cytotoxic and promoted the growth and proliferation of osteoblasts (Vazquez-Vazquez et al. 2019).

For dental pulp regeneration (Fig. 3.2), hyaluronic acid sponges are able to promote cell proliferation, cell-organization and blood vessel invasion (Inuyama et al. 2010). The presence of defects in interdental papilla is an esthetic concern for patients and can lead to plaque accumulation and generate periodontal diseases. So, the use of minimally invasive strategies for gingiva augmentation has been explored (Becker et al. 2010; Ni et al. 2019). Hyaluronic acid injection has been considered as a minimally invasive treatment and is able to promote fibroblasts proliferation, and formation of collagen fibers induced by hyaluronic acid. PGA matrices have been able to support the growth in high density of pulp-derived fibroblasts which synthesize collagen with morphology similar to native pulp over 60 days (Mooney et al. 1996).

The addition of elastin to electrospun scaffolds based on synthetic polymers allows improving the elasticity of fibers and biochemical signals, facilitating the cell organization to use in epithelial tissues, as salivary glands (Foraida et al. 2017).

In mineralized connective tissues, like bone, cementum and dentin, the organic phase is mainly composed of collagen type I, presented as mineralized collagen fibrils, where hydroxyapatite is found in the gap zones of the collagen fibril with c-axis parallel to long axis of collagen fibril (Sherman et al. 2015). In this perspective, the use of collagen for mineralized tissues is preferred to mimic the hierarchical structure of mineralized collagen in hard tissues, to direct the mineralization process and confer elasticity to the tissue (Nudelman and Sommerdijk 2012).

Collagen-based materials, especially collagen type I, have been used extensively, for guided tissue and Guided Bone Regeneration (GTR/GBR) to separate the bone from epithelial and connective tissues during the regeneration process (Stoecklin-Wasmer et al. 2013). Collagen membranes have shown that the formation of bone and cementum improved on *in vivo* experiments (Tal et al. 1996). Nanofiber composites of fish collagen with bioactive glass and chitosan have been designed to promote bone regeneration in furcation defects on *in vivo* studies. Results have shown an excellent biocompatibility, meanwhile, the addition of bioactive glass improved the mechanical properties (Zhou et al. 2017). On the other hand, there are reports related with the antimicrobial activity of chitosan, therefore the addition of chitosan to these scaffolds may prevent the adhesion of bacteria by controlling the chitosan concentration without cytotoxic effects. Dense collagen gel scaffolds seeded with dental pulp stem cells have shown that it is a promising strategy for bone tissue regeneration on *in vivo* experiment by inducing osteogenic differentiation of MSCs (Chamieh et al. 2016).

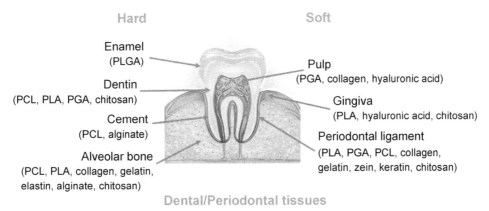

Figure 3.2. Polymers commonly used in dental/periodontal applications.

A composite scaffold of rhBMP-2 loaded zein-based scaffold was synthesized with the addition of SBA-15 nanoparticles with modified chitosan has been an anti-infective composite scaffold for delivery of osteogenic factors for bone tissue regeneration (Zhou et al. 2014). Results showed that the presence of chitosan enabled in to inhibiting bacterial growth. Meanwhile, zein protein provides the proper speed of degradation without immune response or inflammation, with induction of osteogenic differentiation *in vitro* and ectopic ossification *in vivo*.

Furthermore, chitosan has been used to encapsulate molecules as antibiotics for sustained drug release to prevent the bacteria colonization in periodontal treatments. For this propose, electrospinning and electrospray were used to design a platform able to influence the cell adhesion by fiber morphology of PCL fibers, meanwhile chitosan nanoparticles were loaded with amoxicillin with a positive antibacterial effect, this is an attractive approach for periodontal treatments (Guarino et al. 2017a). It is known that PLA nanofiber scaffolds have good biocompatibility and promote cell proliferation (Santoro et al. 2016b; Afanasiev et al. 2017). However, when an appropriate concentration of chitosan is added, cell proliferation and osteogenic differentiation are enhanced (Shen et al. 2018).

Due to the high hydrophobicity of zein, it has been used in combination with other polymers to improve cell adhesion. Co-electrospinning fibers of zein and gelatin have shown to enhance the biocompatibility of zein, meanwhile, the *in vitro* degradation showed a gradual change in fiber morphology, indicating that can be used as a scaffold for periodontal tissue regeneration (Yang et al. 2017). Electrospun fibers have shown to have a high capability to encapsulate propolis to work as patches for oral mucosa in case of abrasions and inflammation by releasing propolis gradually (Bonadies et al. 2019).

It is known that chitosan has anti-inflammatory effects, so the combination of chitosan with atorvastatin is able to enhance the anti-inflammatory effect (Özdoğan et al. 2018). The results have shown an improvement of the anti-inflammatory effect with sustained release of drug. Additionally, bone and tissue healing was induced *in vivo* in periodontitis induced rats. By improving mechanical properties with the incorporation of other natural or synthetic polymers, or inorganic compounds, as hydroxyapatite, chitosan has been proposed for bone tissue regeneration (Jiang et al. 2006; Wang and Stegemann 2010; Van Hong Thien et al. 2013).

Hydrogels of alginate have been proposed as injectable scaffolds with the periodontal ligament and gingival stem cells to direct them to an osteogenic tissue (Moshaverinia et al. 2012). The encapsulated stem cells were viable and proliferated after their encapsulation. After four weeks they differentiated into osteo- and adipo-like cells, thus, encapsulated stem cells can be a promising strategy for bone tissue engineering.

Scaffolds of hyaluronic acid have shown to promote a faster mineralization process (Manferdini et al. 2010). The osteoinductive property of hyaluronic acid has led to study its influence on the osseointegration of dental implant (Yazan et al. 2019).

In this regard, other natural polymers such as chitosan have been used to mimic the composition and morphology of ECM, resulting in improved biocompatibility on *in vitro* and *in vivo* studies and formation of vasculature (Ding et al. 2017). A porous scaffold of silk fibroin for maxillofacial bone defects have been modified with decellularized pulp and fibronectin to improve the cell behavior's adhesion, proliferation and calcium synthesis (Sangkert et al. 2017).

Titanium implants are a current strategy to replace teeth loss, however, there is a lack in the osseointegration, and restoration of surrounding tissues. There are different approaches to overcome these disadvantages. A promising strategy to improve the osseointegration is the immobilization of molecules as sericin (Nayak et al. 2013). In combination with RGD sequence, adhesion and proliferation of osteoblast-like cells were improved, and the expression of osteoblast cells markers was upregulated. To improve the interface interaction between titanium surfaces and host tissues, keratin fibers have been deposited onto titanium surfaces to improve the cellular respond of fibroblast (Ferraris et al. 2017). Moreover, to provide a morphological signal, alignment of keratin fibers has been proposed to stimulate fibroblasts to restore the soft tissue around transmucosal dental implants and seal the interface to avoid bacterial penetration (Ferraris et al. 2018).

Conclusions

The use of polymers currently represents a valid practice to design scaffolds for hard and soft tissues replacement in oral implants. In this perspective, bioinspired approaches based on the use of biopolymers (i.e., proteins, polysaccharides) naturally present in specific compartments of the biological tissues are tracing new routes for the restoration/regeneration of natural tissues located in the oral cavity—from hard (i.e., bone, dentin) to softer ones (i.e., pulp, gum).

Acknowledgements

INCIPIT project COFUND European Union's Horizon2020 research under the Marie Skłodowska-Curie project (Grant n. 665403).

References

Afanasiev, S.A., Muslimova, E.F., Nashchekina, Y.A., Nikonov, P.O., Rogovskaya, Y.V., Bolbasov, E.N. et al. 2017. Peculiarities of cell seeding on polylactic acid-based scaffolds fabricated using electrospinning and solution blow spinning technologies. Bull. Exp. Biol. Med. 164: 281–284.

Aguirre-Chagala, Y.E., Altuzar-Aguilar, V.M., León-Sarabia, E., Tinoco-Magaña, J.C., Yañez-Limón, J.M. and Mendoza-Barrera, C.O. 2017. Physicochemical properties of polycaprolactone/collagen/elastin nanofibers fabricated by electrospinning. Mater. Sci. Eng. C 76: 897–907.

Ahmad, S., Minhas, M.U., Ahmad, M., Sohail, M., Abdullah, O. and Badshah, S.F. 2018. Preparation and evaluation of skin wound healing chitosan-based hydrogel membranes. AAPS PharmSciTech.

Aldana, A.A. and Abraham, G.A. 2016. Current advances in electrospun gelatin-based scaffolds for tissue engineering applications. Int. J. Pharm.

Ali, S., Khatri, Z., Oh, K.W., Kim, I.S. and Kim, S.H. 2014. Zein/cellulose acetate hybrid nanofibers: Electrospinning and characterization. Macromol. Res. 22: 971–977.

Aurrekoetxea, M., Garcia-Gallastegui, P., Irastorza, I., Luzuriaga, J., Uribe-Etxebarria, V., Unda, F. et al. 2015. Dental pulp stem cells as a multifaceted tool for bioengineering and the regeneration of craniomaxillofacial tissues. Front. Physiol. 6: 1–10.

Bai, S., Han, H., Huang, X., Xu, W., Kaplan, D.L., Zhu, H. et al. 2015. Silk scaffolds with tunable mechanical capability for cell differentiation. Acta Biomater. 20: 22–31.

Barron, C. and He, J.Q. 2017. Alginate-based microcapsules generated with the coaxial electrospray method for clinical application. J. Biomater. Sci. Polym. Ed. 28: 1245–1255.

Bartold, P.M., McCulloch, Ch.A.G., Narayanan, A.S. and Pitaru, S. 2000. Tissue engineering: a new paradigm for periodontal regeneration based on molecular and cell biology. Periodontol. 24: 253–269.

Becker, W., Gabitov, I., Stepanov, M., Kois, J., Smidt, A. and Becker, B.E. 2010. Minimally invasive treatment for papillae deficiencies in the esthetic zone: A pilot study. Clin. Implant. Dent. Relat. Res. 12: 1–8.

Berahim, Z., Moharamzadeh, K., Rawlinson, A. and Jowett, A.K. 2011. Biologic interaction of three-dimensional periodontal fibroblast spheroids with collagen-based and synthetic membranes. J. Periodontol. 82: 790–797.

Bonadies, I., Cimino, F. and Guarino, V. 2019. *In vitro* degradation of zein nanofibres for propolis release in oral treatments. Mater Res. Express. 6: 075407.

Bonino, C.A., Krebs, M.D., Saquing, C.D., Jeong, S.I., Shearer, K.L., Alsberg, E. et al. 2011. Electrospinning alginate-based nanofibers: From blends to crosslinked low molecular weight alginate-only systems. Carbohydr. Polym. 85: 111–119.

Carrasco, F., Pagès, P., Gámez-Pérez, J., Santana, O.O. and Maspoch, M.L. 2010. Processing of poly(lactic acid): Characterization of chemical structure, thermal stability and mechanical properties. Polym. Degrad. Stab. 95: 116–125.

Chamieh, F., Collignon, A.-M., Coyac, B.R., Lesieur, J., Ribes, S., Sadoine, J. et al. 2016. Accelerated craniofacial bone regeneration through dense collagen gel scaffolds seeded with dental pulp stem cells. Sci. Rep. 6: 38814.

Chanes-Cuevas, O.A., Perez-Soria, A., Cruz-Maya, I., Guarino, V. and Antonio Alvarez-Perez, M. 2018. Macro-, micro- and mesoporous materials for tissue engineering applications. AIMS Mater. Sci. 5: 1124–1140.

Cirillo, V., Guarino, V., Alvarez-Perez, M.A. Marrese, M. and Ambrosio, L. 2014. Optimization of fully aligned bioactive electrospun fibers for "*in vitro*" nerve guidance. J. Mat. Sci. Mat. Med. 25: 2323–2332.

Cruz-Maya, I., Guarino, V. and Alvarez, M.A. 2018. Protein based devices for oral tissue repair and regeneration. AIMS Mater. Sci. 5: 156–170.

Cruz-Maya, I., Guarino, V., Almaguer-Flores, A., Alvarez-Perez, M.A., Varesano, A. and Vineis, C. 2019. Highly polydisperse keratin rich nanofibers: Scaffold design and *in vitro* characterization. J. Biomed. Mater. Res. Part A. jbm.a.36699.

Danhier, F., Ansorena, E., Silva, J.M., Coco, R., Breton, A. Le and Préat, V. 2012. PLGA-based nanoparticles: An overview of biomedical applications. J. Control. Release. 161: 505–522.

Ding, Z.Z., Ma, J., He, W., Ge, Z.L., Lu, Q. and Kaplan, D.L. 2017. Simulation of ECM with silk and chitosan nanocomposite materials. J. Mater. Chem. B 5: 4789–4796.

Dong, J., Sun, Q. and Wang, J.-Y. 2004. Basic study of corn protein, zein, as a biomaterial in tissue engineering, surface morphology and biocompatibility. Biomaterials 25: 4691–4697.

Felt, O., Buri, P. and Gurny, R. 1998. Chitosan: a unique polysaccharide for drug delivery. Drug. Dev. Ind. Pharm. 24: 979–993.

Ferraris, S., Truffa Giachet, F., Miola, M., Bertone, E., Varesano, A., Vineis, C. et al. 2017. Nanogrooves and keratin nanofibers on titanium surfaces aimed at driving gingival fibroblasts alignment and proliferation without increasing bacterial adhesion. Mater. Sci. Eng. C 76: 1–12.

Ferraris, S., Guarino, V., Cochis, A., Varesano, A., Cruz-Maya, I., Vineis, C. et al. 2018. Aligned keratin submicrometric-fibers for fibroblasts guidance onto nanogrooved titanium surfaces for transmucosal implants. Mater. Lett. 229: 1–4.

Foraida, Z.I., Kamaldinov, T., Nelson, D.A., Larsen, M. and Castracane, J. 2017. Elastin-PLGA hybrid electrospun nanofiber scaffolds for salivary epithelial cell self-organization and polarization. Acta Biomater. 62: 116–127.

Fukushima, K. and Kimura, Y. 2008. An efficient solid-state polycondensation method for synthesizing stereocomplexed poly(lactic acid)s with high molecular weight. J. Polym. Sci. Part A Polym. Chem. 46: 3714–3722.

Granados-Hernández, M.V., Serrano-Bello, J., Montesinos, J.J., Alvarez-Gayosso, C., Medina-Velázquez, L.A., Alvarez-Fregoso, O. et al. 2018. *In vitro* and *in vivo* biological characterization of poly(lactic acid) fiber scaffolds synthesized by air jet spinning. J. Biomed. Mater. Res.—Part B Appl. Biomater. 106: 2435–2446.

Grover, C.N., Cameron, R.E. and Best, S.M. 2012. Investigating the morphological, mechanical and degradation properties of scaffolds comprising collagen, gelatin and elastin for use in soft tissue engineering. J. Mech. Behav. Biomed. Mater. 10: 62–74.

Guarino, V., Alvarez-Perez, M., Cirillo, V. and Ambrosio, L. 2011. hMSC interaction with PCL and PCL/gelatin platforms: A comparative study on films and electrospun membranes. J. Bioact. Compat. Polym. 26: 144–160.

Guarino, V., Altobelli, R., Cirillo, V., Cummaro, A. and Ambrosio, L. 2015. Additive electrospraying: a route to process electrospun scaffolds for controlled molecular release. Polym. Adv. Technol. 26: 1359–1369.

Guarino, V., Cruz-Maya, I., Altobelli, R., Abdul Khodir, W.K., Ambrosio, L., Alvarez-Perez, M.A. et al. 2017a. Electrospun polycaprolactone nanofibres decorated by drug loaded chitosan nano-reservoirs for antibacterial treatments. Nanotechnology 28: 505103.

Gupta, B., Revagade, N. and Hilborn, J. 2007. Poly(lactic acid) fiber: An overview. Prog. Polym. Sci. 32: 455–482.

Gupte, M.J. and Ma, P.X. 2012. Nanofibrous scaffolds for dental and craniofacial applications. J. Dent. Res. 91: 227–34.

Hong Thien, D. Van, Hsiao, S.W., Ho, M.H., Li, C.H. and Shih, J.L. 2013. Electrospun chitosan/hydroxyapatite nanofibers for bone tissue engineering. J. Mater. Sci. 48: 1640–1645.

Horst, O., Chavez, M.G., Jheon, A.H., Desai, T. and Klein, O. 2012. Stem cell and biomaterilas reserch in dental tissue engineering and regeneration. NIH Public Acces. 56: 495–520.

Inuyama, Y., Kitamura, C., Nishihara, T., Morotomi, T., Nagayoshi, M., Tabata, Y. et al. 2010. Effects of hyaluronic acid sponge as a scaffold on odontoblastic cell line and amputated dental pulp. J. Biomed. Mater. Res. Part B Appl. Biomater. 92B: 120–128.

Jiang, T., Abdel-Fattah, W.I. and Laurencin, C.T. 2006. *In vitro* evaluation of chitosan/poly(lactic acid-glycolic acid) sintered microsphere scaffolds for bone tissue engineering. Biomaterials 27: 4894–4903.

Kim, J.H., Park, C.H., Perez, R.a., Lee, H.Y., Jang, J.H., Lee, H.H. et al. 2014. Advanced biomatrix designs for regenerative therapy of periodontal tissues. J. Dent. Res. 1203–1212.

Kim, Y., Ko, H., Kwon, I.K. and Shin, K. 2016. Extracellular matrix revisited: Roles in tissue engineering. Int. Neurourol. J. 20: S23–S29.

Kolambkar, Y.M., Dupont, K.M., Boerckel, J.D., Huebsch, N., Mooney, D.J., Hutmacher, D.W. et al. 2011. An alginate-based hybrid system for growth factor delivery in the functional repair of large bone defects. Biomaterials 32: 65–74.

Kunz, R.I., Brancalhão, R.M.C., Ribeiro, L. de F.C. and Natali, M.R.M. 2016. Silkworm sericin: Properties and biomedical applications. Biomed. Res. Int. 2016: 1–19.

Lam, C.X.F., Hutmacher, D.W., Schantz, J.-T., Woodruff, M.A. and Teoh, S.H. 2009. Evaluation of polycaprolactone scaffold degradation for 6 months *in vitro* and *in vivo*. J. Biomed. Mater. Res. A 90: 906–919.

Lan Levengood, S. and Zhang, M. 2015. Chitosan-based scaffolds for bone tissue engineering Sheeny. J. Mater. Chem. B Mater. Biol. Med. 2: 3161–3184.

Lan, W., He, L. and Liu, Y. 2018. Preparation and properties of sodium carboxymethyl cellulose/sodium alginate/chitosan composite film. Coatings 8: 291.

Lee, P., Bax, D.V., Bilek, M.M.M. and Weiss, A.S. 2014. A novel cell adhesion region in tropoelastin mediates attachment to integrin $\alpha v\beta 5$. J. Biol. Chem. 289: 1467–1477.

Liu, M., Zeng, X., Ma, C., Yi, H., Ali, Z., Mou, X. et al. 2017. Injectable hydrogels for cartilage and bone tissue engineering. Bone Res. 5.

Lopes, M., Jardini, A. and Filho, M.R. 2014. Synthesis and characterizations of poly(lactic acid) by ring-opening polymerization for biomedical applications. Chem. Eng. 38: 331–336.

Luong-Van, E., Grøndahl, L., Chua, K.N., Leong, K.W., Nurcombe, V. and Cool, S.M. 2006. Controlled release of heparin from poly(epsilon-caprolactone) electrospun fibers. Biomaterials 27: 2042–2050.

Ma, D., Wang, Y. and Dai, W. 2018. Silk fibroin-based biomaterials for musculoskeletal tissue engineering. Mater. Sci. Eng. C 89: 456–469.

Ma, P.X. and Langer, R. 1995. Degradation, structure and properties of fibrous nonwoven poly(glycolic acid) scaffolds for tissue engineering. MRS Proc. 394: 99–104.

Magin, T.M., Vijayaraj, P. and Leube, R.E. 2007. Structural and regulatory functions of keratins. Exp. Cell Res. 313: 2021–2032.

Makadia, H.K. and Siegel, S.J. 2011. Poly Lactic-co-Glycolic Acid (PLGA) as biodegradable controlled drug delivery carrier. Polymers (Basel). 3: 1377–1397.

Manferdini, C., Guarino, V., Zini, N., Raucci, M.G., Ferrari, A., Grassi, F. et al. 2010. Mineralization behavior with mesenchymal stromal cells in a biomimetic hyaluronic acid-based scaffold. Biomaterials 31: 3986–3996.

Milleret, V., Hefti, T., Hall, H., Vogel, V. and Eberli, D. 2012. Influence of the fiber diameter and surface roughness of electrospun vascular grafts on blood activation. Acta Biomater. 8: 4349–4356.

Minardi, S., Taraballi, F., Wang, X., Cabrera, F.J., Eps, J.L. Van Robbins, A.B. et al. 2016. Biomimetic collagen/elastin meshes for ventral hernia repair in a rat model. Acta Biomater. Mar. 1; 50: 165–177.

Moioli, E.K., Clark, P.A., Xin, X., Lal, S. and Mao, J.J. 2007. Matrices and scaffolds for drug delivery in dental, oral and craniofacial tissue engineering. Adv. Drug. Deliv. Rev. 59: 308–324.

Mooney, D.J., Powell, C., Piana, J. and Rutherford, B. 1996. Engineering dental pulp-like tissue *in vitro*. Biotechnol. Prog. 12: 865–868.

Moshaverinia, A., Chen, C., Akiyama, K., Ansari, S., Xu, X., Chee, W.W. et al. 2012. Alginate hydrogel as a promising scaffold for dental-derived stem cells: An *in vitro* study. J. Mater. Sci. Mater. Med. 23: 3041–3051.

Nayak, S., Dey, T., Naskar, D. and Kundu, S.C. 2013. The promotion of osseointegration of titanium surfaces by coating with silk protein sericin. Biomaterials 34: 2855–2864.

Ni, J., Shu, R. and Li, C. 2019. Efficacy evaluation of hyaluronic acid gel for the restoration of gingival interdental papilla defects. J. Oral. Maxillofac. Surg. 1–8.

Nudelman, F. and Sommerdijk, N.A.J.M. 2012. Biomineralization as an inspiration for materials chemistry. Angew. Chemie-Int. Ed. 51: 6582–6596.

Ozdil, D. and Aydin, H.M. 2014. Polymers for medical and tissue engineering applications. J. Chem. Technol. Biotechnol. 89: 1793–1810.

Özdoğan, A.I., İlarslan, Y.D., Kösemehmetoğlu, K., Akca, G., Kutlu, H.B., Comerdov, E. et al. 2018. *In vivo* evaluation of chitosan based local delivery systems for atorvastatin in treatment of periodontitis. Int. J. Pharm. Oct 25; 550(1-2): 470–476.

Pamuła, E., Błazewicz, M., Paluszkiewicz, C. and Dobrzyński, P. 2001. FTIR study of degradation products of aliphatic polyesters-carbon fibres composites. J. Mol. Struct. 596: 69–75.

Panilaitis, B., Altman, G.H., Chen, J., Jin, H.-J., Karageorgiou, V. and Kaplan, D.L. 2003. Macrophage responses to silk. Biomaterials 24: 3079–3085.

Parenteau-Bareil, R., Gauvin, R. and Berthod, F. 2010. Collagen-based biomaterials for tissue engineering applications. Materials (Basel) 3: 1863–1887.

Rodríguez-Cabello, J.C., González de Torre, I., Ibañez-Fonseca, A. and Alonso, M. 2018. Bioactive scaffolds based on elastin-like materials for wound healing. Adv. Drug Deliv. Rev. 129: 118–133.

Sangkert, S., Kamonmattayakul, S., Chai, W.L. and Meesane, J. 2017. Modified porous scaffolds of silk fibroin with mimicked microenvironment based on decellularized pulp/fibronectin for designed performance biomaterials in maxillofacial bone defect. J. Biomed. Mater Res.—Part A 105: 1624–1636.

Santoro, M., Shah, S.R., Walker, J.L. and Mikos, A.G. 2016a. Poly(lactic acid) nanofibrous scaffolds for tissue engineering. Adv. Drug Deliv. Rev.

Santoro, M., Shah, S.R., Walker, J.L. and Mikos, A.G. 2016b. Poly(lactic acid) nanofibrous scaffolds for tissue engineering. Adv. Drug Deliv. Rev. 107: 206–212.

Şenel, S. 2010. Potential applications of chitosan in oral mucosal delivery. J. Drug Deliv. Sci. Technol. 20: 23–32.

Shen, R., Xu, W., Xue, Y., Chen, L., Ye, H., Zhong, E. et al. 2018. The use of chitosan/PLA nano-fibers by emulsion eletrospinning for periodontal tissue engineering. Artif Cells, Nanomedicine Biotechnol. 46: 419–430.

Sherman, V.R., Yang, W. and Meyers, M.A. 2015. The materials science of collagen. J. Mech. Behav. Biomed. Mater. 52: 22–50.

Smith, A.M., Moxon, S. and Morris, G.A. 2016. 13—Biopolymers as wound healing materials. pp. 261–287. *In*: M.S.B.T.-W.H.B. Ågren, (ed.). Woodhead Publishing.

Srinivasan, B., Kumar, R., Shanmugam, K., Sivagnam, U.T., Reddy, N.P. and Sehgal, P.K. 2010. Porous keratin scaffold-promising biomaterial for tissue engineering and drug delivery. J. Biomed. Mater. Res.—Part B Appl. Biomater. 92: 5–12.

Stoecklin-Wasmer, C., Rutjes, A.W.S., Costa, B.R., da Salvi, G.E., Jüni, P. and Sculean, A. 2013. Absorbable collagen membranes for periodontal regeneration: a systematic review. J. Dent. Res. 92: 773–781.

Su, K. and Wang, C. 2015. Recent advances in the use of gelatin in biomedical research. Biotechnol. Lett. 37: 2139–2145.

Suarez-Franco, J.L., Vázquez-Vázquez, F.C., Pozos-Guillen, A., Montesinos, J.J., Alvarez-Fregoso, O. and Alvarez-Perez, M.A. 2018. Influence of diameter of fiber membrane scaffolds on the biocompatibility of hpdl mesenchymal stromal cells. Dent. Mater. J. 37: 465–473.

Susarla, S.M., Swanson, E. and Gordon, C.R. 2011. Craniomaxillofacial reconstruction using allotransplantation and tissue engineering: Challenges, opportunities, and potential synergy. Ann. Plast. Surg. 67.

Tal, H., Pitaru, S., Moses, O. and Kozlovsky, A. 1996. Collagen gel and membrane in guided tissue regeneration in periodontal fenestration defects in dogs. J. Clin. Periodontol. 23: 1–6.

Tiwari, S. and Bahadur, P. 2019. Modified hyaluronic acid based materials for biomedical applications. Int. J. Biol. Macromol. 121: 556–571.

Vazquez-Vazquez, F.C., Chanes-Cuevas, O.A., Masuoka, D., Alatorre, J.A., Chavarria-Bolaños, D., Vega-Baudrit, J.R. et al. 2019. Biocompatibility of developing 3D-printed tubular scaffold coated with nanofibers for bone applications. J. Nanomater. 2019: 1–13.

Vepari, C. and Kaplan, D.L. 2007. Silk as a biomaterial. Prog. Polym. Sci. 32: 991–1007.

Wang, H.-L., Greenwell, H., Fiorellini, J., Giannobile, W., Offenbacher, S., Salkin, L. et al. 2005. Periodontal regeneration. J. Periodontol. 76: 1601–22.

Wang, L. and Stegemann, J.P. 2010. Thermogelling chitosan and collagen composite hydrogels initiated with β-glycerophosphate for bone tissue engineering. Biomaterials 31: 3976–3985.

Wang, S., Wang, Z., Foo, S.E.M., Tan, N.S., Yuan, Y., Lin, W. et al. 2015. Culturing fibroblasts in 3D human hair keratin hydrogels. ACS Appl. Mater. Interfaces 7: 5187–5198.

Westhrin, M., Xie, M., Olderøy, M., Sikorski, P., Strand, B.L. and Standal, T. 2015. Osteogenic differentiation of human mesenchymal stem cells in mineralized alginate matrices. PLoS One 10: 1–16.

Xu, T., Miszuk, J.M., Zhao, Y., Sun, H. and Fong, H. 2015. Electrospun polycaprolactone 3D nanofibrous scaffold with interconnected and hierarchically structured pores for bone tissue engineering. Adv. Healthc. Mater. 4: 2238–2246.

Yang, C., Lee, J.S., Jung, U.W., Seo, Y.K., Park, J.K. and Choi, S.H. 2013. Periodontal regeneration with nano-hyroxyapatite-coated silk scaffolds in dogs. J. Periodontal. Implant. Sci. 43: 315–322.

Yang, F., Miao, Y., Wang, Y., Zhang, L.-M. and Lin, X. 2017. Electrospun zein/gelatin scaffold-enhanced cell attachment and growth of human periodontal ligament stem cells. Materials (Basel) 10: 1168.

Yao, R., Zhang, R., Luan, J. and Lin, F. 2012. Alginate and alginate/gelatin microspheres for human adipose-derived stem cell encapsulation and differentiation. Biofabrication 4.

Yazan, M., Kocyigit, I.D., Atil, F., Tekin, U., Gonen, Z.B. and Onder, M.E. 2019. Effect of hyaluronic acid on the osseointegration of dental implants. Br. J. Oral Maxillofac. Surg. 57: 53–57.

Zhang, H.L., Wang, J., Yu, N. and Liu, J.S. 2014. Electrospun PLGA/multi-walled carbon nanotubes/wool keratin composite membranes: Morphological, mechanical, and thermal properties, and their bioactivities *in vitro*. J. Polym. Res. 21.

Zhang, Q., Lv, S., Lu, J., Jiang, S. and Lin, L. 2015. Characterization of polycaprolactone/collagen fibrous scaffolds by electrospinning and their bioactivity. Int. J. Biol. Macromol. 76: 94–101.

Zhang, Y., Cui, L., Li, F., Shi, N., Li, C., Yu, X. et al. 2016. Design, fabrication and biomedical applications of zein-based nano/micro-carrier systems. Int. J. Pharm. 513: 191–210.

Zhou, P., Xia, Y., Cheng, X., Wang, P., Xie, Y. and Xu, S. 2014. Enhanced bone tissue regeneration by antibacterial and osteoinductive silica-HACC-zein composite scaffolds loaded with rhBMP-2. Biomaterials 35: 10033–10045.

Zhou, T., Liu, X., Sui, B., Liu, C., Mo, X. and Sun, J. 2017. Development of fish collagen/bioactive glass/chitosan composite nanofibers as a GTR/GBR membrane for inducing periodontal tissue regeneration. Biomed. Mater. 12: 055004.

4

Calcium Phosphate and Bioactive Glasses

Osmar A. Chanes-Cuevas,[1] *José L. Barrera-Bernal,*[1]
Iñigo Gaitán-S.[1] and *David Masuoka*[2,*]

Introduction

The focus of biomaterials has been evolving in the last 50 years, leading to the design of bioactive materials. Bioactive material is defined as a material that has the potential effect on or inducing a response from living tissues, organisms or cells (Hench 1991; Salinas and Vallet-Regí 2007; Hench and Polak 2002).

Bone defects are among the most common diseases in clinical orthopedics and are mainly caused by infections, defects, tumors and congenital diseases. Generally, these defects need bone grafts since they cannot heal by themselves, and inappropriate treatment can lead to death or invalidity. Researchers have, therefore, tried to find novel materials for bone repair and substitution. Ceramic based materials are important sources of biomaterials for the biomedical field. The ceramics intended to be in contact with living tissues are called bioceramics (Hench 1991; Salinas and Vallet-Regí 2007). These types of bioceramics have been used to repair and replace diseased and damaged parts of musculoskeletal systems. Its biocompatibility varies from ceramic oxides, which are bioinerts in the body, to the other end of biodegradable bioactive materials that allow the stimulation of specific cellular responses at the molecular level like any other biomaterial. Bioceramics can be manufactured in porous or dense, amorphous or crystalline forms and applied as coatings, cements, scaffolds and nanoparticles (Yilmaz et al. 2019). The most widely used biodegradable bioactive ceramics include calcium phosphates (CaPs), they are the main constituents of bones, teeth and tendons of mammals, giving these organs hardness and stability (Šupová 2015). Biomineralization can be described as a phenomenon in which a mineral is integrated as a functional and often structural part of living organisms, often in direct and close contact to a matrix forming protein or carbohydrate structure. Among biominerals found in nature, the most common are calcium carbonate-based biominerals

[1] Tissue Bioengineering Laboratory, Division of Graduate Studies and Research, Faculty of Dentistry, National Autonomous University of Mexico (UNAM), Circuito Exterior s/n. Col. Copilco el Alto, Alcaldía de Coyoacán, C.P. 04510, CDMX, México.

[2] Biomedical Technology Research and Develop Laboratory, Health Science Center, Stomatology Department, Autonomous University of Aguascalientes (UAA), Unidad Medico Didáctica, Av. Universidad 940, Ciudad Universitaria, C.P. 20131 Aguascalientes, Mexico.

* Corresponding author: david.masuoka@edu.uaa.mx, david.masuoka@gmail.com

like aragonite (nacre) and calcite (mussels, exoskeletons of crayfish, etc.), CaPs (in vertebrate bone and teeth) and silicates (plants, sea sponges) but much rarer natural minerals also exist (Habraken et al. 2016). They can be produced in large quantities, against relatively low cost, are stable and therefore available off-the-shelf. Recent knowledge from the field of biomineralization has clearly indicated that the architecture of natural composites such as bone and nacre is responsible for their high tensile and toughness properties. Also, various authors have proposed innovative methods to produce architecturally complex structures with a high level of control and have shown that outstanding properties can be achieved (Habraken et al. 2016).

The medical needs of an increasingly aging population have prompted a great deal of research in search of new biomaterials for the manufacture of products for many uses. These biomaterials are used to regenerate and repair living tissues damaged by disease or trauma. For specific clinical applications, mainly in orthopedics and dentistry, such as drug delivery agents, biocomposites filler, as coatings in metals and in bone implants, bioceramics play a key role (Aparicio 2014).

Calcium Phosphate

Calcium phosphate (CaP) compounds have been investigated as bone repair materials since 1920. However, they were of little use in clinical applications until the 1970s, when CaP materials were used as bone substitutes in the form of porous blocks and granules (Vallet-Regí and Salinas 2009). The clinical potential of CaP materials increased further in the early 1980s when LeGeros postulated the concept of calcium phosphate-based cements, and which was concretized by Brown and Chow in 1983 when publishing the first study on a bone cement that was set at a physiological temperature and meant the possibility of obtaining HA in monolithic form (Szurkowska and Kolmas 2017). The cement was obtained by mixing two salts of calcium phosphate, specifically a basic phase teracálcico phosphate and an acid phase, dicalcium phosphate dihydrated with an aqueous solution, as a result a paste was set that set at room temperature (O'Neill et al. 2017; Brown 1987).

Cement setting is the result of a process of dissolving the reagents and precipitation of a new phase, which takes place at room or body temperature. The entanglement between the precipitated crystals is responsible for the setting and hardening of the cement. CaP cements are produced by two types of setting reaction. The first occurs with an acid-base reaction; a relatively acidic calcium phosphate phase reacts with a relatively alkaline phase producing a neutral calcium phosphate phase. The second setting reaction occurs when the initial calcium phosphate and the final product of the setting reaction share the same Ca/P molar ratio. CaP cements have a microporous structure after setting, with quite high porosities that can vary from 30 to 50%, depending on various processing factors such as the powder-liquid ratio used (Brown 1987; Regí et al. 2000). In addition, CaP cements exhibit unique bioactive and osteoconductive properties that make them a viable alternative for bone regeneration, have excellent handling properties (i.e., injectability, moldability and cohesion). This gives them the additional advantage of being molded at irregularly shaped defect sites (O'Neill et al. 2017; Kucko et al. 2019). Compared with grafts of spongy bone or other bone substitutes, calcium phosphate has a much higher compressive strength (4 to 10 times greater than spongy bone once hardened) (Hak 2007). Calcium phosphate cements have been used clinically as bone grafts in dental and orthopedic applications (Dorozhkin 2015). In general, the calcium phosphate powder component is kneaded with setting liquid to produce a paste-like fluid, which is injected or filled by hand to cover bone defects with complicated shapes. Therefore, the use of CaP cements makes minimally invasive operations possible (Unuma and Matsushima 2013).

Some clinical studies have reported the use of CaP cement in fractures of the distal radius; in a prospective study 110 patients over 50 years with fracture of the distal radius were randomized to form two groups; a group with conventional immobilization treatment and another group with CaP cement treatment where patients treated with calcium phosphate, 82% had a satisfactory result, compared with a 56% result in the conventional immobilization group (Sanchez-Sotelo et al. 2000). Another small prospective randomized study compared calcium phosphate cement (BoneSource) with an autograft in the treatment of fractures with metaphyseal defects. Twenty-nine patients with 30 defects were evaluated for reduction radiographic loss and clinical outcome at 6 months follow-up. The majority of patients had sustained fracture of the tibial plateau or distal radius. Joint reduction was maintained in 83% of patients treated with calcium phosphate cement (Bone-Source®), compared with 67% of those treated with autograft (Dickson et al. 2002). Despite the progress made towards the manufacture of CaP cements that possesses a variety of surface and chemical characteristics, the influence of

Table 4.1. Summary of the uses of calcium phosphate.

Ceramic	Application	Reference
Calcium Phosphate	o Bone graft to restore bone defects caused by periodontal disease o Scaffolding for bone or dentin regeneration o Orthodontic cement o Coating material for the surface of dental implants	(Samavedi et al. 2013)

material properties in the organization of cellular events such as adhesion and differentiation is still poorly understood (Samavedi et al. 2013).

Calcium Sulfate

The hemihydrated form of calcium sulfate is more commonly known by orthopedic surgeons as plaster of Paris (Fig. 4.1), a material used for splinting and plastering (Hak 2007). It has long been used to immobilize fracture areas, and has also been widely used as dental impression material (Tang et al. 2010).

Calcium sulfate has been used since 1892 to fill bone defects and to act as a bone graft substitute. During the last 75 years, several authors reported variability on its use (Table 4.2), although in general with good results as a graft of bone defects, without complications attributed to calcium sulfate. The

Figure 4.1. Calcium sulfate. The different morphology of the dust particles, in general irregular prisms of 50–80 μm, is observed by SEM micrograph. In addition, a composition of 48% O, 27% Ca and 14% S, the main components of $CaSO_4$, was identified by EDS analysis, and traces of typical plaster contaminants were found, such as 9% Si, 1% Na and 0.3% Mg.
Own source.

Table 4.2. Some clinical uses of calcium sulfate.

Ceramic	Application	Reference
Calcium Sulfate	o Fracture zone immobilizer o Dental impression material o Bone substitute	(Balhaddad et al. 2019)

first results were variable, probably due to the inconsistent crystalline structure, purity and quality of calcium sulfate (Kelly et al. 2001). Some histological studies on the use of calcium sulfate in animals have shown that the trabecular bone obtained in surgically created bone defects is qualitatively similar to that observed after using an autogenous bone graft (Turner et al. 2001). Currently calcium sulfate is available in individual pellets or as a powder that can be mixed in solution to form an injectable paste or molded to form the desired shapes; its compressive strength is similar to that of a spongy bone (Pietrzak and Ronk 2000). Of the available bone graft substitutes, calcium sulfate is the one that reabsorbs more rapidly (4–6 weeks) (Hak 2007). Several studies have reported the use of calcium sulfate; During the 50s and 60s, Peltier et al. described data on their clinical experience in the use of calcium sulfate to fill bone defects (Peltier et al. 1957); in the 70s Peltier and Jones reported the results of 26 patients who had been treated by unicameral bone cysts with calcium sulfate with a long-term follow-up (1 to 20 years), where 92.3% of the patients had a successful cure of the defect with bone formation in the cyst, without complication or need for additional surgery (Peltier et al. 1978). In the 1980s, Coetzee reported that 110 patients treated with calcium sulfate mainly for bone defects in the skull and facial bones, where he concluded that calcium sulfate was an excellent substitute for bone grafting (Coetzee 1980). By the 2000s Watson reported promising results on the use of injectable calcium sulfate in five patients with tibial plateau fracture and in three patients with tibial pylon fracture (Watson 2004). On the other hand, Moed et al. reported the use of calcium sulfate implanted during acetabular fracture surgery in 32 patients, whereby monitoring CT scans they showed good bone growth (> 90%) in 22 of the 31 patients (Moed et al. 2003).

Tricalcium Phosphate

It has been known for more than 250 years that calcium phosphates are a component of the human body. Therefore, it was logical to assume that this material was thought of as a natural substitute for bone. Within calcium phosphates, one of the most used for this purpose is tricalcium phosphate (TCP). The first reported case of a tricalcium phosphate used as a graft dates back to 1920 by Albee who reported that a TCP graft promoted the formation of new bone tissue (Moed et al. 2003). Tricalcium phosphate is a bioceramics of formula $Ca_3(PO_4)_2$, which is divided into two main groups, the form α and β, both with the same Ca/P ratio of 1.5 but with different crystallographic, monoclinic and rhombohedral structures, respectively (Samavedi et al. 2013; Barrere et al. 2006; Jeong et al. 2019). This structure gives them different properties, since it depends on their stability, solubility, mechanical resistance and even their biological properties, and therefore their final applicability. The α form is generally obtained by two methods, the first is by means of a thermal transformation in which the temperature rises above 1125°C of some precursor with a Ca/P ratio of 1.5, such as hydroxyapatite deficient in calcium, amorphous calcium phosphate or the β-TCP form. The second method of obtaining it is by the so-called solid state reaction, in which precursors are mixed and the temperature is raised in the same way. Some authors recognize an α' form that occurs only at temperatures greater than 1430°C and returns to the α form when cooled below its transition temperature, and therefore lacks clinical interest (Carrodeguas and De Aza 2011). In general, the α-TCP form is more soluble and less stable, and also has less mechanical resistance than the β-TCP form, so it is not used as a bone graft, but rather as cement, which can be accompanied by a polymer that functions as a vehicle for α-TCP particles. Colpo and collaborators developed an α-TCP cement using an acrylamide-based polymer as a vehicle, which was also used to release drugs in a controlled manner (Colpo et al. 2018). An and collaborators tested an α-TCP cement together with carboxymethylcellulose and hyaluronic acid, achieving values of up to 10 MPa, which shows that polymers also help improve their mechanical properties (An et al. 2016). The α-TCP cements set by a reaction in which three α-TCP molecules incorporate hydrogen from water to form an acidic phosphate, also adding a hydroxyl group to the resulting molecule (Gildenhaar et al. 2012).

The β form can be obtained by various methods, the most common is by calcination at a temperature of 900 to 1100°C, it is more stable at room temperature and has been reported to have osteoconductive and osteoinductive properties, it has also been shown to stimulate the differentiation to preosteoblastic cells, therefore, is mainly used as a bone graft. Guillaume proposes that β-TCP can be used to fill in different types of bone defects in maxillofacial surgery such as: post extraction alveoli, apicectomy defects or prior to the placement of an implant to increase bone volume in different areas, etc. (Guillaume 2017).

When β-TCP is in the presence of water (physiological fluids) at 37°C, a reaction occurs in which hydroxyapatite is formed with the release of calcium and acid phosphates to the surrounding environment, which lowers the pH of the area, and increases its reabsorption, it is important to note that both the α and β forms have different dissolution kinetics at different pH, the α form being more soluble, as mentioned earlier (Carrodeguas and De Aza 2011; Ratner et al. 2013).

The different dissolution and absorption rates are important for these types of materials, and depend on their physicochemical properties, their porosity, their defects in the crystallographic network, and the processes to which they are subjected, for example, if they are sintered. It is well known that hydroxyapatite is not completely reabsorbed, hindering the formation of new bone tissue. In general, β-TCP is reabsorbed for 6 to 9 months, which makes it an ideal candidate for use as scaffolding (Guillaume 2017). While α-TCP is dissolved by the physiological medium, in the β form, in addition to the physiological medium, osteoclasts appear to be involved, since, when these are experimentally inhibited, the remodeling process becomes slower (Tanaka et al. 2017).

However, the process by which graft remodeling occurs is more complex than it might seem at first instance, although β-TCP has an acceptable solubility rate, it can release excess calcium ions that prevent osteoclast activity (Yamada et al. 1997), and it is important to remember that for efficient bone remodeling, a coordinated mechanism between osteoclasts and osteoblasts is needed.

Through a sintering process the degree of graft porosity can be controlled, it has been shown that porosities less than 100 μm influence the rate of reabsorption of the material, facilitating the apposition of osteoblasts on the material (Chan et al. 2015), in addition, a porosity insufficiency does not allow the formation of blood vessels, however, excessive porosity considerably decreases the mechanical strength of the material.

Although both forms of TCP are reabsorbed faster than hydroxyapatite, they lack some of its advantages, which is why it has been proven to form composite ceramics between these materials, giving rise to the so-called Biphasic Calcium Phosphate (BCP), with the intention to control the rate of graft resorption, and take advantage of each of its elements (Samavedi et al. 2013). Chan and colleagues conducted a study to compare commercial hydroxyapatite against a BCP formed by α-TCP and hydroxyapatite, and demonstrated that the BCP had better osteoinductivity *in vitro* (Chan

Table 4.3. Clinical applications of TCP.

Ceramic	Application	Reference
α-TCP	Bone cement	(Carrodeguas and De Aza 2011)
α-TCP	Bone cement	(Colpo et al. 2018)
β-TCP	Bone graft	(Gildenhaar et al. 2012)
BCP	Bone graft	(Yamada et al. 1997)
α-TCP and β-TCP	Graft in Stereolithograph	(Chan et al. 2015)
β-TCP	Graft in Stereolithograph	(Klammert et al. 2010)

et al. 2015). One of the most innovative applications of tricalcium phosphates is in stereolithography, which is an additive-type 3-D printing technique based on a powder-liquid system, in the dental area it is used in prosthetics and maxillofacial surgery to develop models of study, however, it has been proposed that by this 3-D printing technique grafts based on calcium phosphates can be made with the advantage that they will fill the defect perfectly. For example, klammert and collaborators performed a mixture of 45% α-TCP and 55% β-TCP (Klammert et al. 2010). Khalyfa et al., tested a combination of tetra calcium phosphate and tricalcium phosphate to perform models on a stereolithograph, showing good results in cell proliferation (Khalyfa et al. 2007).

Bioactive Glasses (45S5)

Some types of bioactive glasses or bioglasses have shown the ability to bind to the bone, these have also been considered as a type of bioceramics. From the point of view of materials science, the difference between glasses and ceramics is that glasses are amorphous materials, that is, their atoms do not have a long-range three-dimensional arrangement, as if they have crystalline materials.

Bioglasses are mainly used in tissue bioengineering to promote bone regeneration by the ease with which they bind to the bone by forming a layer of hydroxyapatite substituted with carbonate or hydroxycarbonate apatite (HCA). There are three main types of bioglass that are, made of silicate, of phosphate and of borate, in addition to the combinations that occur between these and among other bioceramics (Rahaman et al. 2011; Fu et al. 2011; Jones 2015). Although the chemical mechanism by which the transformation of the bioglass into the HCA layer is well known, the process by which it joins the bone and favors the formation of new bone tissue is less studied. When placed in an organism, the bioglass will interact with physiological fluids, producing an ionic exchange that raises the pH of the medium, which dissolves the silicate and forms a silanol layer on which phosphate groups and calcium ions are added to form an amorphous calcium phosphate (APC), to which OH groups are attached, and CO_3 to finally form the HCA layer (Rahaman et al. 2011; Fu et al. 2011; Jones 2015). It is believed that the formed layer of HCA, as well as the ions released during its formation, has important roles in the regeneration of bone tissue, functioning as chemoattractants of stem cells and favoring their differentiation towards an osteoblastic phenotype.

The bioactive glass called 45S5 under the trade name BIOGLASS® is the most widely used, and was the first to be discovered in 1969 by Larry L. Hench. The name 45S5 comes from its composition since it is formed by 45% SiO_2, and a 5:1 molar ratio of calcium and phosphorus, it also has other components such as Na_2O, CaO and P_2O_5. Another silicate-based bioactive glass is called 13–93, which is similar in composition to 45S5, however, it differs in its concentration of SiO_2 and other fillers, which makes it easier to sinter for obtaining porous 3-D scaffolding, and gives a slower absorption rate.

In addition, there are bioactive phosphate glasses that by their composition are more bone-like, and therefore are biocompatible and non-toxic, are made from P_2O_5, using CaO and Na_2O as fillers, they are reabsorbed faster than glasses. However, this process can be controlled by altering the composition of silicate. Due to the elements that make it up, when degraded, phosphate groups are released that will serve as a substrate for the formation of new bone tissue (Ahmed et al. 2004a; Ahmed et al. 2004b).

The third group of bioactive glasses are those of borate, the process by which they become HCA is similar to 45S5 and 13–93, with faster degradation and a higher conversion rate. It has also been tried to join two types of bioactive glasses, silicates and borates producing borosilicates, controlling the proportions of these glasses has enabled to regulate the degradation rate of these materials to more closely resembling the bone. In addition, it has been proven that they can function as vehicles for the controlled release of important elements for bone tissue formation (Zn, Cu, Mg, Sr), as well as for drug release. Among the disadvantages shared by the different types of bioactive glasses are the lack of porosity of the three-dimensional scaffolding and the low mechanical resistance thereof. Unlike the TCP, bioglasses cannot be sintered to obtain grafts of specific shapes and porosities because this process dramatically reduces their mechanical resistance, so, historically, their use as a graft was reduced to areas without significant mechanical loads. At the same time, the potential risk of release of toxic byproducts, especially borates, is still being investigated. Different techniques have been tested to synthesize bioactive glasses, which have achieved scaffolding with mechanical properties with potential application in bone load tissue, with values ranging from 0.2 to 150 MPa with different ranges of porosity, from 30 to 95% (Fu et al. 2011), where as the porosity increases, the strength of the material decreases. Another important factor seems to be the microstructure of the scaffolding, the more that the order and similar orientation of the porosity is achieved, the better resistance values are obtained.

One of the main dental applications for bioglasses is its use in the coating of dental implants, to take advantage of its bioactivity and improve the osseointegration of the implant, since titanium alone is inert. Mistry and collaborators followed 62 titanium implants placed in 31 patients for 12 months who were placed with a hydroxyapatite and bioglasses coating, finding successful osseointegration and good prosthetic support for both groups (Mistry et al. 2011). Kirsten et al. managed to coat zirconia implants with 45S5 added with alkali metals, making the sintering process compatible between both components and demonstrating that the bioactive behavior in fetal bovine serum was maintained (Kirsten et al. 2015). At the same time it has been studied that bioactive glasses favor the secretion of endothelial vascular growth factor (VEGF) by surrounding cells, which contributes to the formation of new blood vessels, it has also been proven to use bioglasses as vehicles of the same VEGF and some metal ions such as Cu^{2+}, Co^{2+}, to favor neovascularization (Kargozar et al. 2018). Li et al. manufactured a membrane with bioglasses, Cu and other components, managing to promote angiogenesis and the formation of an epithelium layer *in vivo* with potential application for skin regeneration (Li et al. 2016). With all of the above it is concluded that, although bioactive

Table 4.4. Summary of the clinical applications of the bioglasses.

Ceramic	Application	Reference
Bioglass silicate	Ti implants coating	(Ahmed et al. 2004b)
45S5	Zirconia implant coating	(Mistry et al. 2011)
Bioactive glass	Neovascularization	(Kirsten et al. 2015)
Bioactive glass	Epithelial formation	(Kargozar et al. 2018)

glasses were previously thought of as a material of the past and with few applications in the field of tissue regeneration, today it has proven to be a material with a lot of potential and with different applications in this field of study.

Hydroxyapatite (HA)

Most of the biomaterials currently available for hard tissue replacement and regeneration therapies are synthetic bone graft osteoconductive materials. Hydroxyapatite (HA) (Fig. 4.2) $[Ca_{10}(PO_4)_6OH_2]$ is well known as a valuable material for bone replacement due to its osteogenic, osteoconductive and osteoinductive properties (Rabiee et al. 2010).

It is known that during biomineralization processes living organisms can crystallize and deposit a wide range of minerals. Among them are calcium phosphates that occur in vertebrates not only in a normal way (bones, teeth), but also in a pathological way (calculation and dental and urinary stones, atherosclerotic lesions). Different types of calcium phosphates vary in chemical formulas (Table 4.5).

Bones are a complex system composed of hydroxyapatite (HA) and type I collagen fibrils. HA represents 70% of the bones, while collagen constitutes 20% and water about 10%. In mammals, the

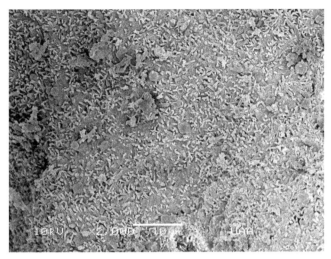

Figure 4.2. SEM micrograph of HA deposited crystals are observed.

Table 4.5. Main calcium phosphates.

Relation Ca/P	Name	Symbol	Formula
0.5	Monocalcium phosphate monohydrate	MCPM	$Ca(H_2PO_4)_2 \cdot H_2O$
1.0	Dicalcium phosphate dihydrate (brushite)	DCPD	$CaHPO_4 \cdot 2H_2O$
1.0	Dicalcium phosphate anhydrous (monetite)	DCPA	$CaHPO_4$
1.33	Octacalcium phosphate	OCP	$Ca_8(HPO_4)_2(PO_4)_4 \cdot 5H_2O$
1.5	β-Tricalcium phosphate	(β-TCP)	$Ca_3(PO_4)_2$
1.2–2.2	Amorphous calcium phosphate	ACP	$Ca_x(PO_4)_y \cdot nH_2O$
1.67	Hydroxyapatite	HA	$Ca_{10}(PO_4)_6(OH)_2$

main mineral form is called biological apatite, which is an apatite rich in carbonates and hydroxyl deficiency with a Ca/P ratio of less than 1.67 (Szcześ et al. 2017).

HA has found wide use due to its chemical and crystallographic similarity with carbonated apatite in human bones and teeth. Among the clinical applications of HA are: as a coating material for implants, as a bone substitute in the hip and knee; another particularly popular application of HA is as a carrier of chemotherapy drugs and antibiotics (Yilmaz et al. 2019). The HA can be synthesized using different techniques such as precipitation, hydrolysis and hydrothermal synthesis or extracted from natural resources, for example, from seashells, eggshells, bovine bones, among others. Other non-clinical uses of HA have also been as a scaffold for bone regeneration (Szcześ et al. 2017), studies such as Kim et al. reported that a higher content of hydroxyapatite (up to 70% by weight) also helps develop greater differentiation and mineralization of MC3T3-E1 cells in compound scaffolds (Kim et al. 2015); while another study by Juhasz reported that the hydrogel of poly-2-hydroxyethylmethacrylate/ polycaprolactone (PHEMA/PCL) incorporated with 10% by weight of HA of nanometric size and HA substituted with carbonate (CHA) exhibited significantly higher cellular activity in cells of human osteoblasts in the compound obtained than pure hydrogel (Juhasz et al. 2010).

Biocompatibility

In the context of bone regeneration, bioceramics have been developed with the aim of overcoming the possible disadvantages of bone grafts. These materials must have mechanical properties similar to those of the bone and reflect the anatomical geometry of the tissue without causing any immune response in the host during regeneration. These bioceramics are fragile materials that have low load properties, however, they have mineral components similar to natural bone and show good biocompatibility; some uses include:

Bone cell transport and growth factors.
Implant coating to improve integration.
Combined use with bone cements (Kankilic et al. 2016).

The most commonly used trial to report biocompatibility is the MTT trial. The MTT or Bromide 3-(4,5-dimethylthiazol-2-yl)-2,5-diphenyltetrazo test is a colorimetric technique that is used to assess cell density (Stockert et al. 2018; Carmichael et al. 1987). This test is based on the fact that a tetrazole salt (yellow color) is reduced to formazan crystals (purple color) in living cells (Mosmann 1983). This reaction occurs because the NAD(P)H-dependent cellular oxidoreductases enzymes can, under defined conditions, reflect the number of viable cells present, by the activity of dehydrogenases (Stockert et al. 2018). After the reduction of MTT, the culture medium is currently removed, the formazan deposits are extracted and the DMSO (Mosmann 1983) is evaluated colorimetrically, usually used for extraction.

Several studies report the usefulness of the MTT test, for example (Wang et al. 2007) manufactured scaffolds composed of Nano HAp/polyamide, which were cultured with Mesenchymal Stem Cells (MSC) derived from the bonemarrow of neonatal rabbits. They used the MTT test where they obtained results on adhesion, proliferation and differentiation of MSC. Thanks to these results, they concluded that the Nano HAp/polyamide compound scaffolds were biocompatible and showed no cytotoxic effect. Subsequently, scaffolds were implanted in rabbit jaws, obtaining good osteoconductivity, as well as good osteogenesis. Another study by Liuyun et al. 2009 by lyophilization, synthesized a scaffold composed of Nano HAp, chitosan and biodegradable carboxymethylcellulose. Using MG63 and MSC cells, they evaluated the biocompatibility by the MTT assay, finding that the compound scaffold was not toxic and biocompatible with both cell types.

Bioceramics can also be characterized by other methods. For example, it is possible to characterize by X-ray diffraction, used to determine the crystalline structure of biomaterials based on nanocrystalline apatite. On the other hand, characterization by Fourier Transformed Infrared Transmission Spectroscopy (FTIR) is also possible, which characterizes the molecular bonds of the material. The specific area of the powdered material, the porosity or the particle size are known because of the analysis of the surface area of Brunauer-Emmett-Teller. Finally, in the field of microscopy techniques, both Scanning Electron Microscopy (SEM) and Transmission Electron Microscopy (TEM) are techniques that allow imaging with greater increases in the structure of the material (Phillips et al. 2003).

There are numerous studies on applications for bone regeneration that demonstrate its biocompatibility. Zandi et al. (2010) conducted research on the biocompatibility of nanoHap-coated hydroxyapatite (HAp)/gelatin scaffolds with Mesenchymal Stem Cells (MSC) obtained from bone marrow from Wistar rats. They used the techniques of SEM-EDS and FTIR for the characterization of the biomaterial. Coated scaffolds were cultured with MSC and analyzed with SEM and MTT assay. They concluded that the samples coated with Nano HAp showed more bioactivity and biocompatibility than the uncoated ones.

The biological activity of biomaterials can also be reflected by osteoinduction, which is defined as the ability of the material to induce bone formation, osteoinductive properties are also attributed to biomaterials capable of stimulating the specific cellular response, that is, activating genes which stimulate the tissue regeneration (Łączka et al. 2016). There are a limited number of reports on osteoinduction in polymeric and metallic materials. In contrast, an increasing number of studies report the *in vivo* osteoinductive properties of bioceramics, particularly those based on calcium phosphate, among which are: HA, β-TCP, biphasic calcium phosphate, that is, a mixture of HA and TCP, anhydrous calcium dihydrophosphate (DCPA), dehydrated calcium di-phosphate (DCPD), carbonated apatite and calcium pyrophosphates (CPP) (Łączka et al. 2016).

Studies such as that of (Xie et al. 2016) describe osteoconductive properties due to the ability to bind live bone through the formation of a layer of apatite on its surfaces, in addition, through the SEM analysis they reported that the dissolution and precipitation process allowed the formation of calcium phosphate globules, and consequently the formation of a new layer of apatite after immersion of the samples in simulated body fluid (FCS); this confirmed that the porous hybrid bioceramics α5/β5-TCP had good *in vitro* bioactivity, which is considered as one of the necessary conditions for biomedical applications.

Many studies suggest that osteoinductive properties are attributed to the fact that bioceramics containing calcium phosphate (Ca-P) allow a superficial layer of Ca-P to be formed in a biological environment, a condition that is essential for them to adhere to the bone being alive. Therefore, the source of Ca-P in a biomaterial is a prerequisite for its potential for heterotopic bone formation (Łączka et al. 2016; Ohtsuki et al. 1992).

Biomedical Applications

There is a need for new biomaterials that can substitute damaged tissues, stimulate the body's own regenerative mechanisms and promote tissue healing. Porous templates referred to as 'scaffolds' are thought to be required for three-dimensional tissue growth. Bioceramics, a special set of fully, partially or non-crystalline ceramics (e.g., calcium phosphates, bioactive glasses and glass-ceramics) that are designed for the repair and reconstruction of diseased parts of the body, have high potential as scaffold materials. Traditionally, bioceramics have been used to fill and restore bone and dental defects (repair of hard tissues). More recently, this category of biomaterials has also revealed promising applications in the field of soft-tissue engineering. Therefore, the use of bioceramics in biomedical applications is on the increase.

Ceramics have always been considered for the repair of hard tissues in clinical applications due to the unique composition and mechanical properties (Hench 1991), depending on the application, the choice of a particular bioceramic for a given application will depend on the type of bioceramic/ tissue attachment required. Bioceramics can directly interact with the surrounding tissue, either supporting tissue growth or inducing new tissue regeneration for bioactive ceramics (Zhou et al. 2019). It can also remain inactive at the application site serving the purpose of mechanical load carrier as in the case of bioinert ceramics. Resorbable bioceramics are designed to degrade gradually over time and be replaced by the natural host tissue. In fact, this could be seen as the optimal solution to biomaterials problems. Osteo-inductive stimulation of ceramics creates an ideal environment for bone regeneration, and bioactive ions released during ceramic degradation play an important role in the activation of cells to produce a new tissue (Salinas and Vallet-Regí 2007). Developments of resorbable ceramics present complications, mainly due to problems related to matching the rate of resorption with the replacement by the natural host tissue and maintenance of strength and stability of interfaces during the degradation period. The rate of tissue growth varies from patient to patient and with tissue type. There are different configurations and shapes of ceramics such as: powders, particles, granules, dense blocks, porous scaffolds, suspensions, implant coatings that are used depending on the biomedical application. According on their tissue response, bioceramics can be categorized in bioactive, bioresorbable and bioinert (Table 4.6).

Tissue engineering has taken a new direction, to take advantage of the combined use of living cells and tri-dimensional ceramic scaffolds to deliver vital cells to the damaged site of the patient. Feasible and productive strategies have been aimed at combining a relatively traditional approach, such as bioceramics implants, with the acquired knowledge applied to the field of cell growth and differentiation of osteogenic cells.

There are numerous bioceramics currently used in both dentistry and medicine. Bioceramics are widely used for orthopedic applications, for coatings to improve the compatibility of metal implants, and can function as resorbable lattices which provide a framework that is eventually dissolved as the body rebuilds tissue (Driscoll et al. 2019; Nabiyouni et al. 2018). The field of dentistry is constantly

Table 4.6. Bioceramics category according to tissue response.

Type	Application	Description
Bioactive	Bioglass/glass ceramic	Eventually replaced or incorporated into tissue.
Bioresorbable	Calcium phosphate	Durable tissues that can undergo interfacial interactions with surrounding tissue.
Bioinert	Alumina, zirconia, carbon	Non interactive with biological systems.

changing due to introduction of new techniques and technological advances (Dorozhkin 2013). Advances in dentistry material sciences contribute significantly to study and find new potential materials. Basically, in dentistry, they are used for prosthetic dentures, filling bone defects, root repair and for apical retro fills (LeGeros 1988; LeGeros n.d.).

Conclusion

The development of bioactive ceramics at the regeneration of tissues has created an important field of interest, since their characteristics are those converted into the application of routine use in various medical areas. Here of some advances achieved in the field of bioactive ceramics are described. Not only did the knowledge of fundamental processes governing biomineralization grow tremendously, but their application as targeted delivery vehicles and as synthetic bone graft substitutes has demonstrated very important successes. They were present and have been used for many years clinically, these successes have been somewhat downplayed among other developments in the field of biomaterials. This, however, does not reflect the enormous diversity they have to offer both in terms of products and their applications. And these have not yet been explored to their maximum extent. Recent technological developments will bring bioactive ceramics research and development to another step further, that fits well in the search for largely available and affordable strategies for damaged and diseased tissues. For this reason the biomedical industry has great potential for better understanding these types of materials while supporting the continuous development of their potential biomaterials.

References

Ahmed, I., Lewis, M., Olsen, I. and Knowles, J.C. 2004a. Phosphate glasses for tissue engineering: Part 1. Processing and characterisation of a ternary-based P_2O_5–CaO–Na_2O glass system. Biomaterials 25: 491.

Ahmed, I., Lewis, M., Olsen, I. and Knowles, J.C. 2004b. Phosphate glasses for tissue engineering: Part 2. Processing and characterisation of a ternary-based P_2O_5–CaO–Na_2O glass fibre system. Biomaterials 25: 501.

An, J., Wolke, J.G.C., Jansen, J.A. and Leeuwenburgh, S.C.G. 2016. Influence of polymeric additives on the cohesion and mechanical properties of calcium phosphate cements. J. Mater. Sci. Mater. Med. 27: 58.

Balhaddad, A.A., Kansara, A.A., Hidan, D., Weir, M.D., Xu, H.H.K. and Melo, M.A.S. 2019. Toward dental caries: Exploring nanoparticle-based platforms and calcium phosphate compounds for dental restorative materials Bioact. Mater. 4: 43–55.

Barrere, F., van Blitterswijk, C.A. and de Groot, K. 2006. Bone regeneration: Molecular and cellular interactions with calcium phosphate veramics. Int. J. Nanomedicine 1: 317–32.

Brown, W.E. and Chow, L.C. 1986. A new calcium phosphate, water-setting cement. Cem. Res. Prog. 352–379.

Carmichael, J., DeGraff, W.G., Gazdar, A.F., Minna, J.D. and Mitchell, J.B. 1987. Evaluation of a tetrazolium-based semiautomated colorimetric assay: assessment of chemosensitivity testing. Cancer Res. 47: 936–42.

Carrodeguas, R.G. and De Aza, S. 2011. α-Tricalcium phosphate: Synthesis, properties and biomedical applications. Acta Biomater. 7: 3536.

Chan, Y.-H., Chang, Y.-S., Shen, Y.-D., Yang, T.-S., Ou, S.-F., Hsu, Y.-J. et al. 2015. Comparative *in vitro* osteoinductivity study of HA and α-TCP /HA bicalcium phosphate. Int. J. Appl. Ceram. Technol. 12: 192–8.

Coetzee, A.S. 1980. Regeneration of bone in the presence of calcium sulfate. Arch. Otolaryngol. 106: 405–9.

da Silva, R.V., Bertran, C.A., Kawachi, E.Y. and Camilli, J.A. 2007. Repair of cranial bone defects with calcium phosphate ceramic implant or autogenous bone graft. J. Craniofac. Surg. 18: 281–6.

Dickson, K.F., Friedman, J., Buchholz, J.G. and Flandry, F.D. 2002. The use of bone source hydroxyapatite cement for traumatic metaphyseal bone void filling. J. Trauma. 53: 1103–8.

Dorozhkin, S.V. 2013. Calcium orthophosphates in dentistry. J. Mater. Sci. Mater. Med. 24: 1335–63.

Dorozhkin, S.V. 2015. Calcium orthophosphate bioceramics. Ceram. Int. 41: 13913–66.

Driscoll, J.A., Lubbe, R., Jakus, A., Chang, K., Haleem, M., Yun, C. et al. 2019. 3D-printed ceramic-demineralized bone matrix hyperelastic bone composite scaffolds for spinal fusion. Tissue Eng. Part A 26: 3–4.

Fu, Q., Saiz, E., Rahaman, M.N. and Tomsia, A.P. 2011. Bioactive glass scaffolds for bone tissue engineering: state of the art and future perspectives. Mater. Sci. Eng. C 31: 1245–56.

Gildenhaar, R., Knabe, C., Gomes, C., Linow, U., Houshmand, A. and Berger, G. 2012. Calcium alkaline phosphate scaffolds for bone regeneration 3D-fabricated by additive manufacturing. Key Eng. Mater. 493: 849–54.

Guillaume, B. 2017. Filling bone defects with β-TCP in maxillofacial surgery: A review. Morphologie 101: 113–9.

Habraken, W., Habibovic, P., Epple, M. and Bohner, M. 2016. Calcium phosphates in biomedical applications: Materials for the future. Mat. Today 19: 69–87.

Hak, D.J. 2007. The use of osteoconductive bone graft substitutes in orthopaedic trauma. J. Am. Acad. Orthop. Surg. 15: 525–36.

Hench, L.L. 1991. Bioceramics: From concept to clinic. J. Am. Ceram. Soc. 74: 1487–510.

Hench, L.L. and Polak, J.M. 2002. Third-generation biomedical materials. Science 295: 1014–7.

Jeong, J., Kim, J.H., Shim, J.H., Hwang, N.S. and Heo, C.Y. 2019. Bioactive calcium phosphate materials and applications in bone regeneration. Biomater. Res. 23: 4.

Jones, J.R. 2015. Reprint of review of bioactive glass: from hench to hybrids. Acta Biomater. 23: S53–82.

Juhasz, J.A., Best, S.M. and Bonfield, W. 2010. Preparation of novel bioactive nano-calcium phosphate–hydrogel composites. Sci. Technol. Adv. Mater. 11: 14103.

Kankilic, B., Köse, S., Korkusuz, P., Timuçin, M. and Korkusuz, F. 2016. Mesenchymal stem cells and nano-bioceramics for bone regeneration. Curr. Stem Cell Res. Ther. 11: 487–93.

Kargozar, S., Baino, F., Hamzehlou, S., Hill, R.G. and Mozafari, M. 2018. Bioactive glasses: sprouting angiogenesis in tissue engineering. Trends Biotechnol. 36: 430–44.

Kelly, C.M., Wilkins, R., Gitelis, S., Hartjen, C., Tracy Watson, J. and Taek Kim, P. 2001. The use of a surgical grade calcium sulfate as a bone graft substitute: results of a multicenter trial. Clin. Orthop. Relat. Res. 382: 42–50.

Khalyfa, A., Vogt, S., Weisser, J., Grimm, G., Rechtenbach, A., Meyer, W. et al. 2007. Development of a new calcium phosphate powder-binder system for the 3D printing of patient specific implants. J. Mater. Sci. Mater. Med. 18: 909–16.

Kim, H.-L., Jung, G.-Y., Yoon, J.-H., Han, J.-S., Park, Y.-J. Kim, D.-G. et al. 2015. Preparation and characterization of nano-sized hydroxyapatite/alginate/chitosan composite scaffolds for bone tissue engineering. Mater. Sci. Eng. C 54: 20–5.

Kirsten, A., Hausmann, A., Weber, M., Fischer, J. and Fischer, H. 2015. Bioactive and thermally compatible glass coating on zirconia dental implants. J. Dent. Res. 94: 297–303.

Klammert, U., Gbureck, U., Vorndran, E., Rödiger, J., Meyer-Marcotty, P. and Kübler, A.C. 2010. 3D powder printed calcium phosphate implants for reconstruction of cranial and maxillofacial defects. J. Cranio-Maxillofacial Surg. 38: 565–70.

Kucko, N.W., Herber, R.-P., Leeuwenburgh, S.C.G. and Jansen, J.A. 2019. pp. 591–611. In: Atala, A., Lanza, R., Mikos, A.G. and Nerem, R. (eds.). Principles of Regenerative Medicine. Amsterdam: Elsevier Academic Press.

Łączka, M., Cholewa-Kowalska, K. and Osyczka, A.M. 2016. Bioactivity and osteoinductivity of glasses and glassceramics and their material determinants. Ceram. Int. 42: 14313–25.

LeGeros, R.Z. 1988. Calcium phosphate materials in restorative dentistry: A review. Adv. Dent. Res. 2: 164.

Li, J., Zhai, D., Lv, F., Yu, Q., Ma, H., Yin, J. et al. 2016. Preparation of copper-containing bioactive glass/eggshell membrane nanocomposites for improving angiogenesis, antibacterial activity and wound healing. Acta Biomater. 36: 254–66.

Liuyun, J., Yubao, L. and Chengdong, X. 2009. Preparation and biological properties of a novel composite scaffold of nano-hydroxyapatite/chitosan/carboxymethyl cellulose for bone tissue engineering. J. Biomed. Sci. 16: 65.

Mistry, S., Kundu, D., Datta, S. and Basu, D. 2011. Comparison of bioactive glass coated and hydroxyapatite coated titanium dental implants in the human jaw bone. Aust. Dent. J. 56: 68–75.

Moed, B.R., Willson Carr, S.E., Craig, J.G. and Watson, J.T. 2003. Calcium sulfate used as bone graft substitute in acetabular fracture fixation. Clin. Orthop. Relat. Res. 410: 303–9.

Mosmann, T. 1983. Rapid colorimetric assay for cellular growth and survival: Application to proliferation and cytotoxicity assays. J. Immunol. Methods 65: 55–63.

Nabiyouni, M., Brückner, T., Zhou, H., Gbureck, U. and Bhaduri, S.B. 2018. Magnesium-based bioceramics in orthopedic applications. Acta Biomater. 66: 23–43.

O'Neill, R., McCarthy, H.O., Montufar, E.B., Ginebra, M.-P., Wilson, D.I., Lennon, A. et al. 2017. Critical review: Injectability of calcium phosphate pastes and cements. Acta Biomater. 50: 1–19.

Ohtsuki, C., Kokubo, T. and Yamamuro, T. 1992. Mechanism of apatite formation on CaOSiO2P2O5 glasses in a simulated body fluid. J. Non. Cryst. Solids 143: 84–92.

Peltier, L.F., Bickell, E.Y., Lillo, R. and Thein, M.S. 1957. The use of plaster of paris to fill defects in bone. Ann. Surg. 146: 61–9.

Peltier, L.F. and Jones, R.H. 1978. Treatment of unicameral bone cysts by curettage and packing with plaster-of-paris pellets. J. Bone Joint Surg. Am. 60: 820–2.

Phillips, M.J., Darr, J.A., Luklinska, Z.B. and Rehman, I. 2003. J. Mater. Sci. Mater. Med. 14: 875–82.

Pietrzak, W.S. and Ronk, R. 2000. Calcium sulfate bone void filler: a review and a look ahead. J. Craniofac. Surg. 11: 327–33.

Planell, J.A., Best, S.M., Lacroix, D. and Merolli, A. 2009. Bone Repair Biomaterials. Cambridge, UK: Woodhead Publishing Limited, CRC Press LLC.

Rabiee, S.M., Moztarzadeh, F. and Solati-Hashjin, M. 2010. Synthesis and characterization of hydroxyapatite cement. J. Mol. Struct. 969: 172–5.

Rahaman, M.N., Day, D.E., Bal, B.S., Fu, Q., Jung, S.B., Bonewald, L.F. et al. 2010. Bioactive glass in tissue engineering. Acta Biomater. 7: 2355–73.

Ratner, B.D. 2004. Biomaterials Science: An Introduction to Materials in Medicine. 2nd ed. Amsterdam: Elsevier Academic Press.

Salinas, A.J. and Vallet-Regí, M. 2007. Evolution of ceramics with medical applications. Zeitschrift für Anorg. und Allg. Chemie 633: 1762–73.

Salinas, A.J., Esbrit, P. and Vallet-Regí, M. 2012. A tissue engineering approach based on the use of bioceramics for bone repair. Biomater. Sci. 1: 40–51.

Samavedi, S., Whittington, A.R. and Goldstein, A.S. 2013. Calcium phosphate ceramics in bone tissue engineering: A review of properties and their influence on cell behavior. Acta Biomater. 9: 8037–45.

Sanchez-Sotelo, J., Munuera, L. and Madero, R. 2000. Treatment of fractures of the distal radius with a remodellable bone cement: A prospective, randomised study using Norian SRS. J. Bone Joint Surg. Br. 82: 856–63.

Stockert, J.C., Horobin, R.W., Colombo, L.L. and Blázquez-Castro, A. 2018. Tetrazolium salts and formazan products in cell biology: Viability assessment, fluorescence imaging, and labeling perspectives. Acta Histochem. 120: 159–167.

Šupová, M. 2015. Substituted hydroxyapatites for biomedical applications: A review. Ceram. Int. 41: 9203–31.

Szcześ, A., Hołysz, L. and Chibowski, E. 2017. Advances in structural design of lipid-based nanoparticle carriers for delivery of macromolecular drugs, phytochemicals and anti-tumor agents. Adv. Colloid Interface Sci. 249: 331–45.

Szurkowska, K. and Kolmas, J. 2017. Hydroxyapatites enriched in silicon—Bioceramic materials for biomedical and pharmaceutical applications. Prog. Nat. Sci. Mater. Int. 27: 401–9.

Tanaka, T., Komaki, H., Chazono, M., Kitasato, S., Kakuta, A., Akiyama, S. et al. 2017. Basic Research and clinical application of beta-tricalcium phosphate (β-TCP). Morphologie 101: 164–72.

Tang, M., Shen, X. and Huang, H. 2010. Influence of α-calcium sulfate hemihydrate particle characteristics on the performance of calcium sulfate-based medical materials. Mater. Sci. Eng. C 30: 1107–1111.

Turner, T.M., Urban, R.M., Gitelis, S., Kuo, K.N. and Andersson, G.B. 2001. Radiographic and histologic assessment of calcium sulfate in experimental animal models and clinical use as a resorbable bone-graft substitute, a bone-graft expander, and a method for local antibiotic delivery. One Institution's Experience. J. Bone Joint Surg. Am. 83(A Suppl.): 8–18.

Unuma, H. and Matsushima, Y. 2013. Preparation of calcium phosphate cement with an improved setting behavior. J. Asian Ceram. Soc. 1: 26–9.

Vallet-Regí, M. 2001. Ceramics for medical applications. J. Chem. Soc. Dalton. Trans. 2: 97–108.

Wang, H., Li, Y., Zuo, Y., Li, J., Ma, S. and Cheng, L. 2007. Biocompatibility and osteogenesis of biomimetic nano-hydroxyapatite/polyamide composite scaffolds for bone tissue engineering. Biomaterials 28: 3338–48.

Watson, J.T. 2004. The use of an injectable bone graft substitute in tibial metaphyseal fractures. Orthopedics 27: s103–7.

Xie, L., Yu, H., Deng, Y., Yang, W., Liao, L. and Long, Q. 2016. Preparation, characterization and *in vitro* dissolution behavior of porous biphasic α/β-tricalcium phosphate bioceramics. Mater. Sci. Eng. C 59: 1007–15.

Yamada, S., Heymann, D., Bouler, J.-M. and Daculsi, G. 1997. Osteoclastic Resorption of calcium phosphate ceramics with different hydroxyapatite/beta-tricalcium phosphate ratios. Biomaterials 18: 1037–41.

Yilmaz, B. and Alshemary, A.Z. 2019. Co-doped hydroxyapatites as potential materials for biomedical applications. Z. Evis. Microchem. J. 144: 443–53.

Zandi, M., Mirzadeh, H., Mayer, C., Urch, H., Eslaminejad, M.B., Bagheri, F. et al. 2010. Biocompatibility evaluation of nano-rod hydroxyapatite/gelatin coated with nano-HAp as a novel scaffold using mesenchymal stem cells. J. Biomed. Mater. Res. 92: 1244–55.

Zhou, Y., Wu, C. and Chang, J. 2019. Bioceramics to regulate stem cells and their microenvironment for tissue regeneration. Mater. Today 24: 41–56.

5

From Conventional Approaches to Sol-gel Chemistry and Strategies for the Design of 3D Additive Manufactured Scaffolds for Craniofacial Tissue Engineering

Gloria A.,[1,*] *Russo T.,*[1] *Martorelli M.*[2] and *De Santis R.*[1]

Introduction

Over the past two decades, great interest has been focused on craniofacial tissue engineering (Langer and Vacanti 1993; Hollister 2005; Thrivikraman et al. 2017). In particular, a significant contribution was clearly due to the research and development in the field of bone augmentation, thus leading to tissue engineering as an alternative treatment option in dentistry and medicine (Bertassoni and Coelho 2015; Thrivikraman et al. 2017).

In this context, biomaterials play an important role for tissue regeneration as it is widely recognized how their properties may be suitably tailored by varying the composition and, in many cases, the architecture, in order to regulate the cell microenvironment during the tissue formation process, also controlling the rate of regeneration (Bertassoni and Coelho 2015; Thrivikraman et al. 2017).

Generally, biomaterials are used as scaffold systems which should allow cell migration, proliferation and differentiation, thus promoting new tissue formation.

With regard to craniofacial bone regeneration and augmentation, a wide range of organic and inorganic biomaterials have been considered. As an example, bioceramics such as calcium phosphate (CaP) have been used to develop 'inorganic' scaffolds, whereas synthetic and natural biopolymers have been taken into account to manufacture 'organic' scaffolds.

---

[1] Institute of Polymers, Composites and Biomaterials—National Research Council of Italy, V.le J.F. Kennedy 54–Mostra d'Oltremare Pad. 20, 80125 Naples, Italy.
[2] Department of Industrial Engineering, Fraunhofer JL IDEAS—University of Naples Federico II, P.le Tecchio 80, 80125 Naples, Italy.
* Corresponding author: antonio.gloria@cnr.it

The great effort to mimic the organic-inorganic composition and the structure of the natural bone, in which crystallites of a natural bioceramic (hydroxyapatite) reinforce the fibrils of a natural organic polymer (collagen), has led research towards the design of strong and durable biomaterials (Thrivikraman et al. 2017).

In the current chapter, an overview of conventional and recent developments in the field of craniofacial bone regeneration and augmentation will be first reported. Bioceramics, bioglasses, biopolymers and biocomposites will be described together with their peculiar features. Furthermore, starting from experimental and theoretical analysis performed on the material, a special focus will be devoted to the potential combination of sol-gel chemistry and CAD-based approach to design additively manufactured scaffolds for craniofacial tissue engineering.

Materials: Bioceramics, Bioglasses, Biopolymers and Biocomposites

Most candidate materials for bone regeneration are represented by bioceramics such as calcium phosphates (CaP), calcium carbonates, calcium sulfates, bioactive glasses and composites consisting of biodegradable polymers and bioactive inorganic materials (Gerhardt and Boccaccini 2010; Thrivikraman et al. 2017).

The ability of CaP bioceramics to stimulate bone growth is well documented as in the 1920s an effort to improve bone formation was made using an aqueous slurry of 'triple calcium phosphate' (Albee and Morrison 1920). In the late 90s, CaP materials started to receive great attention in the field of bone tissue engineering, as a consequence of their ability to rapidly promote bone formation (Langer and Vacanti 1993; Thrivikraman et al. 2017).

Many reasons have been frequently reported to justify their interesting properties: in terms of composition, the similarities to hydroxyapatite, which is the main constituent of the bone; osteoprogenitor cells are able to process and resorb them; the possibility to trigger the intracellular signaling of osteogenesis through the presence of inorganic phosphates and soluble calcium (Thrivikraman et al. 2017) derived from byproducts of CaP crystal dissolution (Thrivikraman et al. 2017).

Many commercial CaP materials are currently available for bone regeneration, generally including one or more phases of CaP in different mineral phases, processing conditions and crystal structures.

Experimental studies have also demonstrated that the specific material phase which constitutes the CaP materials strongly determines the efficacy of the osteogenesis process. A key factor which regulates osteoinduction is represented by the solubility of the CaP mineral phase (Kamakura et al. 2002; Thrivikraman et al. 2017).

The synthesis of materials with a composition similar to the mineral phase of the natural bone was a starting point towards the design of scaffolds consisting of CaP and possessing high mechanical properties (Galea et al. 2008; Thrivikraman et al. 2017). Thus, sintered hydroxyapatite (HA) or sintered β-tricalcium phosphate (β-TCP) were initially obtained by CaP graft sintering.

Anyway, scaffolds fabricated by the sintering process have low injectability and are generally too brittle for load bearing applications. In addition, the process of scaffold remodeling and osteoclast-driven biodegradation is hindered by their low solubility.

In fabricating novel CaP scaffolds for bone regeneration, a wide range of possibilities was offered by the development of bone cements consisting of CaP phases which may be formed at room temperature (i.e., calcium deficient HA, octacalcium phosphate, monetite and brushite) (Munting et al. 1993; Steffen et al. 2001; Thrivikraman et al. 2017).

Interestingly, such materials with less crystalline phases possess a much lower solubility rates and may transform into a more stable HA phase upon implantation (Suzuki et al. 1991; Constantz et al. 1998; Bohner et al. 2003; Thrivikraman et al. 2017), also showing better biological performances if compared to sintered CaP (Thrivikraman et al. 2017).

From a historical point of view, CaP materials have been processed in different forms, including particulates, solid and porous blocks, and this led to a limitation for their clinical use in dentistry, if compared to injectable materials with enhanced handling characteristics and less invasive features.

Furthermore, in comparison to soft and moldable sponges and polymers, it is often difficult to strictly control the shape and architecture for particulate materials. Thus, their applications are largely limited.

Although many efforts have been made to properly understand the mechanisms related to CaP-induced bone formation, pre-clinical results have shown interesting results (Thrivikraman et al. 2017).

In this context, it was demonstrated that similar results were obtained when using an osteoinductive TCP ceramic, an autograft scaffold and scaffolds loaded with growth factors such as rhBMP-2 or recombinant human BMP (Yuan et al. 2010; Thrivikraman et al. 2017). Moreover, the use of CaP materials results in very promising craniofacial bone regeneration, taking into consideration the interesting results obtained from clinical trials on commercially available products.

However, many analyses performed on HA, TCP and biphasic calcium phosphate have generally shown how the presence of a more soluble phase may improve bone formation, even if the ability to induce bone formation has already been displayed by several compositions and phases of CaP (Habibovic et al. 2005; Yuan et al. 2006; Thrivikraman et al. 2017). Grain size, specific surface area and microporosity of CaP are features playing a key role in osteoinduction (Hulbert et al. 1970; Klawitter and Hulbert 1971; Klein et al. 1985; Thrivikraman et al. 2017).

Methodologies involving the combination of CaP materials with pre-fabricated blood vessels and capillaries, as well as a vascularized collagen-β-tricalcium phosphate graft, would represent novel strategies for the development of advanced scaffolds (Kang et al. 2013; 2015; Thrivikraman et al. 2017).

Bio-inorganic substitution of CaP materials and bioglasses represent further interesting approaches. It is well known that the mineral phase of the natural bone is characterized by many trace elements, and a cationic substitution with Sr of Mg on CaP-based scaffolds may alter the biological and mechanical performances, as they modify the physiochemical properties of CaP (i.e., microstructure, crystallinity, solubility) (Tarafder et al. 2013; Thrivikraman et al. 2017). A study in an animal model reported the effect of SrO and MgO doping β-TCP scaffolds resulting in an increase of the early bone formation in the doped structures, in comparison to the undoped ones (Tarafder et al. 2013; Thrivikraman et al. 2017).

Like the CaP materials, bioactive glass and glass-ceramics have also been widely investigated and proposed as potential candidates for bone regeneration (Kaur et al. 2014; Thrivikraman et al. 2017). The ability of bioglasses to induce the formation of a hydroxycarbonate apatite layer, when exposed to body fluids, has been discussed in the literature (Gerhardt and Boccaccini 2010; Thrivikraman et al. 2017). Even though bioactive glasses and glass ceramics are commercially available in particulate form, their use for load-bearing devices is strongly limited by their low strength and low fracture toughness.

Taking into account the potential to promote the osteogenic cell response, bioactive glasses are generally employed as fillers or coatings for polymer-based scaffolds. BonAlive® (BonAlive Biomaterials Ltda), Perioglass® (NovaBone), GlassBone® (Noraker) and Vitoss® (Stryker) are examples of current materials for bone regeneration available for clinical applications. However, despite the interesting results obtained from these kinds of materials in comparison to the typical CaP sintered ceramics, studies are still ongoing to assess their long term clinical performance and their efficiency for craniofacial applications (Thrivikraman et al. 2017).

With regard to biopolymers, gels and hydrogels have been studied for cell and growth factor delivery. The attention on natural polymers (i.e., gelatin, collagen, silk, chitosan) has been focused with the aim to mimic the chemical composition, structure and biochemical properties of the organic matrix of the natural bone, considering their low immunogenic properties, as well as their ability to stimulate cell response and function, and to support tissue remodeling (Russo et al. 2013; Thrivikraman et al. 2017).

In the field of tissue regeneration, collagen represents an immediate choice because of its presence in the extracellular matrix (ECM) of vertebrates as the most abundant protein.

The low mechanical properties of collagen limit its applications in the case of bone regeneration. In addition, it cannot be mass produced and risks of allergic reactions and infections may be related to the use of animal tissues, even if the use of recombinant collagen may reduce these drawbacks (Lynn et al. 2004; Adachi et al. 2006; Thrivikraman et al. 2017). Alginate, chitosan and agarose are natural polysaccharides which can be considered to develop polymeric scaffolds. The presence of positively charged amino groups enables the interactions with DNA, proteins, lipids and cell membranes.

In the case of chitosan, cytokines and growth factors may be stimulated by its cationic nature which favors the interactions with proteoglycans and glycosaminoglycans, playing an important role in the tissue regeneration process (Costa-Pinto et al. 2011; Thrivikraman et al. 2017). Silk fibroin has also provided interesting results in terms of cell proliferation and *in vivo* bone formation (Meechaisue et al. 2007; Zhang et al. 2010; Uebersax et al. 2013; Thrivikraman et al. 2017). The low mechanical strength obviously limits its applications in the biomedical field (Thrivikraman et al. 2017).

The Demineralized Bone Matrix (DBM) is a further kind of bone substitute which has been clinically studied (Sawkins et al. 2013; Kolk et al. 2012; Thrivikraman et al. 2017).

It mainly consists of type I collagen (about 90%) with several growth factors (i.e., osteopontin, bone sialoprotein, IGF1, BMPs, as well as some inorganic phosphates, calcium-based particles and cellular debris (Gruskin et al. 2012; Holt and Grainger 2012; Thrivikraman et al. 2017).

Even if different DBM products are available in the market, they show osteoinductive and osteoconductive properties, but differ from each other in terms of sterilization methods, process conditions, donor specifications (Kinney et al. 2010; Thrivikraman et al. 2017). As for the osteoinductive ability, different DBM products generally provide different results. Surgical handling also affects the efficacy and, for this reason, DBM is often mixed with binders (i.e., hyaluronan, glycerol, bovine collagen with sodium alginate, gelatin, calcium sulfate, lecithin, carboxymethyl cellulose).

DBM may be used as a bone graft extender as it possesses low mechanical strength. It has also been considered better than standard autografts due to its ability to revascularize and locally induce the release of agents and growth factors promoting bone formation.

Regarding synthetic polymers, polylactic acid (PLA), polyglycolide (PGA), poly(lactide-co-glycolide) (PLGA), poly(ε-caprolactone) (PCL), polyhydroxyalkanoates (PHAs) and (propylene fumarate) (PPF). Such polymers have been developed to tailor the mechanical properties of the scaffolds by varying the degrees of cross-linking and concentration or by means of copolymerization strategies (Thrivikraman et al. 2017).

In particular, over the past years a great deal of attention has been focused on PCL, which is a biodegradable aliphatic polyester (Gloria et al. 2010; 2012; 2019; Thrivikraman et al. 2017).

Generally, neat synthetic polymers also do not properly satisfy the mechanical requirements for bone tissue engineering due to their flexibility and weakness. Thus, polymer-based 'biocomposites', consisting of biopolymers reinforced with inorganic fillers, have been considered an intriguing alternative for hard-tissue engineering.

If compared to neat polymers, polymer-based composites should possess enhanced mechanical properties and better structural integrity and flexibility than brittle ceramics. Porous composite scaffolds with improved bioactivity and tailored resorption rate were obtained by properly combining polymers and inorganic fillers, showing a combination of the best features of their constituents, and, in many cases, further properties which the single constituents generally do not show (Gloria et al. 2010).

Sol-gel Chemistry and CAD-based Approach Towards the Design of Additive Manufactured Scaffolds: An Intriguing Strategy

In the field of tissue regeneration, the design of multifunctional substrates and scaffolds is challenging.

Several kinds of polymers and polymer-based composites have been considered together with different approaches to enhance their biological and mechanical features (Gloria et al. 2010; Gloria et al. 2013; De Santis et al. 2013; De Santis et al. 2015; Thrivikraman et al. 2017).

In designing substrates and scaffolds for craniofacial tissue engineering, mechanical performances play a key role, as higher strength and higher elastic modulus are clearly required, if compared to the case of soft tissue regeneration.

Pore size, geometry and interconnectivity influence the functional properties of the 3D porous scaffolds and, consequently, the tissue regeneration process (Hollister 2005; Gloria et al. 2012; De Santis et al. 2015).

In general, scaffold manufacturing processes are divided into two groups, 'conventional' and 'advanced' methods (Gloria et al. 2012; De Santis et al. 2015; Martorelli et al. 2016; Lanzotti et al. 2018). In the field of tissue engineering, different from conventional methods, the introduction of additive manufacturing techniques as advanced fabrication methods, based on Computer-Aided Design (CAD)/Computer-Aided Manufacturing (CAM), and the reverse engineering approach have allowed the potential to design customized scaffolds characterized by complex shapes, reproducible architectures and morphology, as well as by tailored and improved mechanical and mass transport properties (De Santis et al. 2013; De Santis et al. 2015; Domingos et al. 2017; Gloria et al. 2019).

Technical considerations related to the stress transfer mechanism and stress concentration, as well as the design criteria for composite materials provide the possibility to develop devices where a polymer matrix may be reinforced with an optimized amount of micro/nano-particles, avoiding weakness in the structure.

As an example, the effect of the inclusion of HA nanoparticles, morphological and structural features (i.e., lay-down pattern, filament distance, pore size and geometry) was investigated in a study on nanocomposite PCL/HA scaffolds obtained by additive manufacturing (fused deposition modeling/3D fiber deposition technique) in order to define a strategy for the design of customized structures for craniofacial tissue regeneration (De Santis et al. 2010). The obtained results demonstrated that the presence of HA nanoparticles enhanced the mechanical properties and biological performances of the neat PCL scaffolds. Taking into account the experimental findings, together with the reverse engineering approach and additive manufacturing, the potential to design customized scaffolds for mandibular defect regeneration (i.e., symphysis and ramus) was also shown (De Santis et al. 2010) (Fig. 5.1).

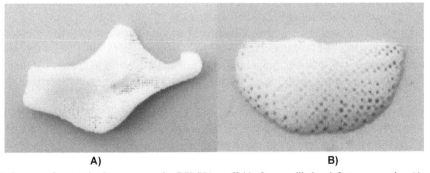

A)   B)

Figure 5.1. Images of customized nanocomposite PCL/HA scaffolds for mandibular defect regeneration (A-ramus and B-symphysis) fabricated by additive manufacturing (fused deposition modeling/3D fibre deposition technique). The image was adapted from De Santis et al. 2010.

However, in this context, organic–inorganic hybrids have also been developed as biomaterials with improved properties (Catauro et al. 2010; 2017).

Starting from the hydrolysis and polycondensation of metal alkoxides $(M(OR)_x$, with M = Si, Sn, Zr, Ti, Al, Mo, V, W, Ce, and so forth) (Catauro et al. 2010; 2017; Russo et al. 2010), the sol–gel method can be used to synthesize most hybrid organic–inorganic materials, trying to combine the best features of polymers and inorganic materials. Among the advantages, such a process can be carried out at low temperatures (i.e., room temperature) (Catauro et al. 2010; 2017).

The introduction of polymers into inorganic networks led to the development of hybrid materials showing several morphological and physical features (Catauro et al. 2010; 2017).

Based on the nature of the chemical interactions between the organic and inorganic phases, organic–inorganic hybrids are generally classified into Class I and Class II, the latter being characterized by stronger bonds (covalent or ionic-covalent bonds) (Catauro et al. 2010; 2017).

Among synthetic polymers, PCL has been widely proposed for biomedical applications, such as tissue engineering and drug delivery, the bioactivity of TiO_2 and ZrO_2 glasses was already demonstrated by the formation of a bone-like apatite layer on the surfaces (Catauro et al. 2010; 2017).

Consequently, PCL-based organic–inorganic hybrids (PCL/TiO_2 and PCL/ZrO_2) were synthesized by the sol–gel method, incorporating PCL into the network via the hydrogen bonds between the hydroxyl groups of the inorganic phase and the carboxylic groups of the polymer.

The bioactivity of PCL/TiO_2, PCL/ZrO_2 and other organic–inorganic hybrids was assessed (Catauro et al. 2010; 2017).

The properties of PCL were improved by embedding the bioactive PCL/ZrO_2 and PCL/TiO_2 organic–inorganic hybrid microparticles (Russo et al. 2010). Several compositions were used for the organic–inorganic hybrid microparticles, but, in terms of mechanical and biological performances, small punch tests and cell viability/proliferation analyses provided better results for substrates consisting of PCL loaded with Class I Zr2 (PCL = 12, ZrO_2 = 88 wt%) and Ti2 (PCL = 12, TiO_2 = 88 wt%) hybrid microparticles, in comparison to the other particle composition (Russo et al. 2010).

Accordingly, 3D additive manufactured composite scaffolds for hard-tissue engineering were then fabricated and analyzed, first benefiting from the optimization of 2D substrates (De Santis et al. 2013).

As a first step, composite pellets consisting of PCL and organic–inorganic hybrid particles were developed. Successively, the composite pellets were processed by additive manufacturing (fused deposition modeling/3D fiber deposition technique). PCL pellets (M_w = 65,000) were dissolved in tetrahydrofuran (THF). Organic–inorganic hybrid microparticles (Class I Ti2 and Zr2, diameter < 38 μm) and ethanol were then added to the solution during stirring. A PCL/filler weight ratio (wt/wt) of 80/20 was considered. An ultrasonic bath was also used to optimize the dispersion of the hybrid particles (De Santis et al. 2013).

Cylindrical scaffolds (diameter of 6 mm, height of 3 mm) were built by extruding and depositing the material according to the selected lay-down pattern (0°/0°/90°/90°).

A 3D Bioplotter (Envisiontec GmbH, Gladbeck, Germany) was used and the composite pellets were placed in a stainless steel syringe. The pellets were then heated to 140°C and a suitable pressure was applied. A nozzle with an inner diameter of 500 μm and a deposition speed of 30 mm/min were used to process the material. The additive manufactured scaffolds were characterized by the fiber diameter, fiber spacing and layer thickness.

In terms of biological performance, with regard to composite scaffolds consisting of PCL loaded with Ti2 and Zr2, human Bone Marrow-derived Mesenchymal Stem Cells (BMSCs) and Dental Pulp Stem Cells (DPSCs) were viable over time (De Santis et al. 2013).

On the other hand, results from compression tests performed on the 3D additive manufactured scaffolds showed stress-strain curves similar to those found in literature (De Santis et al. 2013) (Fig. 5.2).

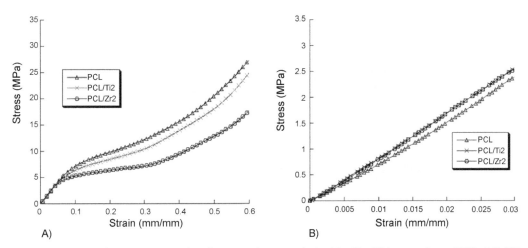

Figure 5.2. Compression tests: (A) examples of stress–strain curves obtained for 3D additive manufactured PCL, PCL/Ti2, and PCL/Zr2 scaffolds tested at a rate of 1 mm/min up to a strain of 0.6 mm/mm; (B) initial linear region of stress–strain curves up to a strain of 0.03 mm/mm. The image was adapted from De Santis et al. 2013.

The composite scaffolds displayed a mean value of compressive modulus (90 MPa) which was greater than that achieved for the 3D PCL structures (79 MPa), even if at a strain of 0.6 mm/mm the latter provided a higher maximum stress especially in comparison to the PCL/Zr2 (De Santis et al. 2013).

Although the small punch test may be considered a useful test method for the evaluation of mechanical properties, if only small quantities of material are available, this test method cannot be used to calculate the Young's modulus. As consequence, a numerical simulation is usually needed (De Santis et al. 2013; Maietta et al. 2018). For this reason, Finite Element Analysis (FEA) was carried out on 2D substrates (PCL loaded with sol–gel-synthesized organic-inorganic hybrid fillers) (De Santis et al. 2013; Maietta et al. 2018). The adopted design strategy provided the possibility to correlate the small punch test data and the Young's modulus of the materials.

Specifically, the theoretical analysis involved different values of Young's modulus (from 200 to 5000 MPa) and disk specimens. The Poisson's ratio was set to 0.40.

For the different investigated composites, FEA provided similar findings in terms of the displacement contour plot when different loads were applied according to the Young's modulus of each specimen (200–5000 MPa) to achieve a final displacement of 0.2 mm (Fig. 5.3) (Maietta et al. 2018).

FEA was also used to determine the force-displacement curves (Fig. 5.4A) for all the materials, together with a normalized force-displacement curve (Fig. 5.4B). The normalized curve was determined dividing the values of the force by the Young's modulus of the material.

The computational findings together with the obtained equation for the curve represent interesting results, as they provided the possibility to evaluate the Young's modulus of a material when subjected to the small punch test.

A modulus of 193 MPa was calculated for PCL (Maietta et al. 2018), which was higher than that found in a previous theoretical analysis (144 MPa) (De Santis et al. 2013). However, the obtained values were still lower in comparison to the neat PCL (i.e., 343.9–571.5 MPa). This was probably due to the intrinsic porosity of the substrates as they were manufactured via molding and solvent casting techniques (Russo et al. 2010; De Santis et al. 2013; Maietta et al. 2018).

Furthermore, Young's moduli of 378 MPa and 415 MPa were computed for PCL/Ti2 and PCL/Zr2, respectively (Maietta et al. 2018). These values were also higher than those evaluated using a previous theoretical model (282 MPa and 310 MPa) (De Santis et al. 2013). In comparison to a previous theoretical analysis (De Santis et al. 2013), as higher values of the Young's modulus were

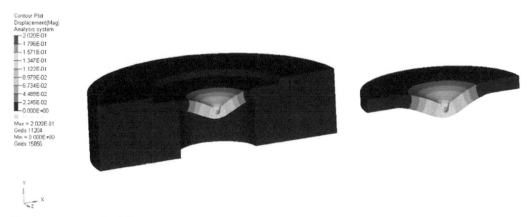

Figure 5.3. An example of displacement contour plot for a composite disk obtained from FEA. The image was adapted from Maietta et al. 2018.

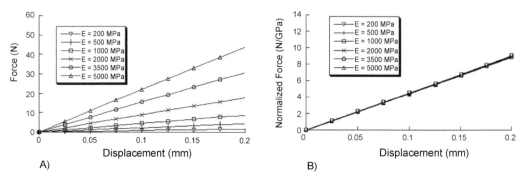

A)   B)

Figure 5.4. Force-displacement curves (A) and normalized force-displacement curve (B). The image was adapted from Maietta et al. 2018.

found for PCL, PCL/Ti2 and PCL/Zr2, the results would suggest a better approximation of the experimental data (Maietta et al. 2018).

Young's moduli reported above were significantly higher than those in the experimental study on the design of 3D additive manufactured scaffolds (De Santis et al. 2013), where mean values of compressive modulus of 90 MPa and 79 MPa were obtained for composite (PCL loaded with sol-gel synthesized organic-inorganic hybrid fillers—PCL/Ti2 and PCL/Zr2) and neat PCL structures, respectively. These differences are related to the controlled architectural features and porosity of the 3D additive manufactured scaffolds, where 'apparent' stress and strain are generally defined. Accordingly, considering a methodology already described for 3D porous scaffolds with controlled architectures (Gloria et al. 2012), the apparent stress (σ) and strain (ε) can be calculated as $\sigma = F/A_0$ and $\varepsilon = \Delta H/H_0$, when the values of the Force (F) measured by the load cell, the apparent initial cross-sectional area (A_0), the initial height (H_0) and the height variation (ΔH) of the scaffold are known.

For this reason, it is worth noting that although in the case of porous structures the compressive modulus provides information on the stiffness of the structures, it does not represent the elastic modulus, which must be considered as an intrinsic property of the material.

The obtained correlation would be an important design step in the determination of the Young's modulus of the material from small punch test data, as well as a starting point for the design of 3D additive manufactured scaffolds.

Finally, the strategy based on the experimental and theoretical analysis performed on the material may be considered as a promising step towards the development of 3D customized advanced scaffolds for craniofacial tissue regeneration, combining the reverse engineering approach with the recent advances in the design for additive manufacturing and sol-gel chemistry.

Conclusions

The understanding of physiology, complex processes, molecular pathways and remodeling features are crucial for the regeneration of craniofacial tissues (Tevlin et al. 2014; Thrivikraman et al. 2017).

Even though it results in difficulty to reproduce the nature and structure of complex tissues, current scientific and technological advances (i.e., sol-gel chemistry, CAD-based approach, additive manufacturing and reverse engineering techniques) provide a great potential for the design 3D customized scaffolds for craniofacial tissue regeneration, which are able to guide systemic and local biological processes.

Materials, geometry, architectural features and porosity, together with the ability to release specific biomolecules and to tailor drug release kinetics, clearly play a key role in the design of advanced and multifunctional scaffolds for craniofacial tissue regeneration. As an example, the possibility of combining growth factors or stem cell-based therapy with 3D porous scaffolds characterized by controlled architectural features could become one of the most promising strategies (Tevlin et al. 2014; Thrivikraman et al. 2017).

However, in this context 3D cell printing technology would also be an interesting approach enabling researchers to suspend and position cells embedded in materials such as hydrogels (Tao et al. 2019). 3D bioprinting could allow in obtaining specific mechanical properties, cell interactions and desired distribution of growth factors. The possibility of printing blood capillaries has been already reported and cell printing for craniofacial tissue regeneration would seem feasible, even if many studies are still ongoing (Tao et al. 2019).

Accordingly, although in recent years much progress has already been made in terms of interdisciplinary collaboration, a continuous and collaborative research among engineers, chemists, biologists and surgeons would be necessary.

References

Adachi, T., Tomita, M., Shimizua, K., Ogawa, S. and Yoshizato, K. 2006. Generation of hybrid transgenic silkworms that express *Bombyx mori* prolylhydroxylase alpha-subunits and human collagens in posterior silk glands: Production of cocoons that contained collagens with hydroxylated proline residues. J. Biotechnol. 126(2): 205–19.

Albee, F.H. and Morrison, H.F. 1920. Studies in bone growth triple calcium phosphate as a stimulus to osteogenesis. Ann. Surg. 71: 32–39.

Bertassoni, L.E. and Coelho, P.G. 2015. Engineering Mineralized and Load Bearing Tissues. Springer, New York.

Bohner, M., Theiss, F., Apelt, D., Hirsiger, W., Houriet, R., Rizzoli, G. et al. 2003. Compositional changes of a dicalcium phosphate dihydrate cement after implantation in sheep. Biomaterials 24(20): 3463–3474.

Catauro, M., Verardi, D., Melisi, D., Belotti, F. and Mustarelli, P. 2010. Novel sol-gel organic-inorganic hybrid materials for drug delivery. J. Appl. Biomater. Biomech. 8: 42–51.

Catauro, M., Tranquillo, E., Illiano, M., Sapio, L., Spina, A. and Naviglio, S. 2017. The influence of the polymer amount on the biological properties of PCL/ZrO$_2$ hybrid materials synthesized via sol-gel technique. Materials 10: 1186.

Constantz, B.R., Barr, B.M., Ison, I.C., Fulmer, M.T., Baker, J., McKinney, L. et al. 1998. Histological, chemical, and crystallographic analysis of four calcium phosphate cements in different rabbit osseous sites. J. Biomed. Mater Res. 43(4): 451–61.

Costa-Pinto, A.R., Reis, R.L. and Neves, N.M. 2011. Scaffolds based bone tissue engineering: The role of chitosan. Tissue Eng. Part B Rev. 17(5): 331–4.

De Santis, R., Gloria, A. and Ambrosio, L. 2010. Materials and technologies for craniofacial tissue repair and regeneration. Top. Med. 16(1-4).

De Santis, R., Gloria, A., Russo, T., D'Amora, U., D'Antò, V., Bollino, F. et al. 2013. Advanced composites for hard-tissue engineering based on PCL/organic-inorganic hybrid fillers: From the design of 2D substrates to 3D rapid prototyped scaffolds. Polym. Compos. 34: 1413–141.

De Santis, R., D'Amora, U., Russo, T., Ronca, A., Gloria, A. and Ambrosio, L. 2015. 3D fibre deposition and stereolithography techniques for the design of multifunctional nanocomposite magnetic scaffolds. J. Mater. Sci. Mater. Med. 26(10): 250. DOI: 10.1007/s10856-015-5582-4.

Domingos, M., Gloria, A., Coelho, J., Bartolo, P. and Ciurana, J. 2017. Three-dimensional printed bone scaffolds: The role of nano/micro-hydroxyapatite particles on the adhesion and differentiation of human mesenchymal stem cells. Proc. Inst. Mech. Eng. H 231: 555–564.

Galea, L.G., Bohner, M., Lemaître, J., Kohler, T. and Müller, R. 2008. Bone substitute: Transforming beta-tricalcium phosphate porous scaffolds into monetite. Biomaterials 29(24-25): 3400–7.

Gerhardt, L.-C. and Boccaccini, A.R. 2010. Bioactive glass and glass-ceramic scaffolds for bone tissue engineering. Materials 3(7): 3867.

Gloria, A., De Santis, R. and Ambrosio, L. 2010. Polymer-based composite scaffolds for tissue engineering. J. Appl. Biomater. Biomech. 8(2): 57–67.

Gloria, A., Causa, F., Russo, T., Battista, E., Della Moglie, R., Zeppetelli, S. et al. 2012. Three-dimensional poly(ε-caprolactone) bioactive scaffolds with controlled structural and surface properties. Biomacromolecules 13(11): 3510–21.

Gloria, A., Russo, T., D'Amora, U., Zeppetelli, S., D'Alessandro, T., Sandri, M. et al. 2013. Magnetic poly(ε-caprolactone)/iron-doped hydroxyapatite nanocomposite substrates for advanced bone tissue engineering. J. R. Soc. Interface 10: 20120833.

Gloria, A., Frydman, B., Lamas, M.L., Serra, A.C., Martorelli, M., Coelho, J.F.J. et al. 2019. The influence of poly(ester amide) on the structural and functional features of 3D additive manufactured poly(ε-caprolactone) scaffolds. Mater Sci. Eng. C Mater. Biol. Appl. 98: 994–1004.

Gruskin, E., Doll, B.A., Futrell, F.W., Schmitz, J.P. and Hollinger, J.O. 2012. Demineralized bone matrix in bone repair: History and use. Adv. Drug Deliv. Rev. 64(12): 1063–77.

Habibovic, P., Yuan, H., van der Valk, C.M., Meijer, G., van Blitterswijk, C.A. and de Groot, K. 2005. 3D microenvironment as essential element for osteoinduction by biomaterials. Biomaterials 26(17): 3565–3575.

Hollister, S.J. 2005. Porous scaffold design for tissue engineering. Nat. Mater. 4(7): 518–24.

Holt, D.J. and Grainger, D.W. 2012. Demineralized bone matrix as a vehicle for delivering endogenous and exogenous therapeutics in bone repair. Adv. Drug Deliv. Rev. 64(12): 1123–8.

Hulbert, S.F., Young, F.A., Mathews, R.S., Klawitter, J.J., Talbert, C.D. and Stelling, F.H. 1970. Potential of ceramic materials as permanently implantable skeletal prostheses. J. Biomed. Mater. Res. 4(3): 433–456.

Kamakura, S., Sasano, Y., Shimizu, T., Hatori, K., Suzuki, O., Kagayama, M. et al. 2002. Implanted octacalcium phosphate is more resorbable than beta-tricalcium phosphate and hydroxyapatite. J. Biomed. Mater. Res. 59(1): 29–34.

Kang, Y., Kim, S., Fahrenholtz, M., Khademhosseini, A. and Yang, Y. 2013. Osteogenic and angiogenic potentials of monocultured and co-cultured human-bone marrow-derived mesenchymal stem cells and human-umbilical-vein endothelial cells on three-dimensional porous beta-tricalcium phosphate scaffold. Acta Biomater. 9(1): 4906–15.

Kang, Y., Mochizuki, N., Khademhosseini, A., Fukuda, J. and Yang, Y. 2015. Engineering a vascularized collagen-β-tricalcium phosphate graft using an electrochemical approach. Acta Biomater. 11: 449–458.

Kasten, P., Luginbühl, R., van Griensven, M., Barkhausen, T., Krettek, C., Bohner, M. et al. 2003. Comparison of human bone marrow stromal cells seeded on calcium-deficient hydroxyapatite, β-tricalcium phosphate and demineralized bone matrix. Biomaterials 24(15): 2593–2603.

Kaur, G., Pandey, O.P., Singh, K., Homa, D., Scott, B. and Pickrell, G. 2014. A review of bioactive glasses: Their structure, properties, fabrication and apatite formation. J. Biomed. Mater Res. A 102(1): 254–74.

Kinney, R.C., Ziran, B.H., Hirshorn, K., Schlatterer, D. and Ganey, T. 2010. Demineralized bone matrix for fracture healing: Fact or fiction? J. Orthop. Trauma 24(Suppl. 1): S52–5.

Klawitter, J.J. and Hulbert, S.F. 1971. Application of porous ceramics for the attachment of load bearing internal orthopedic applications. J. Biomed. Mater. Res. 5(6): 161–229.

Klein, C.P.A.T. et al. 1985. Interaction of biodegradable β-whitlockite ceramics with bone tissue: An *in vivo* study. Biomaterials 6(3): 189–192.

Kolk, A., Handschel, J., Drescher, W., Rothamel, D., Kloss, F., Blessmann, M. et al. 2012. Current trends and future perspectives of bone substitute materials—from space holders to innovative biomaterials. J. Craniomaxillofac. Surg. 40(8): 706–18.

Langer, R. and Vacanti, J.P. 1993. Tissue engineering. Science 260(5110): 920–6.

Lanzotti, A., Martorelli, M., Russo, T. and Gloria, A. 2018. Design of additively manufactured lattice structures for tissue regeneration. Mater. Sci. Forum 941: 2154–2159.

Lynn, A.K., Yannas, I.V. and Bonfield, W. 2004. Antigenicity and immunogenicity of collagen. J. Biomed. Mater. Res. B Appl. Biomater. 71(2): 343–54.

Maietta, S., Russo, T., Santis, R., Ronca, D., Riccardi, F., Catauro, M. et al. 2018. Further theoretical insight into the mechanical properties of polycaprolactone loaded with organic-inorganic hybrid fillers. Materials 11(2): pii: E312. doi: 10.3390/ma11020312.

Martorelli, M., Maietta, S., Gloria, A., De Santis, R., Pei, E. and Lanzotti, A. 2016. Design and analysis of 3D customized models of a human mandible. Procedia CIRP 49: 199–202.

Meechaisue, C., Wutticharoenmongkol, P., Waraput, R., Huangjing, T., Ketbumrung, N., Pavasant, P. et al. 2007. Preparation of electrospun silk fibroin fiber mats as bone scaffolds: a preliminary study. Biomed. Mater. 2(3): 181–8.

Munting, E., Mirtchi, A.A. and Lemaitre, J. 1993. Bone repair of defects filled with a phosphocalcic hydraulic cement: an *in vivo* study. J. Mater. Sci. Mater. Med. 4(3): 337–344.

Russo, T., Gloria, A., D'Antò, V., D'Amora, U., Ametrano, G., Bollino, F. et al. 2010. Poly(ε-caprolactone) reinforced with sol-gel synthesized organic-inorganic hybrid fillers as composite substrates for tissue engineering. J. Appl. Biomater. Biomech. 8(3): 146–152.

Russo, T., D'Amora, U., Gloria, A., Tunesi, M., Sandri, M. and Rodilossi, S. 2013. Systematic analysis of injectable materials and 3D rapid prototyped magnetic scaffolds: From CNS applications to soft and hard tissue repair/regeneration. Procedia Eng. 59: 233–239.

Sawkins, M.J., Bowen, W., Dhadda, P., Markides, H., Sidney, L.E., Taylor, A.J. et al. 2013. Hydrogels derived from demineralized and decellularized bone extracellular matrix. Acta Biomater. 9(8): 7865–7873.

Steffen, T., Stoll, T., Arvinte, T. and Schenk, R.K. 2001. Porous tricalcium phosphate and transforming growth factor used for anterior spine surgery. Eur. Spine J. 10(2): S132–S140.

Suzuki, O., Nakamura, M., Miyasaka, Y., Kagayama, M. and Sakurai, M. 1991. Bone formation on synthetic precursors of hydroxyapatite. Tohoku J. Exp. Med. 164(1): 37–50.

Tao, O., Kort-Mascort, J., Lin, Y., Pham, H.M., Charbonneau, A.M. and ElKashty, O.A. 2019. The applications of 3D printing for craniofacial tissue engineering. Micromachines 10: 480.

Tarafder, S., Davies, N.M., Bandyopadhyay, A. and Bose, S. 2013. 3D printed tricalcium phosphate scaffolds: Effect of SrO and MgO doping on *in vivo* osteogenesis in a rat distal femoral defect model. Biomater. Sci. 1(12): 1250–1259.

Tevlin, R., McArdle, A., Atashroo, D., Walmsley, G.G., Senarath-Yapa, K. and Zielins, E.R. 2014. Biomaterials for craniofacial bone engineering. J. Dent. Res. 93(12): 1187–1195.

Thrivikraman, G., Athirasala, A., Twohig, C., Kumar Boda, S. and Bertasson, L.E. 2017. Biomaterials for craniofacial bone regeneration. Dent. Clin. North Am. 61(4): 835–856.

Uebersax, L., Apfel, T., Nuss, K.M., Vogt, R., Kim, H.Y., Meinel, L. et al. 2013. Biocompatibility and osteoconduction of macroporous silk fibroin implants in cortical defects in sheep. Eur. J. Pharm. Biopharm. 85(1): 107–18.

Yuan, H., van Blitterswijk, C.A., de Groot, K. and de Bruijn, J.D. 2006. Cross-species comparison of ectopic bone formation in biphasic calcium phosphate (BCP) and hydroxyapatite (HA) scaffolds. Tissue Eng. 12(6): 1607–15.

Yuan, H., Fernandes, H., Habibovic, P., de Boer, J., Barradas, A.M.C., de Ruiter, A. et al. 2010. Osteoinductive ceramics as a synthetic alternative to autologous bone grafting. Proc. Natl. Acad. Sci. 107(31): 13614–13619.

Zhang, Y., Wu, C., Friis, T. and Xiao, Y. 2010. The osteogenic properties of CaP/silk composite scaffolds. Biomaterials 31(10): 2848–56.

6

Mesenchymal Stem Cells from Dental Tissues

Febe Carolina Vázquez Vázquez,[1] *Jael Adrián Vergara-Lope Núñez,*[1]
Juan José Montesinos[2] and *Patricia González-Alva*[1,*]

Introduction

Stem cells biology has become a relevant field for the understanding of tissue regeneration and implementation of regenerative medicine. The ongoing research on stem cells continues to advance the knowledge on the development of living organisms, and also on the processes of how healthy cells replace damaged cells in adult organisms.

Moreover, most of the knowledge regarding Mesenchymal Stem Cells (MSCs) comes from bone marrow cells. In this chapter, we will define the nomenclature for mesenchymal stem cells (MSCs), their morphology and immunophenotype, the biological markers used to identify them, the plasticity of these cells in forming different types of cells and tissues, and their high proliferative ability.

We will outline the current understanding of the differences between Bone Marrow Mesenchymal Stem Cells (BM-MSCs), and Dental Mesenchymal Stem Cells (D-MSCs). Since their isolation, D-MSCs properties have been studied for different applications including dental therapy, craniofacial regeneration and also for immunomodulatory capacity, these features will be discussed in the present chapter.

All stem cells, regardless of their source, share general properties. One is that they are unspecialized cells capable of self-renewing through cell division, even after long periods of inactivity. The second property is that, under certain experimental conditions, they can be induced to become a specific type of cell or tissue (multipotency) (Dominici et al. 2006; Sedgley and Botero 2012).

[1] Tissue Bioengineering Laboratory, Division of Graduate Studies and Research, Faculty of Dentistry, National Autonomous University of Mexico (UNAM), Circuito Exterior s/n. Col. Copilco el Alto, Alcaldía de Coyoacán, C.P. 04510, CDMX, México.
Emails: fcarolina.vazquez@gmail.com; adrianvelonu@gmail.com
[2] Mesenchymal Stem Cells Laboratory, Oncology Research Unit, Oncology Hospital, National Medical Center, Mexican Institute of Social Security. Avenida Cuauhtémoc 330. Col. Doctores. C.P. 06720 Ciudad de México, México.
Email: montesinosster@gmail.com
* Corresponding author: pgonzalezalva@comunidad.unam.mx, goap.unam@gmail.com

The essential property of stem cells for self-renewal, i.e., their ability to go through numerous cycles of cell division while maintaining an undifferentiated state represents opportunities for the treatment of diseases such as diabetes and heart diseases (Tatullo et al. 2017; Sedgley and Botero 2012).

Despite the general recognition of stem cell properties, the definition of their characteristics is inconsistent among investigators (Dominici et al. 2006). Because, many laboratories have developed different methodologies to isolate and expand MSCs, which results in different findings.

In this regard, the International Society for Cellular Therapy (ISCT) has stated that "*multipotent mesenchymal stromal cells (MSCs) is the currently recommended designation for plastic-adherent cells isolated from bone marrow and other tissues that have often been labeled mesenchymal stem cell*" (Zuk et al. 2001; Dominici et al. 2006).

Also, in 2006, the ISCT proposed the minimum characterization criteria for human MSCs. Other than their propensity for adherence to plastic when maintained under standard culture conditions, MSCs must be able to differentiate into osteoblasts, adipocytes and chondroblasts *in vitro* differentiating conditions (Yildirim et al. 2016).

Additionally, $\geq 95\%$ of the MSCs population must positively express a cluster of differentiation superficial antigenic markers of hematopoietic differentiation called CDs (for cluster of differentiation) such as CD105 (endoglin), CD73 (ecto-5'-nuclease), and CD90 (Thy1), as measured by flow cytometry. At the same time, these cells must lack expression (52% positive) of CD45, CD34, CD14 or CD11b, C79α or CD19 and HLA-DR (Dominici et al. 2006).

Stem cells can either be embryonic or adult (postnatal); however, a significant challenge in regenerative cell therapy is to find the most appropriate cell source (Sedgley and Botero 2012; Mason et al. 2014).

The discovery and characterization of BM-MSCs, require the search for more accessible MSCs, as a consequence, this has driven interest in dental tissues, which are rich sources of stem cells (Sedgley and Botero 2012).

The present chapter summarizes the properties of different stem cells from dental sources, called D-MSCs populations, such as defining the cluster of differentiation marker of each population of D-MSCs, *in vitro* multipotency, population doubling and immunomodulatory properties. The D-MSCs were also grouped according to their position in the tooth/dental tissue.

Moreover some ethical questions raised by recent advances in biological research regarding D-MSCs will included here.

Dental Mesenchymal Stem Cells (D-MSCs)

The discovery and characterization of BM-MSCs led to the characterization of different MSCs populations from other tissues, all based on the standard criteria for BM-MSCs (Liu et al. 2015).

Sources of MSCs, other than BM-MSCs, include a variety of tissues, such as umbilical cord blood, synovium, liver, adipose tissue, lungs, amniotic fluid, tendons, placenta, skin and breast milk (Fig. 6.1) (Liu et al. 2015; Miura et al. 2003).

As for the maxillocraniofacial complex, the search for MSC-like cells led to the discovery of a unique population of MSCs from human dental tissues (Liu et al. 2015). Dental Mesenchymal Stem Cells (D-MSCs) have a variety of clinical benefits, and are expected to be applied in different areas of regenerative medicine.

Next the different sources of D-MSCs and their characteristics will be described, based on their position in the tooth/dental tissue.

Dental Pulp Stem Cells (DPSC)

At the beginning of 2000, several human dental stem/progenitor cells were isolated and characterized. The D-MSCs populations come from a variety of dental tissues; however, they share common

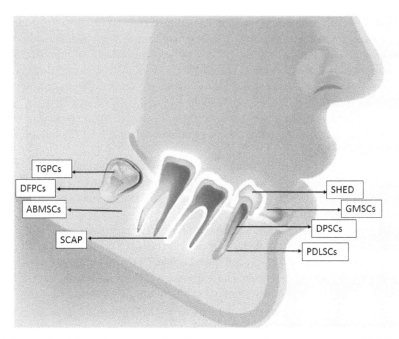

Figure 6.1. Human dental tissue-derived mesenchymal stem cells come from different sources. AB-MSCs, Alveolar Bone-derived Mesenchymal Stem Cells; DFPCs, Dental Follicle Progenitor Cells; DPSCs, Dental Pulp Stem Cells; G-MSCs, Gingiva-derived MSCs; PDLSCs, periodontal ligament stem cells; SCAP, Stem Cells from the Apical Papilla; SHED, Stem Cells from Exfoliated Deciduous teeth; TGPCs, Tooth Germ Progenitor Cells.

characteristics, such as the capacity for self-renewal and the ability to differentiate into at least three distinct lineages (Fig. 6.1) (Sedgley and Botero 2012). Table 6.1 summarizes surface markers that have been identified in D-MSCs.

In the last few decades, several studies have reported that human tooth germs contain multipotent cells that give rise to dental and periodontal tissues (Yalvac et al. 2010). The dental pulp, particularly from third molars, have shown to be a significant stem cell source (Arthur et al. 2009).

Dental Pulp Stem Cells (DPSC) were first isolated from human permanent third molars in 2000, and the cell population was clonogenic and highly proliferative (Gronthos et al. 2000; Gronthos et al. 2002).

The cell population isolated by enzymatic digestion from dental pulp tissue was found to express several surface markers, such as CD73, CD90, and CD105, but not CD14, CD34, or CD45 (Liu et al. 2015). Table 6.2 summarizes the surface markers found in the D-MSCs discussed here.

Later on, studies made by Gronthos et al. confirmed that MSCs derived from dental pulp fulfilled the criteria needed for stem cells, including their ability to differentiate into adipocytes, neural cells and odontoblasts; together with their ability of self-renewal capabilities (Gronthos et al. 2002).

Besides, DPSCs have been reported to differentiate into osteoblasts, chondrocytes and myoblast-like cells, and demonstrate axon guidance (Arthur et al. 2009). Fast population doubling time, immunosuppressive properties and the ability to form a dentin-pulp-like complex have also been reported for DPSC (Liu et al. 2015).

Also, our research group has isolated and characterized DPSCs from third molars. The cells can differentiate into adipogenic, osteogenic and chondrogenic lineages, and were found to express surface markers, such as CD150, CD90, CD4 and HLA-ABC. This is in concordance with the minimum criteria to be considered and classified as mesenchymal stem cells (Fig. 6.2).

Table 6.1. Features identified in dental mesenchymal stem cells.

	DPSC	SHED	PDLSC	DFPC	ABMSCs	SCAP	TGPC	GMSC
Adherence to plastic in standard culture conditions	+	+	+	+	+	+	+	+
Phenotype	CD105‡, CD73‡, CD90(+)‡, CD14, CD19, CD34‡, CD45‡⁻	CD105‡, CD73‡, CD90(+)‡, CD14, CD19, CD34‡, CD45‡, CD11b⁻	CD105‡, CD73‡, CD90(+)‡, CD14, CD34‡, CD45(⁻)‡	CD105‡, CD73‡, CD90(+)‡, CD34‡, CD45(⁻)‡	CD105‡, CD73‡, CD90(+)‡, CD14, CD19, CD34‡, CD45‡, CD11b⁻	CD105‡, CD73‡, CD90(+)‡, CD14, CD34‡, CD45(⁻)‡	CD105‡, CD73‡, CD90(+)‡, CD14, CD34‡, CD45(⁻)‡	CD105‡, CD73‡, CD90‡ (+), CD34‡, CD45‡(+)
***In vitro* differentiation to adipocytes, osteoblast, chondroblasts**	Yes	Yes	Yes	Yes	Yes	Yes	Yes	Yes

+, adherent to plastic wells; (+): positive; (−): Negative, (≤ 2%); ‡, common markers among different population of Dental Mesenchymal Stem Cells.

Table 6.2. Surface markers of human dental mesenchymal stem cells.

Cell Type	CD105	CD90	CD73	CD45	CD34	CD31	HLA-ABC	HLA-DR
DPSC	+	+	+	−	−	−	+	−
PDLSC	+	+	+	−	−	−	+	−
GNSC	+	+	+	−	−	−	+	−

CD105 Endoglin, TGFB receptor; **CD90** Thy-1, T cells activator; **CD73** Ecto-5' nucleotidase, hydrolyses AMP to adenosine and phosphate; **CD45** lymphocyte common antigen, lymphocyte signaling; **CD34** gp 105–120, transmembrane phosphoglycoprotein; **CD31** PECAM-1, leukocyte transendothelial migration; **HLA-ABC** major histocompatibility antigen, L-selectin; **HLA-DR** major histocompatibility antigen, VCAM-1.

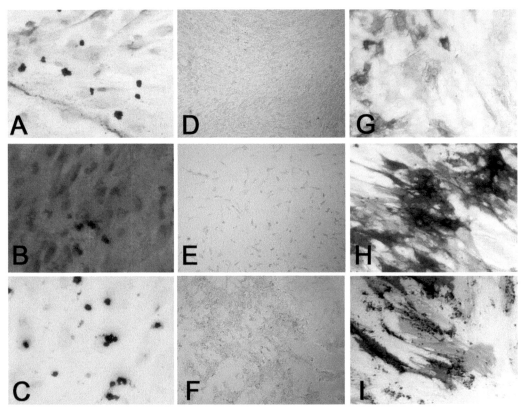

Figure 6.2. Adipogenic, chondrogenic and osteogenic differentiation of human D-MSCs. D-MSCs showed positive staining for Oil red (A, D, and G); Alcian blue (B, E, and H); or Alkaline phosphatase stain (C, F, and I); D-MSCS were either culture with regular medium, adipogenic, chondrogenic or osteogenic medium. Magnification 5X.

Stem Cells from Human Exfoliated Deciduous Teeth (SHED)

Tooth development is a well-orchestrated process, which includes shedding of the deciduous teeth, followed by the eruption of permanent teeth. The transition from deciduous teeth to permanent adult teeth is a dynamic process, that coordinates development and eruption of permanent teeth with the resorption of the deciduous teeth roots (Liu et al. 2015).

Researchers took advantage of such a process isolating a distinct population of multipotent stem cells from the remnant pulp of exfoliated deciduous teeth and expanding them *ex vivo*; unexpectedly providing a unique and accessible tissue source of MSCs (Miura et al. 2003).

Stem Cells from Human Exfoliated Deciduous Teeth (SHEDs) are highly proliferative stem cells isolated from exfoliated deciduous teeth, and are capable of differentiating into a variety of cell types under defined culture conditions, including osteoblasts, odontoblast, adipocytes and neural cells (Morsczeck et al. 2005; Miura et al. 2003). It has also been demonstrated that SHED can induce dentin and bone formation (Morsczeck et al. 2005).

Furthermore, SHEDs express a variety of neural cell markers, such as Nestin and GFAP; and can form sphere-like clusters, and multicytoplasmic processes when cultured under neurogenic conditions (Miura et al. 2003; Sedgley and Botero 2012).

SHEDs have a higher proliferation rate than DPSCs and BM-MSCs, suggesting that they represent a more immature population of multipotent stem cells (Morsczeck et al. 2005). As for their gene expression profile, SHEDs are also different from DPSCs and BM-MSCs. Genes related to cell proliferation and extracellular matrix formation such as, Transforming Growth Factor (TGF-β), Fibroblast Growth Factor (FGF2), TGF-β2, collagen type I (Col I), and Col III, show increased expression in SHEDs compared to DPSCs (Sedgley and Botero 2012).

The continuously accumulating evidence suggests that SHEDs from exfoliated deciduous teeth might be an excellent source for stem cell-based therapies, including autologous stem cell transplantation and tissue engineering (Miura et al. 2003).

Periodontal Ligament Stem Cells (PDLSC)

The periodontal ligament (PDL) is a soft connective tissue embedded between the cementum and the alveolar bone socket. Early evidence showed that PDL not only played a vital role in supporting teeth, but it also contributed to tooth nutrition, homeostasis and the regeneration of periodontal tissue (Liu et al. 2015).

McCulloch reported the presence of progenitor/stem cells in the periodontal ligament of mice in 1985. Subsequently, the isolation and identification of multipotent MSCs in human periodontal ligaments were first reported in 2004 (Seo et al. 2004).

In this regard, explants cultures or enzymatic digestion treatment of the PDL made it possible to obtain a population of PDLSCs; and the postnatal multipotent stem cells can be expanded *in vitro* to generate a cementum/PDL-like complex (Liu et al. 2015).

In their study, Seo et al. (Seo et al. 2004) demonstrated that PDLSCs were similar to other MSCs regarding their expression of STRO-1/CD146, and suggested that PDLSCs might also be derived from a population of perivascular cells. Moreover, later works showed that PDLSCs' differentiation could be promoted by Hertwig's epithelial root sheath cells *in vitro* (Sonoyama et al. 2007). Besides, the lineages of differentiation for PDLSCs are cementoblast-like cells, adipocytes and fibroblasts that secrete collagen type I (Sedgley and Botero 2012).

Similar to BM-MSCs, PDLSCs can undergo osteogenic, adipogenic and chondrogenic differentiation (Sedgley and Botero 2012; Sonoyama et al. 2007).

PDLSCs, like DPSCs, also show a higher number of populations doubling than those of BM-MSCs; however, the mechanisms contributing to the long lifespan of PDLSCs and DPSCs is still unclear (Seo et al. 2004).

We have been able to isolate and characterize PDLSCs, the cells can differentiate into adipogenic, osteogenic and chondrogenic lineages; and expressed surface markers, such as, CD150, CD90, CD4, and HLA-ABC in concordance with the minimum criteria to be considered and classified as MSCs (Fig. 6.2).

Dental Follicle Stem Cells (DFPCs)

Tooth development represents a unique and temporal series of events; in which the dental follicle is an ectomesenchymal tissue that surrounds the developing tooth germ preceding its eruption. This tissue is thought to contain stem cells, and lineage-committed progenitor cells form cementoblasts, periodontal ligament cells and osteoblasts (Morsczeck et al. 2005).

MSCs can be isolated from the dental follicle of human third molars (Liu et al. 2015). Like other dental stem cells, Dental Follicle Stem Cells (DFPCs) have an extensive proliferative potential, express similar cell surface antigens, and are capable of forming hard tissue both *in vitro* and *in vivo*. They also express the recognized stem cell markers Notch-1 and Nestin and form the tissue of the periodontium, including alveolar bone, PDL and cementum (Morsczeck et al. 2005).

At the early stage of tooth development, a condense ectomensenchyme progenitor cells limiting the dental papilla and encapsulating the enamel organ is formed, that is, the dental follicle or sac (Nanci 2008). Dental follicles are loose vascular connective tissues that surround the developing tooth germ, and progenitors for periodontal ligament cells, cementoblasts and osteoblasts.

The first isolation of DFPCs was from the dental follicle of human third molars (Morsczeck et al. 2005). DFPCs come from developing tissues; and therefore, it was considered that they might exhibit greater plasticity than other DSCs. In this regard, different cloned DFPC lines have demonstrated considerable heterogeneity (Sedgley and Botero 2012; Botelho et al. 2017; Steimberg et al. 2018).

Although DFPCs and SHED cells can differentiate into neural cells, reports are inconsistent even with similar culture conditions in these cells (Morsczeck et al. 2005).

Alveolar Bone Mesenchymal Stem Cells (AB-MSC)

The alveolar bone is derived from the dental follicle and includes a condensed edge containing the tooth sockets in the bones that hold teeth. Recently, the isolation and culture of human AB-MSCs was reported (Liu et al. 2015). The isolated cells exhibited the morphology of spindle-shaped fibroblast-like, accompanied by adherence to plastic plates and colony formation. These cells express the surface markers CD73, CD90, CD105 and STRO-1; however, they showed negative expression of hematopoietic markers, such as CD14, CD34 and CD45 (Liu et al. 2015).

Stem Cells from the Dental Apical Papilla (SCAP)

During tooth development, root formation begins with the apical proliferation of epithelial cells from the cervical loop. Then, proliferating epithelial cells give shape to the apical papilla, a soft tissue found at the apices of developing permanent teeth. The dental papilla contributes to tooth formation and is eventually converted into pulp tissue (Liu et al. 2015; Sonoyama et al. 2006).

A distinctive population of MSCs referred to as Stem Cells from the dental Apical Papilla (SCAPs) was isolated from the apical papilla of human immature permanent teeth. SCAPs showed a higher proliferation rate and mineralization potential than DPSCs; they express typical MSCs markers, including STRO-1, CD73, CD90 and CD105. Since SCAPs represent a population of cells from a developing tissue, they might exhibit greater plasticity than other dental stem cells (Sonoyama et al. 2006).

SCAPs are found in the apical papilla, at the junction of the apical papilla and the dental pulp. Extracted human third molar and their apical papillae were the primary sources of SCAPs. These cell populations are clonogenic and can differentiate to odontoblastic/osteogenic, adipogenic or neurogenic lineage. Compared with DPSCs, SCAPs show higher proliferation rates; interestingly, differentiation of SCAPs decreases CD24 marker expression, while alkaline phosphatase expression increases (Sedgley and Botero 2012; Sonoyama et al. 2006).

Tooth Germ Progenitor Cells (TGPCs)

Tooth Germ Progenitor Cells (TGPCs) are a stem cell population identified in the dental mesenchyme of the third molar tooth germ during the late bell stage. TGPCs can be expanded in plastic culture plates and maintained for approximately 60 populations doublings, during which they retain their spindle-shaped morphology and high proliferation rate. TGPCs express the MSCs associated markers STRO-1 and CD73, CD90, CD105 and CD166, but are negative for CD34, CD45 and CD133 (Yalvac et al. 2010).

TGPCs also demonstrate a tendency for pluripotency-associated gene expression, such as Nanog, Oct4, Sox2, Klf4 and C-myc; indicating mesenchymal phenotype (Liu et al. 2015; Yalvac et al. 2010).

Gingiva-derived Mesenchymal Stem Cells (G-MSC)

The human gingiva is an oral tissue overlaying the alveolar ridge and retromolar region that is recognized as a biological mucosa barrier and a distinct component of the oral mucosa immunity (Wang et al. 2011).

The gingival tissue can often be obtained as a discarded biological sample. Recently, G-MSCs were isolated from human gingiva; the cells exhibited clonogenicity, self-renewal and multipotent differentiation capacity. These cells also possess both stem cell-like and immunomodulatory properties and display positive signals for Oct4, Sox2, Nanog, Nestin, SSEA-4 and Stro-1 (Liu et al. 2015; Wang et al. 2011).

Our research group has successfully isolated and characterized cells of the dental pulp, periodontal ligament and gingival tissue from primary cultures by explant. In all the lines, a high rate of proliferation and differentiation has been observed, as well as positive and negative percentages of the surface markers (Fig. 6.2).

MSC-based Cell Therapy: Potential Clinical Applications

Stem cell-based therapy has been extensively studied and discussed. Stem cell biology has also become an important field for the understanding of tissue regeneration and the implementation of regenerative medicine.

Next some of the recent discoveries in D-MSCs functions and behavior will be reviewed. The current developments for biologically-based dental therapies will also be discussed. D-MSCs in dentistry have been mainly used to repair damage dentin, pulp revascularization and regeneration and for periodontal disorders.

Periodontal Regeneration

Periodontal diseases are a group of infectious diseases that are characterized by the destruction of tooth-supporting tissues, such as, periodontal ligament, cementum, alveolar bone and gingiva (Seo et al. 2004). Considered as the leading cause of tooth loss, periodontal diseases represent a public health burden worldwide (Holmstrup et al. 2017).

Although much work is still needed for the translation of data from *in vitro* and animal studies to clinical applications, there are exciting findings for the use of D-MSCs in tissue engineering and regenerative medicine applications, particularly for root canal treatments and periodontal regeneration (Lymperi et al. 2013).

For example, a surgically created periodontal defect, in a miniature swine model, was treated with autologous transplantation of PDLSCs. The PDLSCs transplantation was found useful for the regeneration of periodontal tissues (Liu et al. 2008).

As for SHED cells, one of their striking features is that they are capable of inducing murine host cells to differentiate them from bone-forming cells, in an immunocompromised mouse model (Seo et al. 2008; Miura et al. 2003). Although SHEDs could not be directly differentiated into osteoblast, they appeared to induce new bone formation through the establishment of an osteoinductive template in order to recruit murine host osteogenic cells (Miura et al. 2003).

The results obtained with SHEDs implied that deciduous teeth might not only provide a guidance for permanent teeth eruption but may also be involved in bone formation during the eruption of permanent teeth (Miura et al. 2003; Seo et al. 2008).

PDLSCs can also differentiate into cementoblast-like cells, adipocytes and fibroblasts that secrete collagen type I (Sedgley and Botero 2012). In this regard, typical cementum/PDL-like architecture can be regenerated after transplantation of *ex vivo*-expanded human PDLSCs into immunocompromised mice (Huang et al. 2009).

Interestingly, collagen fibers generated *in vivo* were able to connect with newly formed cementum-like structures, similar to the physiological attachment of Sharpey's fibers responsible for the functional attachment of cementum/periodontal ligament architecture (Seo et al. 2004; Huang et al. 2009).

The human PDLSCs implanted into periodontal defects of immunocompromised mice could regenerate periodontal ligament-like tissues. These human stem cells were also identified to be closely associated with the trabecular bone next to regenerated periodontal ligament, suggesting their involvement in alveolar regeneration (Seo et al. 2004).

It has been suggested that PDLSCs may contain a subpopulation of cells capable of differentiating into cementoblasts/cementocytes and collagen-forming cells *in vivo* (Huang et al. 2009; Seo et al. 2004).

PDLSCs have already been used in the clinic. In this regard, two patients were treated for intrabony defects with autologous periodontal ligament cells derived from third molars implanted on bone grafting material (Feng et al. 2010).

The combination of different dental mesenchymal stem cells could aid in the regeneration of root/periodontal ligament complex (Lymperi et al. 2013). For example, in a mouse model, SCAPs and PDLSCs along with hydroxyapatite/tricalcium phosphate as the carrier resulted in the formation of dentin and cementum/Sharpey's fibers respectively (Lymperi et al. 2013).

Sonoyama et al. used PDLSCs combined with SCAPs, isolated from the third molar, and seeded them in a scaffold. The scaffold with the cells was then transplanted into the alveolar bone of young pigs, root and periodontal complex formation was observed, and these structures were able to support an artificial ceramic crown (Sonoyama et al. 2006).

Pulp Therapy

The dental pulp provides nutrition and acts as a biosensor to detect pathogenic stimuli, therefore pulp vitality is essential for the tooth's vitality (Zhang and Yelick 2010).

Pulp tissue has the potential to self-regenerate lost dentin (Huang 2011). However, the majority of dental infections is difficult to eliminate, and the entire pulp tissue is often needed to be removed (pulpectomy). Consequently, in order to thoroughly disinfect and fill the canal root with gutta-percha, the root canal space must be significantly enlarged, thereby, causing further tooth structure loss (Huang 2011; Zhang and Yelick 2010).

In the end, the lack of blood supplied and the nervous system leaves the pulp less teeth even more vulnerable to injury. The accumulated knowledge of stem cells, particularly DPSC has rendered promising strategies for dental pulp regeneration and revascularization (Zhang and Yelick 2010).

In this regard, a deeper understanding of the biology of DSCs populations is essential to understanding the extent of their efficacy for regenerative medicine (Huang et al. 2009).

Previously reported research shows that DPSC has the potential to regenerate dentin/pulp-like architecture, and are hypothesized to be progenitor cells activated when the pulp needs to be repaired (Shakoori et al. 2017).

Dental stem cells populations, derived from pulp tissue, for example, DPSCs, SHED and SCAP or precursor of pulp (SCAP) are suitable cell sources for pulp/dentin regeneration (Huang 2011).

One of the early works using DPSCs was by Gronthos et al. (Holmstrup et al. 2017; Gronthos et al. 2000); DPSCs were isolated and transplanted alongside with hydroxyapatite/tricalcium phosphate into immunocompromised mice. They observed a dentin-pulp-like structure 6 weeks after transplantation. Moreover, the collagenous matrix was deposited perpendicular to the odontoblast-like layer, the odontoblast-like cells formed cytoplasmatic processes into the dentinal matrix, and the interfaced with a pulp-like interstitial tissue was infiltrated with blood cells (Gronthos et al. 2000).

SCAP cells can also craft fragments of pulp/dentin complex when transplanted into immunocompromised mice, while SHED form mineralized tissue different from the pulp/dentin complex (Huang 2011).

Interestingly, different findings suggest that the use of MCSs, different from DPSCs, SCAP, or SHED cells used to regenerate pulp/dentin is a less straightforward approach (Huang 2011).

The regeneration of the entire pulp represents a particular challenge in tissue engineering. For example, the dental pulp is encased within the dentin, with a single apical foramen to allow angiogenesis. In this context, the microstructure of the dental pulp includes different type of cells, such as odontoblasts and a complex innervation. Finally, the specific location of the dentin located only at the periphery of the pulp tissue, along with a highly organized structure with aligned dentinal tubes, are circumstances that should be considered to regenerate the pulp (Huang 2011).

Craniofacial Regeneration

Considering the diversity and extent of structural defects in the oral and maxillofacial region; several research groups have aimed to regenerate defects of bone, cartilage and fat using adult BMSCs (Shakoori et al. 2017).

For example, with a single 25-Gy dose of irradiation combined with tooth extraction to generate osteoradionecrosis in a swine model; BM-MSCs were used to treat and rehabilitate the jaw (Xu et al. 2012).

However, the D-MSCs have also shown the potential and ability to regenerate large bone defects (Shakoori et al. 2017). For example, DPSCs from deciduous teeth were used to regenerate critical-size defects in the orofacial bone in a swine model; the results suggested that SHEDs were able to engraft and regenerate bone, with a 6-month follow-up (Zheng et al. 2009).

Moreover, the association of DPSCs with appropriate scaffold have shown promising clinical relevance to regenerate significant craniofacial bone defects (Botelho et al. 2017). In this regard, DPSCs have been used with a collagen scaffold to repair human mandible bone defects, and in the three-year follow-up of the study, the subjects presented alveolar bone, creating a steadier mandible and with positive clinical results (d'Aquino et al. 2009).

SHEDs are also capable of inducing recipient murine cells to differentiate into bone-forming cells, which is not a property attributed to DPSCs following transplantation *in vivo*. However, SHED could not differentiate directly into osteoblasts; they appeared to induce new bone formation by establishing an osteoinductive template to recruit murine host osteogenic cells. The osteoinductive potential of SHEDs suggested that they potentially can repair critical-sized calvaria defects in mice with substantial bone formation (Huang et al. 2009).

SHEDs and DPSCs are embryologically derived from neural crest cells. Therefore, they may share similar tissue origin with the mandibular bone cells, and thus, serve as better cells source for the regeneration of alveolar and orofacial bone defects (Zheng et al. 2009).

Human PDLSCs applied into the periodontal defects of immunocompromised mice also led to the regeneration of periodontal-like tissue, and these human stem cells were identified to be closely associated with the trabecular bone next to the regenerated periodontium, suggesting their involvement in alveolar bone regeneration (Botelho et al. 2017).

Immunomodulatory Properties

The interactions between the immune system and MSCs was first demonstrated by Bartholomew et al., who observed suppression of a mixed lymphocyte response *in vitro* and prevention of rejection in a baboon skin allograft model (Wang et al. 2018).

Subsequent studies have shown that MSCs can mediate immunosuppression in animal and human models. There are two main aspects regarding the immunoregulation of MSCs; the first one involves the immunosuppressive effects of allogeneic MSCs. The second one includes the effects of inflammatory cytokines on MSCs activity and differentiation (Grinnemo et al. 2004).

In this regard, MSCs showed low expression of the Major Histocompatibility Complex (MHC) class I molecules, hence, they are able to evade the host cells immune system. They also lack MHC class II and costimulatory molecules, such as, CD40, CD40L, CD80 and CD86; which are required for immune cell stimulation (Wang et al. 2018; Huang et al. 2009).

MSCs' immunosuppressive effects, such as down-modulate immune reactions executed by T-, denditric, NK and B-cells have also been reported. Moreover, the immunosuppressive abilities of MSCs, have provided new insights for the treatment of immune-mediated diseases (Huang et al. 2009; Wang et al. 2018; Mao et al. 2017).

For example, studies in which MSCs were used for bone and cartilage repair, indicated that the failure of the therapeutic effect of allogeneic MSCs was associated with signs of activated immune responses. However, other studies demonstrated that MSCs loaded into scaffolds showed a similar inflammatory response to that induced by autologous MSCs, and had better outcomes (El-Jawhari et al. 2016).

However, the available information on the effects of pro-inflammatory cytokines in MSCs is still limited. In this regard, a preliminary study demonstrated that interferon might act to differentiate MSCs into osteoblasts. Furthermore, autologous implantation of MSCs for chondrogenesis showed that inflammatory reactions against scaffold materials and serum component led to the production of cytokines, such as interleukin (IL)-1α, and that might inhibit cartilage tissue formation (Huang et al. 2009).

As for D-MSCs, SCAP, PDLSCs and DPSCs have shown immunosuppressive properties *in vitro* (Huang et al. 2009).

For example, DPSCs can induce T-cell apoptosis *in vitro*, and ameliorate inflammation-related tissue injuries in mice with colitis; and this model was associated with the expression of the Fas-ligand (FasL). Moreover, knockdown FasL mice reduced the immunoregulatory properties of DPSCs in the context of inducing T-cell apoptosis. Since DPSCs can suppress T-cell proliferation, they are also suitable for preventing or treating T-cell alloreactivity associated with allogeneic transplants. Besides, *ex vivo*-expanded DPSCs significantly inhibited the proliferation of peripheral blood mononuclear cells (Botelho et al. 2017).

Moreover, Toll-Like Receptors (TLRs), the key molecules in the innate and adaptive immune responses, could be responsible for the immunosuppression of DPSCs; by upregulating the expression of Transforming Growth Factor (TGF)-β and interleukin (IL)-6 (Sedgley and Botero 2012; Botelho et al. 2017). Also, our research group recently demonstrated that PDLSS, DPSCs, and GMSCs have the *in vitro* capacity to decrease the proliferation of activated CD3+ T-cells. Moreover, compared to BM-MSCs, co-culture with T-cells and DMSCs showed differences in the production of cytokines and surface and secreted molecules that may participate in T-cell immunosuppression (De la Rosa-Ruiz et al. 2019).

When it comes to the immune response, macrophage polarization into M1 or M2 phenotype has often been at the origin of immunological paradigms. DPSCs transplantation on diabetic poly-neuropathy, have found that the DPSCs stimulate macrophages polarization towards anti-inflammatory M2 phenotype, therefore, presenting immunosuppressive effects on diabetic polyneuropathy (Botelho et al. 2017).

In the case of G-MSCs, investigators have also shown that they are capable of inducing M2 polarization of tissue macrophages. For example, in the presence of G-MSCs, macrophages exhibit M2-like phenotype, increased expression of mannose receptors and interleukin (IL)-10, and decreased expression of tumor necrosis factor-alpha (TNF)-α (Shakoori et al. 2017).

G-MSCs can also suppress the activation and function of inflammatory Th1 and Th17 cells, and promote the proliferation of regulatory T cells. The combined effects of G-MSCs on both macrophages and T cells suggest that G-MSCs could be used for several inflammatory disease models, such as colitis, contact-allergic-dermatitis, oral mucositis and skin wound healing (Shakoori et al. 2017).

Bone Tissue Engineering Applications of D-MSCs

The extracellular matrix (ECM) has distinctive biochemical and biophysical properties that dictate tissue-specific behavior. In the case of mineralized tissues, their elegant specificity is the result of a unique ECM composition and topographies, including a fibrillar collagenous extracellular matrix infiltrated with additional non-collagenous proteins and proteoglycans which are thought to guide the mineralization process. The process requires assembly and maturation of macromolecules, which demands enzymatic processing of matrix components (Coyac et al. 2013; Winkler et al. 2018; Mouw et al. 2014).

For example, bone, dentin and cementum have a lag time before mineralization occurs, during such time, an unmineralized layer of osteoid, predentin and cementoid is formed. In other words, the understanding of cellular differentiation, tissue morphogenesis and physiological remodeling demand a thoughtful consideration of the EMC components that are produced by cells, as well as the assembly of those macromolecules into a functional three-dimensional structure (Mouw et al. 2014).

Presently, available cell-culture models for bone, cartilage and tooth cells remain limited and not wholly adequate, which makes it unable to replicate 3D matrix environments for cell differentiation and matrix mineralization (Mouw et al. 2014; Coyac et al. 2013).

Some existing cell-culture models are capable of mimicking certain aspects of tooth development and matrix mineralization. However, the amount of *in vitro* experimental data available for D-MSCs is scarce compared with that obtained from osteoblasts and chondroblasts/cytes, and mineralized tissue craniofacial biology research (Coyac et al. 2013).

As previously mentioned, the association of D-MSCs with an appropriate scaffold has shown to be critical to a replicated natural matrix environment in terms of guiding cells differentiation (Botelho et al. 2017).

In this respect, several research groups have used 3D scaffolds with a density approaching that of the native extracellular matrix, similar to osteoid or predentin, to study D-MSCs differentiation. For example, D-MSCs have shown the potential to differentiate into bone, enhancing osteogenesis and mineralization *in vivo* and *in vitro* (Lymperi et al. 2013).

In their study, Cordeiro et al. evaluated morphologic characteristics of tissues formed when SHEDs cells were seeded in poly-L-lactic acid biodegradable scaffolds prepared within human tooth slices, followed by their transplantation into immunodeficient mice. The resulted tissue closely resembles the physiology of the dental pulp (Cordeiro et al. 2008).

Another interesting study demonstrated that 3D dense collagen scaffolds promoted SHED osteo/odontogenic cell differentiation and mineralization. The study demonstrated that SHED cells showed expression of mineralized tissue markers, including tissue-non-specific alkaline phosphatase, dentin matrix protein 1, and osteopontin. The matrix formed by SHEDs was also consistent with the formation of apatite mineral that was frequently aligned along collagen fibrils of the scaffold (Coyac et al. 2013).

Table 6.3 summarizes the advances that have been made in regenerative technologies using D-MSCs.

Table 6.3. Advances in tissue engineering applications with the use of Dental Mesenchymal Stem cells.

Cell	Periodontal regeneration	Pulp therapy	Craniofacial regeneration	Immunomodulatory properties	Mineral formation	References
DPSC	Alveolar bone regeneration. Biotoot formation.*	Dentin/pulp-like complex formation. Angiogenesis.	Regeneration of critical-size orofacial bone defects.*	Immunosuppressive effects. Promotes T-cell apoptosis and decrease their proliferation.	Promote mineralized ECM deposition.*	(Shakoori et al. 2017; Huang 2011; Chalisserry et al. 2017)
SHED	Not reported yet.	Promote dental pulp-like architecture.*	Induce new bone formation. Promotes proliferation of bone forming cells.	Inhibit Th17 activation. Promote M2 macrophage phenotype. Suppress chronic inflammation and liver fibrosis.	Enhance osteogenesis. Promotes differentiation into osteogenic/odontogenic cells.*	(Shakoori et al. 2017; Huang 2011; Volponi and Sharpe 2013; Botelho et al. 2017; Li et al. 2014; Lymperi et al. 2013)
PDLSC	Periodontal like-tissues regeneration.*	Not reported yet.	Promote alveolar bone regeneration.*	Immunosuppressive properties in inflammatory disease models.	Form structures similar to bone and cementum.	(Huang 2011; Chalisserry et al. 2017; Lymperi et al. 2013)
DFPS	Periodontal like-tissues regeneration.*	Not reported yet.	Regeneration of critical-size orofacial bone defects.*	Suppress peripheral blood monocytes proliferation.	Form structures similar to cementum *in vivo*.	(Lymperi et al. 2013; Li et al. 2014)
AB-MSC	Not reported yet.	Not reported yet.	Promote bone formation.*	Not reported yet	Enhance osteogenesis.	(Shakoori et al. 2017)
SCAP	Combine with PDLSCs could form root and periodontal complexes.	Promote dentin/pulp-like architecture. Angiogenesis.	Form bone-like tissue.*	Immunosuppressive properties in inflammatory disease models.	Promotes differentiation into osteogenic/odontogenic cells.*	(Volponi and Sharpe 2013; Huang 2011; Li et al. 2014)
TGPC	Not reported yet.	Not reported yet.	Promote bone formation.*	Not reported yet.	Not reported yet.	(Chalisserry et al. 2017)
G-MSC	Not reported yet.	Not reported yet.	Osteogenic potential.	Immunosuppressive effects on tissue macrophages. Promotes the expression of IL-10. Suppress the activation and function of Th1 and Th17.	Promotes differentiation into osteogenic/odontogenic cells.*	(Shakoori et al. 2017; Li et al. 2014; Chalisserry et al. 2017)

*: Biological effects observed with the use of an appropriate scaffold. DPSC; Dental Pulp Stem Cells. SHED; Stem cells from Human Exfoliated Deciduous Teeth. PDLSC; Periodontal Ligament Stem Cells. FPS; Dental Follicle Stem Cells. AB-MSC; Alveolar Bone Mesenchymal Stem Cells. SCAP; Stem Cells from the dental Apical Papilla. TGPC; Tooth Germ Progenitor Cells. G-MSC; Gingiva-derived Mesenchymal Stem Cells. ECM; extracellular matrix.

However, the translation and implementation of regenerative medicine into clinical practice require accurate research models that enable the production and the standardization of clinical cell-based therapies.

Soon regenerative medicine will be at the core of modern health care; therefore, integration of the discovery, development and delivery of cell-based, acellular and/or biomaterial applications constitute a growing challenge for researchers. These challenges should also consider the ethical implications of new therapies, which includes a consideration of global guidelines, as well as local legislation.

Furthermore, in the present chapter, we will address some ethical consideration regarding D-MSCs-based therapies.

Future Perspectives

D-MSCs therapy represents an innovative approach for the repair of defective tissues or functions through the transplantation of live cells.

The D-MSCs have some advantages over other sources of MSCs, and they represent a reliable, and an accessible source of stem cells, with a wide range of clinical application (Volponi and Sharpe 2013). D-MSCs also have the potential to reduce donor site morbidity compared with other grafting sources for MSCs (Shakoori et al. 2017).

As extensively discussed in the present chapter, D-MSCs have great potential to differentiate into other cell types and aid in the regeneration of oral and craniofacial tissues (Shakoori et al. 2017).

However, the majority of research groups agree that, before using D-MSCs in human patients, an assessment of their safety, and efficacy is required (Shakoori et al. 2017; Xu et al. 2012; Bakopoulou and About 2016; Huang 2011).

In this respect, one can argue that despite the availability of MSCs from adult/newborn tissue, these cells have limited proliferative capacity, a considerable variability because they are derived from different donors, and they lose their differentiation potential when cultured *in vitro* (Huang 2011; Gao et al. 2016).

Therefore, a common concern among investigators is that multiple parameters should be optimized and that many topics need consideration before the development of effective MSC-based cell therapies.

Among such topics, one can talk about the need for maintenance or improvement of D-MSCs stemness during the expansion process. Hence, a deeper understanding of the mechanisms of self-renewal is mandatory. The former would allow clinicians to regulate adult stem cell growth *in vitro*, and generate sufficient cell numbers for clinical applications (Huang et al. 2009).

Moreover, D-MSCs-based therapies could not be considered in the same category of drugs or transplants because they contain viable allogeneic or autologous cells enduring *ex vivo* substantial manipulations, and they may be applied at sites not physiologically present or to perform biological functions they do not usually participate in (Huang et al. 2009; Huang 2011).

As previously mentioned, formation and maturation of the extracellular matrix into specialized tissues, particularly mineralized tissues, involve a sequential activation of signal cascades. Moreover, D-MSCs differentiation depends on their association with an appropriate scaffold. Thus, understanding the regulation of stem cells during differentiation, specific tissue production, and the role of specialized scaffolds is a requirement for tissues such as bone, dentin, cartilage and tendon (Huang et al. 2009).

The interactions between stem cells and the immune system should be mentioned. So far, it has been recognized that MSCs show anti-inflammatory and immune privilege potential, and therefore, they are promising in the treatment of many immune disorders (Huang et al. 2009; Gao et al. 2016).

Furthermore, their immunomodulatory properties, like the reduction of serum antibodies, regulation of Th17 cell, polarization of M2 macrophages phenotype, reduction of inflammation in axonal injury, demyelination or cytokine regulation represent therapeutic opportunities for autoimmune diseases, such as, systemic erythematous lupus, rheumatoid arthritis or autoimmune encephalomyelitis (Volponi and Sharpe 2013).

Several MSCs products have already been approved for clinical applications around the world. For example, Cartistem for degenerative arthritis and Cupistem for anal fistula were approved for the Ministry of Drug and Safety in Korea; and Prochymal for Acute Graft Versus Host Disease in Canada and New Zealand (Gao et al. 2016).

Importantly, the mechanisms responsible for MSCs therapeutic effects in a variety of immune diseases may involve paracrine secretion of several cytokines from the MSCs and their interplay with the immune cells (Mao et al. 2017).

In the case of D-MSCs, it has been demonstrated that SHEDs are capable of reducing inflammation, inducing tissue repair and reducing apoptosis in hepatocytes (Volponi and Sharpe 2013; Huang et al. 2009). Ultimately, their effects over inflammatory processes would eventually lead to the treatment of conditions such as liver fibrosis.

Recent evidence suggests that DPSCs or SHEDs promotes recovery after spinal cord injury due to their anti-inflammatory properties, and their capacity to differentiate into neurons and oligodendrocytes (Huang 2011; Botelho et al. 2017; Volponi and Sharpe 2013).

However, the understanding of the interactions between D-MSCs and the immune system are not fully understood. For example, further research is needed to determine if allogenic D-MSCs suppress recipient host short-term, and the long-term effects of immunological rejection (Huang 2011). Therefore, developing a practical approach to prevent or cure human diseases will be difficult.

Clinical trials for the application of D-MSCs which are used for the regeneration of various tissues are scarce. The former is in contrast to the increased number of clinical trials using other MSCs sources for the treatment of bone/articular, cardiovascular, neurological, immune and blood pathologies, such as bone BM-MSCs (Bakopoulou and About 2016).

During recent decades, the use of MSCs in the fields of tissue engineering and regenerative medicine has been surrounded with controversies regarding ethical issues. Moreover, despite the prevailing principles for research integrity, such as the protection of life, health, privacy and dignity of research participants, donor of cells and organ recipients, among others; guidelines for working with MSCs vary from one country to another (Frati et al. 2017).

Embryonic stem cells used represent an entirely different ethical debate, which involves issues associated with the embryo and its right to live.

As for D-MSCs, it could be argued that most cells obtained from extracted teeth are usually discarded, making them relatively free from an ethical discussion. However, the extensive manipulation required for harvesting and characterizing D-MSCs, also represent some risks that should be considered, such as infection, switched cells, the safety of the tools, drugs and reagents for treating cells. Moreover, the potential secondary effects regarding cells transplantation must be explained to the patients, donors and recipients before obtaining informed consent.

In other words, to achieve accurate and efficient scientific results, rigor is required in all areas of D-MSCs research. As emphasized by Gao et al. (Gao et al. 2016), when working with MSCs, every step of the process must be defined, including the tissue origin, separation or enrichment procedures, cell density and even the medium used.

Conclusions

A profound understanding of the biology of D-MSCs is needed; this, combined with tissue engineering, will provide alternative tools for a broad spectrum of therapeutic strategies.

As shown in the present chapter, D-MSCs can be differentiated into other cell types, and they can aid in the regeneration of tissues composed of different cell types in the oral and craniofacial region.

MSCs obtained from oral and craniofacial regions offer easier accessibility with potentially reduced donor site morbidity compared with other sources (Shakoori et al. 2017).

In this context, ethical concerns are raised regarding D-MSCs therapies, and the need to design appropriate clinical trials is imperative for their success.

Therefore, future research must address the safety of cell-based therapies, as well as the quality control in the production of MSCs, to ensure their reproducible and efficient effects when it is delivered to patients (Gao et al. 2016).

Besides, the type of patients who would benefit the most from MSCs therapies should be identified, and rapid translation of basic science into the clinic must be avoided because it could put the patients at higher risk.

MSCs derived from dental tissues have been intensively studied for a wide range of diseases. Notably, their therapeutic potential for immune disorders have shifted our appreciation for their use.

Rather than unrealistically considering the D-MSCs as an immunological panacea, research has started to define new limits on cell-based therapies and correlating their efficacy. In the future, such restrictions will define the new therapeutic targets for cell-based therapies (English and Mahon 2011).

Although the extensive available research suggests that therapies exist with D-MSCs to treat many intractable conditions is possible, several issues are yet to be resolved. The immune-suppressive functions of D-MSCs and the degree of redundancy that exist among the many suppressive processes is an outstanding and growing field of research (Kichenbrand et al. 2019).

Acknowledgments

We specially thank Dr. Fernando Suaste Olmos, Institute of Physiology, UNAM, for his technical assistance and valuable comments on the manuscript. The research reported in this manuscript, and performed at the Dr. Montesinos' laboratory was supported by grants from the Mexican Institute of Social Security (IMSS, grant no. 1731), and the National Council of Science and Technology (CONACYT, grant no. 258205). Also, research conducted at the Tissue Bioengineering Laboratory, Faculty of Dentistry, UNAM was partially supported by the National Council of Science and Technology (CONACyT-AS-A1-S-9178).

References

Arthur, A., Shi, S., Zannettino, A.C., Fujii, N., Gronthos, S. and Koblar, S.A. 2009. Implanted adult human dental pulp stem cells induce endogenous axon guidance. Stem Cells 27: 2229–37.

Bakopoulou, A. and About, I. 2016. Stem cells of dental origin: Current research trends and key milestones towards clinical application. Stem Cells Int. 2016: 4209891.

Botelho, J., Cavacas, M.A., Machado, V. and Mendes, J.J. 2017. Dental stem cells: Recent progresses in tissue engineering and regenerative medicine. Ann. Med. 49: 644–651.

Chalisserry, E.P., Nam, S.Y., Park, S.H. and Anil, S. 2017. Therapeutic potential of dental stem cells. J. Tissue Eng. 8: 2041731417702531.

Cordeiro, M.M., Dong, Z., Kaneko, T., Zhang, Z., Miyazawa, M., Shi, S. et al. 2008. Dental pulp tissue engineering with stem cells from exfoliated deciduous teeth. J. Endod. 34: 962–9.

Coyac, B.R., Chicatun, F., Hoac, B., Nelea, V., Chaussain, C., Nazhat, S.N. et al. 2013. Mineralization of dense collagen hydrogel scaffolds by human pulp cells. J. Dent. Res. 92: 648–54.

D'aquino, R., De Rosa, A., Lanza, V., Tirino, V., Laino, L., Graziano, A. et al. 2009. Human mandible bone defect repair by the grafting of dental pulp stem/progenitor cells and collagen sponge biocomplexes. Eur. Cell Mater. 18: 75–83.

De la Rosa-Ruiz, M.D.P., Álvarez-Pérez, M.A., Cortés-Morales, V.A., Monroy-García, A., Mayani, H., Fragoso-González, G. et al. 2019. Mesenchymal stem/stromal cells derived from dental tissues: a comparative *in vitro* evaluation of their immunoregulatory properties against T cells. Cells. 22: 8(12). pii.E1491.

Dominici, M., Le Blanc, K., Mueller, I., Slaper-Cortenbach, I., Marini, F., Krause, D. et al. 2006. Minimal criteria for defining multipotent mesenchymal stromal cells. The International Society for Cellular Therapy position statement. Cytotherapy 8: 315–7.

El-Jawhari, J.J., Jones, E. and Giannoudis, P.V. 2016. The roles of immune cells in bone healing; what we know, do not know and future perspectives. Injury 47: 2399–2406.

English, K. and Mahon, B.P. 2011. Allogeneic mesenchymal stem cells: agents of immune modulation. J. Cell Biochem. 112: 1963–8.

Feng, F., Akiyama, K., Liu, Y., Yamaza, T., Wang, T.M., Chen, J.H. et al. 2010. Utility of PDL progenitors for *in vivo* tissue regeneration: a report of 3 cases. Oral Dis. 16: 20–8.

Frati, P., Scopetti, M., Santurro, A., Gatto, V. and Fineschi, V. 2017. Stem cell research and clinical translation: A roadmap about good clinical practice and patient care. Stem Cells Int. 2017: 5080259.

Gao, F., Chiu, S.M., Motan, D.A., Zhang, Z., Chen, L., Ji, H.L. et al. 2016. Mesenchymal stem cells and immunomodulation: current status and future prospects. Cell Death Dis. 7: e2062.

Grinnemo, K.H., Månsson, A., Dellgren, G., Klingberg, D., Wardell, E., Drvota, V. et al. 2004. Xenoreactivity and engraftment of human mesenchymal stem cells transplanted into infarcted rat myocardium. J. Thorac. Cardiovasc. Surg. 127: 1293–300.

Gronthos, S., Mankani, M., Brahim, J., Robey, P.G. and Shi, S. 2000. Postnatal human dental pulp stem cells (DPSCs) *in vitro* and *in vivo*. Proc. Natl. Acad. Sci. USA 97: 13625–30.

Gronthos, S., Brahim, J., Li, W., Fisher, L.W., Cherman, N., Boyde, A. et al. 2002. Stem cell properties of human dental pulp stem cells. J. Dent. Res. 81: 531–5.

Holmstrup, P., Damgaard, C., Olsen, I., Klinge, B., Flyvbjerg, A., Nlelsen, C.H. et al. 2017. Comorbidity of periodontal disease: two sides of the same coin? An introduction for the clinician. J. Oral Microbiol. 9: 1332710.

Huang, G.T., Gronthos, S. and Shi, S. 2009. Mesenchymal stem cells derived from dental tissues vs. those from other sources: Their biology and role in regenerative medicine. J. Dent. Res. 88: 792–806.

Huang, G.T. 2011. Dental pulp and dentin tissue engineering and regeneration: Advancement and challenge. Front. Biosci. (Elite Ed.) 3: 788–800.

Kichenbrand, C., Velot, E., Menu, P. and Moby, V. 2019. Dental pulp stem cell-derived conditioned medium: An attractive alternative for regenerative therapy. Tissue Eng. Part B Rev. 25: 78–88.

Li, Z., Jiang, C.M., An, S., Cheng, Q., Huang, Y.F., Wang, Y.T. et al. 2014. Immunomodulatory properties of dental tissue-derived mesenchymal stem cells. Oral Dis. 20: 25–34.

Liu, J., Yu, F., Sun, Y., Jiang, B., Zhang, W., Yang, J. et al. 2015. Concise reviews: Characteristics and potential applications of human dental tissue-derived mesenchymal stem cells. Stem Cells 33: 627–38.

Liu, Y., Zheng, Y., Ding, G., Fang, D., Zhang, C., Bartold, P.M. et al. 2008. Periodontal ligament stem cell-mediated treatment for periodontitis in miniature swine. Stem Cells 26: 1065–73.

Lymperi, S., Ligoudistianou, C., Taraslia, V., Kontakiotis, E. and Anastasiadou, E. 2013. Dental stem cells and their applications in dental tissue Engineering. Open Dent. J. 7: 76–81.

Mao, X., Liu, Y., Chen, C. and Shi, S. 2017. Mesenchymal stem cells and their role in dental medicine. Dent. Clin. North Am. 61: 161–172.

Mason, S., Tarle, S.A., Osibin, W., Kinfu, Y. and Kaigler, D. 2014. Standardization and safety of alveolar bone-derived stem cell isolation. J. Dent. Res. 93: 55–61.

Miura, M., Gronthos, S., Zhao, M., Lu, B., Fisher, L.W., Robey, P.G. et al. 2003. SHED: Stem cells from human exfoliated deciduous teeth. Proc. Natl. Acad. Sci. USA 100: 5807–12.

Morscheck, C., Götz, W., Schierholz, J., Zeilhofer, F., Kühn, U., Möhl, C. et al. 2005. Isolation of precursor cells (PCs) from human dental follicle of wisdom teeth. Matrix Biol. 24: 155–65.

Mouw, J.K., Ou, G. and Weaver, V.M. 2014. Extracellular matrix assembly: a multiscale deconstruction. Nat. Rev. Mol. Cell Biol. 15: 771–85.

Nanci, A. 2008. Development of the tooth and its supporting tissues. pp. 90–95. *In*: Mosby (ed.). Ten Cate's Oral Histology. Development, Structure, and Function. Seventh Edition. Mosby ELSEVIER, China.

Sedgley, C.M. and Botero, T.M. 2012. Dental stem cells and their sources. Dent. Clin. North Am. 56: 549–61.

Seo, B.M., Miura, M., Gronthos, S., Bartold, P.M., Batouli, S., Brahim, J. et al. 2004. Investigation of multipotent postnatal stem cells from human periodontal ligament. Lancet 364: 149–55.

Seo, B.M., Sonoyama, W., Yamaza, T., Coppe, C., Kikuiri, T., Akiyama, K. et al. 2008. SHED repair critical-size calvarial defects in mice. Oral Dis. 14: 428–34.

Shakoori, P., Zhang, Q. and Le, A.D. 2017. Applications of mesenchymal stem cells in oral and craniofacial regeneration. Oral Maxillofac Surg. Clin. North Am. 29: 19–25.

Sonoyama, W., Liu, Y., Fang, D., Yamaza, T., Seo, B.M., Zhang, C. et al. 2006. Mesenchymal stem cell-mediated functional tooth regeneration in swine. PLoS One 1: e79.

Sonoyama, W., Seo, B.M., Yamaza, T. and Shi, S. 2007. Human Hertwig's epithelial root sheath cells play crucial roles in cementum formation. J. Dent. Res. 86: 594–9.

Steimberg, N., Angiero, F., Farronato, D., Berenzi, A., Cossellu, G., Ottonello, A. et al. 2018. Advanced 3D models cultured to investigate mesenchymal stromal cells of the human dental follicle. Tissue Eng. Part C Methods 24: 187–196.

Tatullo, M., Codispoti, B., Pacifici, A., Palmieri, F., Marrelli, M., Pacifici, L. et al. 2017. Potential use of human periapical cyst-mesenchymal stem cells (hPCy-MSCs) as a novel stem cell source for regenerative medicine applications. Front. Cell Dev. Biol. 5: 103.

Volponi, A.A. and Sharpe, P.T. 2013. The tooth—a treasure chest of stem cells. Br. Dent. J. 215: 353–8.

Wang, F., Yu, M., Yan, X., Wen, Y., Zeng, Q., Yue, W. et al. 2011. Gingiva-derived mesenchymal stem cell-mediated therapeutic approach for bone tissue regeneration. Stem Cells Dev. 20: 2093–102.

Wang, M., Yuan, Q. and Xie, L. 2018. Mesenchymal stem cell-based immunomodulation: properties and clinical application. Stem Cells Int. 2018: 3057624.

Winkler, T., Sass, F.A., Duda, G.N. and Schmidt-Bleek, K. 2018. A review of biomaterials in bone defect healing, remaining shortcomings and future opportunities for bone tissue engineering: The unsolved challenge. Bone Joint Res. 7: 232–243.

Xu, J., Zheng, Z., Fang, D., Gao, R., Liu, Y., Fan, Z. et al. 2012. Mesenchymal stromal cell-based treatment of jaw osteoradionecrosis in Swine. Cell Transplant 21: 1679–86.

Yalvac, M., Ramazanoglu, M., Rizvanov, A., Sahin, F., Bayrak, O., Salli, U. et al. 2010. Isolation and characterization of stem cell derived from human third molar tooth germs of young adults: implications in neo-vascularization, osteoporosis-, adopt- and neurogenesis. Pharmacogenomics J. 10: 105–13.

Yildirim, S., Zibandeh, N., Genc, D., Ozcan, E.M., Goker, K. and Akkoc, T. 2016. The comparison of the immunologic properties of stem cells isolated from human exfoliated deciduous teeth, dental pulp, and dental follicles. Stem Cells Int. 2016: 4682875.

Zhang, W. and Yelick, P.C. 2010. Vital pulp therapy-current progress of dental pulp regeneration and revascularization. Int. J. Dent. 2010: 856087.

Zheng, Y., Liu, Y., Zhang, C.M., Zhang, H.Y., Li, W.H., Shi, S. et al. 2009. Stem cells from deciduous tooth repair mandibular defect in swine. J. Dent. Res. 88: 249–54.

Zuk, P.A., Zhu, M., Mizuno, H., Huang, J., Futrell, J.W., Katz, A.J. et al. 2001. Multilineage cells from human adipose tissue: implications for cell-based therapies. Tissue Eng. 7: 211–28.

7

Composite Materials for Oral and Craniofacial Repair or Regeneration

Teresa Russo, *Roberto De Santis** and *Antonio Gloria*

Introduction

Over the past decade, a wide range of degradable, partially degradable and non-degradable polymer based composites has been investigated to repair or to regenerate hard tissues in oral and craniofacial surgery. These composites can be prepared in the laboratory and then implanted or they can be polymerized *in situ*. For the former approach, the main advancements arise from Additive Manufacturing (AM) technologies, also known as 3D printing, while injectable or spreadable nanocomposites represent the main achievements for *in situ* forming of prostheses, restorative materials and scaffolds for Tissue Engineering (TE). Self-shape adaptation of injectable or spreadable nanocomposites, and the overall reduced time from diagnosis to implantation, is the most convenient approach for dental and cranial bone tissues repair or regeneration. However, several drawbacks such as heat and shrinkage due to the polymerization process and release of unreacted monomers, limit the *in situ* forming approach. On the other hand, patient tailored prostheses and scaffolds, designed and manufactured in the laboratory, involve the use of the Reverse Engineering (RE) applied to organs and hard tissues for defining, *via* Computer Aided Design (CAD), the customized prosthesis or scaffold. 3D imaging clinical tools like X-ray CT, MRI and Laser scanners provide the main data source for developing the digital model. Implant designing, composite materials and engineering technologies, as well as future trends in the field, will be focused.

Surface scanners provide the 3D point cloud representing anatomical features which undergoes an editing process involving: filtering and cleaning of the 3D point cloud data-set (Fig. 7.1a), slicing according to a determined direction (Fig. 7.1b), definition of oriented parametric curves (Fig. 7.1c), and definition of parametric surfaces and mesh (Fig. 7.1d). A variety of software has been developed to edit the 3D point cloud (e.g., RapidForm) (Da Silveira et al. 2003; De Santis et al. 2018).

Volume scanners (e.g., X-ray CT) data undergoes a segmentation process to distinguish the variety of tissues constituting the organ: mandible ramous cross section (Fig. 7.1e), mandible arch

IPCB-CNR Institute of Polymers, Composites and Biomaterials, National Research Council of Italy, Viale J.F. Kennedy, 54 Mostra D'Oltremare Pad. 20, 80125 Naples, Italy.

* Corresponding author: rosantis@unina.it

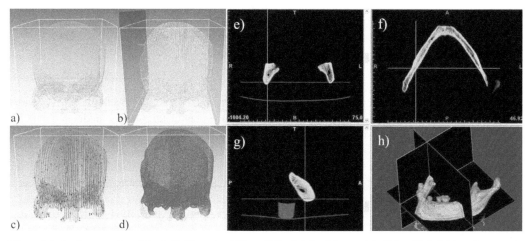

Figure 7.1. Surface scanning of the skull: Cleaned and filtered 3D point cloud data-set (a), slicing (b), definition of parametric curves (c), definition of parametric surfaces and mesh (d), volume scanning of the mandible: mandible ramous cross section (e), mandible arch cross section (f), condyle cross section (g), 3D mandible model (h).

cross section (Fig. 7.1f), condyle cross section (Fig. 7.1g), 3D mandible model (Fig. 7.1h). A variety of software has been developed to edit and process volume data (e.g., Mimics) (Dean et al. 2003; De Santis et al. 2005). The outer surface of anatomical features is often described by an STL format, which is also the common file format in AM.

Cranioplasty

Cranioplasty is a common procedure for correcting bone defects or deformities in the cranium arising from traumatic skull bone fracture, cranial bone deformities and bone cancer and infections. Surgery involving the use of biocompatible materials, and repair or regeneration of large cranial defects is particularly challenging from both the functional and aesthetic point of view (De Santis et al. 2010; Piitulainen et al. 2015; Unterhofer et al. 2017).

At least four approaches involving the use of polymers and composites can be categorized. The first and simplest approach is the *in situ* application of the biomaterial (Costantino et al. 1993; Lee et al. 2009). The second approach is *ex vivo* and considers a plaster impression taken over the defect for realizing a mould into which the prostheses or scaffold is formed (Caro-Osorio et al. 2013; Jaberi et al. 2013). The third *ex vivo* approach uses a 3D scan of the defect for realizing, through AM, a mould into which the prostheses is formed (Saringer et al. 2002; Peltola et al. 2012; Msallem et al. 2017). The fourth approach uses 3D scanning of the defect and RE in conjunction with AM to directly manufacture the prosthesis or scaffold (Espalin et al. 2010; Han et al. 2019).

3D surface imaging through laser scanners and photogrammetry are gaining popularity in this filed as these approaches are less invasive than other clinical imaging tools such as X-ray CT (Arridge et al. 1985; Mertens et al. 2017).

Polymer based Composite Materials in Cranioplasty

Mechanical performances play a crucial role in the design of composite scaffolds for hard TE applications: hard tissues such as bone are stiffer (higher elastic modulus) and stronger (higher strength) than soft tissues. In this context, polymer-based composite materials provide an alternative choice to overcome the problems related to the flexibility and weakness of many polymers. Several criteria have to be considered in designing composite materials. In particular, the stress transfer from polymeric matrix to the reinforcement plays a pivotal role in enhancing composite strength, whilst the different ductility between the polymeric matrix and the reinforcement should create discontinuities in the stress transfer and the generation of stress concentration at the particle/matrix interface.

Poly-Methyl-Methacrylate

Poly-Methyl-Methacrylate (PMMA) is the polymer most widely used for cranioplasty. Its history is longer than half a century and in some instance it showed better long-term outcomes compared to frozen autologous bone (Moreira-Gonzalez et al. 2003; Piitulainen et al. 2015). PMMA can be adopted according to all the four approaches listed earlier. For the *in situ* forming approach, PMMA is provided to surgeons in the form of a solid powder phase made of PMMA and/or copolymers and a liquid monomer component. By mixing the powder and liquid phases the paste is applied *in vivo* and a radical polymerization reaction occurs driven by benzoyl-peroxide and amines (i.e., activators/co-initiators for the formation of radicals) contained in the powder and liquid phase, respectively (Harris et al. 2014; De Santis et al. 2003). Once polymerized, mechanical properties (Table 7.1) of PMMA are in between those of the spongy and cortical bone (McElhaney et al. 1970; Ronca et al. 2014). Heat developed during the exothermic reaction of PMMA limits the direct intra-operative use of this bone cement, especially if large cranial defects need to be restored. However, PMMA can be easily processed through the moulding strategies (van Putten et al. 1992; Cheng et al. 2018; Maricevich et al. 2019). The main advancement of PMMA relies on modifications of the acrylic cement to improving biological and mechanical functions. The loading of antibiotics (i.e., gentamicin)

Table 7.1. Young's modulus of polymer based composite materials.

Composite material	Testing condition	Young's modulus [GPa]	Reference
Cortical/trabecular bone from the skull	Compression	2.41–5.58	McElhaney et al. 1970
Cortical/trabecular bone from the macaca mulatta	Compression	6.47	
Trabecular bone from tibia	Bending	0.05–0.3	Ronca et al. 2014
PMMA (Symplex-P)	Bending	2.6	De Santis et al. 2003
PMMA (CMW1/Gentamicin)	Bending	2.3	
PMMA/HA	Bending	2.0	Zebarjad et al. 2011
PMMA/AgNP	Bending	1.7–2.4	Oei et al. 2012
PEEK 450G	Bending	3.8	Arif et al. 2018
PEEK 450G	Tensile	4.0	
PEI	Bending	9.1	De Santis et al. 2000
PEI	Tensile	14.3	
PEEK/PEI/TiO$_2$	Tensile	5.3	Diez-Pascual and Diez-Vincente 2015
C-FRP-PEEK	Tensile	7.4	Han et al. 2019
G-FRP-PEI	Bending	10.9	De Santis et al. 2000
C-FRP-PEI	Bending	57.7	
PCL	Compression	0.60	Wang et al. 2018
PCL	Tensile	0.59	De Santis et al. 2011
PCL/FeHA	Tensile	0.68	
PLA	Tensile	1.5	Mi et al. 2013
PLA	Compression	0.63	
PGA	Tensile	6.9	Yang et al. 2001
PLGA	Tensile	1.4–2.8	

is suggested to prevent infection (Minelli et al. 2011; Worm et al. 2016). However, a slight reduction of mechanical properties for gentamicin loaded PMMA has been observed (De Santis et al. 2003). The realization of PMMA based nano-composites has been suggested for antimicrobial purposes by incorporating silver (Oei et al. 2012) and gold (Russo et al. 2017) nanoparticles into the polymeric matrix. The integration of phase-change particles has been suggested to reduce temperature levels during polymerization (De Santis et al. 2006). Several PMMA based composites for cranioplasty have been advised for improving osteoconductivity and biocompatibility. Hydroxyapatite (Itokawa et al. 2007; Zebarjad et al. 2011) and bioactive glass (Peltola et al. 2012) represent the most common type of particles functionalizing and reinforcing the acrylic matrix (Table 7.1). Medical-grade PMMA filaments represent the future trends in cranioplasty through AM (Espalin et al. 2010), and gentamicin doped filaments have also been investigated (Mills et al. 2018).

Poly-Ether-Ether-Ketone and Poly-Ether-Imide

A growing interest in alloplastic cranioplasty through high-performance polymers, such as Poly-Ether-Ether-Ketone (PEEK) and Poly-Ether-Imide (PEI), is also recognized. PEEK and PEI are considered expensive thermoplastic semicrystalline polymer showing outstanding mechanical (Table 7.1) and chemical resistance (Arif et al. 2018; Cicala et al. 2018). PEEK cranioplasty is radiolucent and can be sterilized by steam or gamma irradiation (Lethaus et al. 2012; Shah et al. 2014). Short-term follow-up indicates that PEEK is a material of choice especially for patients showing a premorbid neurological status (Lethaus et al. 2011). A large patient series receiving a PEEK temporo-parietal region reconstruction support the high potential of this material for large cranio-facial defects (Ng and Nawaz 2014). A recent ten-year follow up suggests that PEEK cranioplasty leads to a normal progressive healing similar to titanium mesh (Piitulainen et al. 2015). Medical-grade PEEK filaments represent the future trends in cranioplasty through cutting edge technologies of AM (Honigmann et al. 2018), and carbon fibre reinforced PEEK filaments provide further mechanical enhancement (Han et al. 2019). On the other hand, PEI is a well-known polymer in the biomedical field for its biocompatibility (Richardson et al. 1993; Merolli et al. 1999), excellent mechanical properties (De Santis et al. 2000; Cicala et al. 2018) chemical and thermal stability (Johnson and Burlhis 1983). The aforementioned excellent properties, together with the easy sterilization capability (Eltorai et al. 2015), suggested PEI for the manufacturing of surgical guides for cranioplasty (Bekeny et al. 2013; Yang et al. 2018). PEI composites reinforced with continuous carbon and glass fibres have been proposed for the realization of advanced composite bone prostheses (De Santis et al. 2019; Gloria et al. 2008). PEI based composites, incorporating 4%ww titanium dioxide nanoparticles, have further shown antibacterial action versus Gram-positive and Gram-negative (Díez-Pascual et al. 2015). In the last decade, filaments of PEI and PEI blends (also known as ULTEM), are gaining attention for AM processes, representing the future trends in orthopaedics (Parthasarathy 2015; Lee et al. 2017).

Glass or Carbon Fibre Reinforced Composites

Bioactive composite sandwich, consisting of Glass or Carbon Fibre Reinforced Composites (G-FRC or C-FRC) laminates and a core of Bioactive Glass (BG) particles, have also been considered for cranioplasty. G-FRC based on a methacrylic resin matrix coated with BG granules has shown the potential to promote the healing process of calvarial bone defects in rabbits (Tuusa et al. 2008). Soon after, G-FRC based on a dimethacrylic resin matrix incorporating BG granules was used, on humans, for a complex clinical case of a patient undergoing repetitive cranioplasty. The G-FRC/BG explant (after 27 months) showed osteoid formation together with small clusters of mature hard tissue being observed at the margin of the cranioplasty, suggesting that this composite represents a feasible method for calvarial reconstruction (Posti et al. 2016). A G-FRC/BG cranioplasty, based on a photopolymerized dimethacrylic resin matrix and a highly porous BG structure, has also shown antimicrobial and osteoconductivity features (Aitasalo et al. 2014). The advantage of using a photo-cured G-FRC laminate sandwich in conjunction with a BG core is that the material can be precisely layered before polymerization occurs and high load-bearing capability can be expected if the external composite laminates are properly spaced (Piitulainen et al. 2017). A three-year follow-up investigation on paediatric patients with a large cranial defect restored with a thin G-FRC laminate and a core BG scaffold concluded that this composite is a safe and a functional solution for restoring cranial defects in paediatric populations (Piitulainen et al. 2015). C-FRP medical grade implants consist of a woven carbon fabric impregnated with a biocompatible thermoset epoxy matrix. Each C-FRP lamina is laid-up on the positive templates of the cranial defect, and a 3 years follow-up on 27 patients has shown no postoperative complications (Saringer et al. 2002). Although some complications such as infection and adverse reactions may occur, excellent cranioplasty restorations through C-FRP have been observed on a wider sample (Wurm et al. 2004).

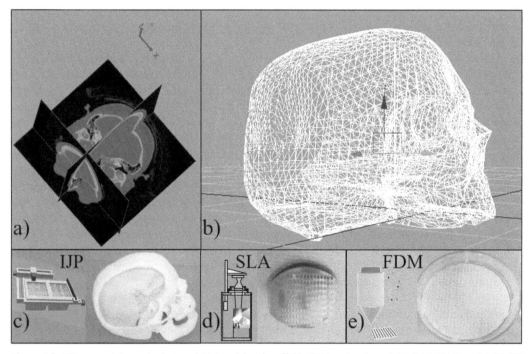

Figure 7.2. RE approach in cranioplasty and AM model and scaffolds. 3D imaging through TAC (a), CAD model (b), IJP skull solid model (c), SLA scaffold model (d), FDM scaffold model (e).

Degradable Polymer based Scaffolds

Degradable scaffolds for cranial TE (Hassan et al. 2019) are particularly important for paediatric patients whose skull growth adds an additional concern in the selection of non-degradable material. Poly-Lactic-Acid (PLA) outer membranes and a HA cement inner core represent a first example for regenerating skull bone defect (Cohen et al. 2004). Tensile and compressive Young's modulus (Table 7.1) of PLA are in between those of a trabecular and cortical bone (Mi et al. 2013). A fully degradable cranioplasty has been achieved following the same strategy but using β-tricalcium phosphate granules enclosed between custom-moulded PLGA mesh (Thesleff et al. 2017). Poly-Lactic-co-Glycolic-Acid (PLGA) membranes have also been used in conjunction with autologous bone according to the *in situ* approach to regenerate skull deformities, suggesting that this approach is safe with tolerable morbidity rates in children (Gephart et al. 2013). Following a similar approach, a PLGA mesh has been layered *in vivo* onto on the dura, a demineralized bone matrix is applied onto the bio-degradable mesh, and a second PLGA mesh is layered to form a sandwich like structure (Chao et al. 2009). This cranioplasty technique is useful if the autologous bone is not available, and it has also promoted bone regeneration with no morbidity. Another interesting approach for cranial bone regeneration has used PLGA meshes contained in the inner core β-tricalcium phosphate granules in conjunction with autologous adipose derived stem cells (Thesleff et al. 2017). PLGA scaffolds with a morphologically controlled micro-architecture, realized through FDM, have been implanted in the parietal skull defect in mice model; with three weeks of *in vivo* observations, angiogenesis originating from desmal bone was evident (Sinikovic et al. 2011). PLGA micro-sphere incorporating bone morphogenetic proteins further enhanced the capability of TE to regenerate cranial bone tissue (Wink et al. 2014). Young's modulus of PGA is close to that of the cortical bone (Table 7.1), while the modulus of PLGA can be tailored over a wide range according to the copolymer composition (Yang et al. 2001). Poly-Capro-Lactone (PCL) is another interesting degradable polyester used for the manufacture of morphologically controlled and full interconnected architectures for bone regeneration (Russo et al. 2013). This polymer has mechanical properties similar to a dense spongy bone (Table 7.1) and its low melting point allows an easy process through AM (De Santis et al. 2011).

Nowadays, the possibility to obtain scaffolds for the repair of skeletal defects in the field of tissue engineering should present a viable alternative to conventional treatment (Hutmacher 2000).

An ideal scaffold should promote bone regeneration and also target the process to certain regions. In particular, it should be advantageous for surgical correction of craniosynostosis in encouraging the formation of cranial bone tissues, while preventing resynostosis (Fedore et al. 2017). Furthermore, medical imaging data would allow for precise and customized replication of the architecture of the skull and bony defect, in a combined approach of RE and AM (Chim and Schantz 2005), thus simplifying surgical placement and retention.

With the latest advancements of 3D scanning, design software and printing technologies, 3D printing of individually customized tissues scaffolds can be created for clinical use (Morrison et al. 2015).

In this scenario, the best combination of materials properties and technologies advancement still represents a challenge in 3D printing research for *in vivo* applications.

PCL scaffolds, manufactured through AM and seeded with mesenchymal progenitor cells and calvarial osteoblasts, have been tested in critical-size defects of the calvarial rabbit model, and results suggest that these 3D constructs have the potential for bone regeneration (Schantz et al. 2003). Short-term human clinical trial suggests that PCL scaffolds realized through AM and coated with bone marrow are well tolerated, have excellent biocompatibility, act as a bone guiding template, and can be firmly anchored to the hosting calvarial tissue (Schantz et al. 2006). A custom-made PCL scaffolds that allow for calvarial osteoblast differentiation and mineralization *in vitro* can be utilized to prevent osteoblast differentiation if coated with polyethylene glycol hydrogel (PEG hydrogel), thus providing a specific control in bone formation during tissue regeneration (Fedore et al. 2017). PCL composites, incorporating calcium phosphate at a weight ratio 1:1, have been implanted in the calvaria of a sheep model, and have shown desirable osteoconductivity and osteointegration (Wang et al. 2018).

A dynamic seeding of osteoblasts and endothelial cells onto a 3D fibre-deposited PCL scaffold may represent a useful approach in order to achieve a functional hybrid in which angiogenesis, furnished by neo-vascular organization of endothelial cells, may further support osteoblasts growth (Kyriakidou et al. 2008).

However, the poor cytocompatibility of the synthetic polymers leads to the inefficiency of the scaffold in obtaining a friendly interface with living cells. Consequently, in order to improve the cell guidance ability of synthetic PCL-based scaffolds (Fig. 7.3a), different modifications have been proposed, the aim being to influence surface properties (i.e., surface charge, wettability, roughness and topography). Surface treatment such as γ-ray irradiation, plasma treatment, endgrafting, ozone oxidization or *in situ* polymerization have been proposed to modify the materials surface properties.

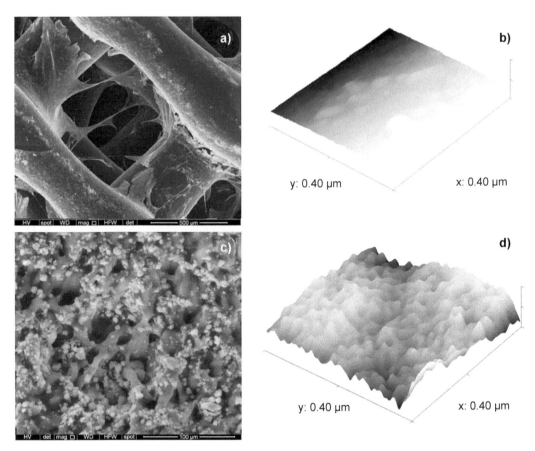

Figure 7.3. Polyester based composite scaffolds. SEM image of PCL scaffold obtained through FDM and seeded with hMSC (a), Atomic force microscopy showing the topography of neat PCL (b), SEM image of PEG/HA nanocomposite scaffold obtained through SLA (c), Atomic force microscopy showing the topography of PEG/HA (d).

However the possibility to introduce amino groups onto the polyester surface through a reaction with diamine will result in decreasing of surface hydrophobicity, neutralization of the acid originated from the scaffold degradation, and the possibility to provide active sites through which other biomolecules such as gelatin, collagen or arginine-glycine-aspartic acid (Arg-Gly-Asp or RGD) peptides can be immobilized. PCL containing laminin-derived peptides sequences IKVAV (Ile-Lys-Val-Ala-Va), YIGSR (Tyr-Ile-Gly-Ser-Arg), or integrin-derived peptides sequence RGD, covalently linked to the polymer surface evidenced an enhancement in cell adhesion and spreading (Santiago et al. 2006; Causa et al. 2010; Gloria et al. 2012).

The conjugation of amine-terminated peptides by means of reductive amination after tether insertion on 3D well-organized scaffolds obtained through AM approach has shown a specific recognition of the solid signal to NIH3T3 integrin cell receptors suggesting a correct presentation of the peptide sequences, taking into consideration its effect on their macromechanical behaviour.

Composite scaffolds based on a PCL matrix have also been investigated for bone TE.

To overcome the limitation of cell penetration and osteoconductivity improvement of the PCL-based implants for craniofacial tissue engineering, composite scaffold of polyvinyl alcohol (PVA), PCL and Bioceramic (HAB) have been proposed. PVA has been adopted since it introduces several free hydroxyl chains, which can be used for scaffold functionalization via linking drugs, biomolecules, or growth factors (Orienti et al. 2001; Prabha et al. 2018), whilst HAB represents a triphasic bioceramic developed by incorporation of hydroxyapatite, beta tricalcium phosphate, calcium silicates and traces of magnesium synergistically acting in improving to produce an osteoconductivity and osteoinductivity (Jones 2013; Prabha et al. 2018). Results indicate that these composite structures should support attachment and growth of stromal stem cells (human bone marrow skeletal (mesenchymal) stem cells (hMSC) and Dental Pulp Stem Cells (DPSC)). In addition, the scaffold supported *in vitro* osteogenic differentiation and *in vivo* vascularized bone formation.

Bioactive PCL and titanium as well as PCL and zirconium organic–inorganic materials have been suitably synthesized by the sol-gel method, highlighting that these organic–inorganic hybrid materials exhibit a hydroxyapatite layer on the surfaces of samples soaked in a fluid simulating the composition of human blood plasma (Russo et al. 2010).

Mechanical performances play a crucial role in the design of composite scaffolds for hard tissue engineering applications: hard tissues such as bone are stiffer (higher elastic modulus) and stronger (higher strength) than the soft tissues. In this context, polymer-based composite materials provide an alternative choice to overcome the problems related to the flexibility and weakness of neat polymers (i.e., PCL). Several criteria have to be considered in designing composite materials. In particular, the stress transfer from polymeric matrix to the reinforcement plays a pivotal role in enhancing composite strength, whilst the different ductility between the polymeric matrix and the reinforcement should create discontinuities in the stress transfer and the generation of stress concentration at the particle/matrix interface.

Thus, composite materials based on PCL matrix reinforced with PCL/TiO_2 or PCL/ZrO_2 hybrid fillers should enhance the mechanical performance of the neat PCL and, at the same time, should improve their bioactivity and biological performances (Russo et al. 2010).

In the innovative concept of magnetic cell guidance, nano-composite polyester substrates incorporating magnetic nanoparticles (MNPs) or iron doped nano-HA particles (Fig. 7.3c) has been also analyzed (Gloria et al. 2013). MNPs may be heated up, allowing their use as hyperthermia agents able to deliver thermal energy to targeted bodies (i.e., tumours) or as elements capable of improving chemotherapy or radiotherapy by providing a degree of tissue warming appropriate for the destruction of malignant cells. The idea of manipulating and controlling specific processes at the cell level by releasing biomolecules and bioactive factors, in turn, linked to magnetic nanocarriers has been proposed. Controlled differentiation of human bone marrow stromal cells using remote magnetic field activation and MNPs has been proposed (Kanczler et al. 2010).

The idea to develop magnetic scaffolds for additionally controlling angiogenesis *in vivo* was first considered by (Bock et al. 2010).

A magnetic scaffold should modify the external magnetic flux distribution, thus causing a much higher concentration of magnetic flux in the vicinity of and inside the scaffold (Gloria et al. 2013). In this context, enhanced cell-material interaction, at least *in vitro* have been proved when PCL scaffolds were reinforced by the inclusion of iron-doped hydroxyapatite (Gloria et al. 2013), while they have promoted, *in vivo* on a rabbit model, new bone formation after 4 weeks from implantation (De Santis et al. 2015a).

Technologies of Composite Materials in Cranioplasty

As reported earlier, technologies of composite materials in cranioplasty can be distinguished according to the surgical procedure. Cranioplasty based on the *in situ* or *in vivo* forming of biomaterials mainly regard the use of PMMA and its composites. It consists of a single step procedure applied intra-operatively (Spence 1954), the liquid and solid PMMA phases are mixed and malleable paste is placed onto the skull defect and shaped for achieving a smooth prosthesis conforming to the normal contours of the patient's skull (Lee et al. 2009; Bloch and McDermott 2011). The use of cold saline solution irrigation or of a damp gauze in saline solution placed between the setting acrylic resin and dura tissue reduce temperature levels due to the heat developed during polymerization (Jaberi et al. 2013; Worm et al. 2016). The polymer based prosthesis is held in place using titanium plates and screws (Caro-Osorio et al. 2013). The *in situ* forming approach can also be applied for regenerative purposes. A PLA or a PLGA membrane is applied on the dura, then autologous, demineralized bone or HA cements is spread over the membrane, and finally a second PLA or PLGA layer is applied to form a degradable sandwich-like structure (Cohen et al. 2004; Chao et al. 2009; Gephart et al. 2013).

Cranioplasty based on moulding technologies, also known as cranial cast fabrication, represent the most common *ex vivo* surgical approach. The first attempts have used the original skull bone flap (Caro-Osorio et al. 2013) or impressions of the defect taken with a gel or wax (Jaberi et al. 2013) to realize the silicone mould into which the prostheses is formed. Interestingly, an approach to realize a 3D solid model of the skull starting from X-ray CT dates back to 1992; polycarbonate layers representing cross-sections of the skull and its defect have been obtained through CAD/CAM milling, and the 3D model has been realized by glueing the milled polycarbonate sheets in a layer by layer fashion typical of AM (van Putten Jr and Yamada 1992). More recently, the RE approach in conjunction with AM (Fig. 7.2), allowed the 3D printing of the cranial defect and hence the realization of the mould into which the prostheses is formed. Fused Deposition Modelling (FDM), Stereo-Lithography (SLA), Ink-Jet Printing (IJP) and Laser Sintering (LS) are the main AM technologies used for the direct or indirect manufacturing of the skull defect (De Santis et al. 2010). Most of the 3D printed skull defect replica have been realized through FDM, however SLA provide higher accuracy in reproducing the solid model (De Santis et al. 2015b). Acrylonitrile butadiene styrene skulls, including bone defects, have been realized through FDM and directly used as moulding template by spreading paraffin oil over the defect to prevent the sticking of the prostheses to the mould (Unterhofer et al. 2017). Using the same material and technology the skull defect has been realized through the indirect moulding approach (Lee et al. 2009; Bloch and McDermott 2011). The IJP of photopolymerizable resin has been used to create the positive shape of the cranial defect around which the silicone mould has been realized (Cheng et al. 2018). A similar indirect method to realize the mould has been achieved through SLA using alginate as final material into which the cranial composite prosthesis is realized (Dean et al. 2003; Peltola et al. 2012). Very recently, polyamide powders have been processed through LS to replicate the positive shape of the cranial defect, hence the silicon mould has been realized around the printed object (Maricevich et al. 2019).

3D printing of biomaterials, without the use of a mould, represent the cutting edge technology in bone cranioplasty (Fig. 7.2). This integrated approach combines RE and AM (Chim and Schantz 2005). Polyester based scaffolds have been realized through AM with a porosity higher than 50% and a mean pore dimension of about 400 μm showed the best bone tissue regeneration (Hassan et al. 2019). FDM is the technology more frequently adopted to manufacture degradable scaffolds for cranial bone regeneration. Through FDM a continuous thermoplastic fibre is deposited from the melt state, and the distance between adjacent fibres and the layer stacking sequence determine the pore size dimension and the scaffold porosity (Gloria et al. 2010). PCL scaffolds have been manufactured through FDM and implanted in critical-size defects of the calvarial rabbit model (Schantz et al. 2003). FDM has also been used to manufacture customized PCL scaffolds for human cranial bone regeneration (Schantz et al. 2006) PLGA scaffolds with a morphologically controlled micro-architecture, realized through FDM (pellets feeding, nozzle diameter 250 μm, distance between fibres 500 μm), have been implanted in the skull defect in mice model; the whole device included a 3D printed PEI chamber designed for both supporting PLGA scaffold implantation and promoting vascularization arising from the dura matter (Sinikovic et al. 2011). Filament fed FDM has instead been used to process PMMA (Espalin et al. 2010) gentamicin doped PMMA (Mills et al. 2018) and PEEK filaments (Honigmann et al. 2018).

Future Trends

Different scenarios may be foreseen in relation to the development of advanced cranioplasty including, but not limited to polymer based composite design, additive manufacturing techniques and the tissue regeneration approach. The emerging additive manufacturing approach is invading all the engineering fields, including biomaterials, showing a very high potential especially as customization, complexity and low weight design are concerned. It is not surprising that this approach has recently been applied to cranioplasty. 3D printing of biomaterials, without the use of a mould, represents the cutting edge technology in bone cranioplasty. This integrated approach combines reverse engineering and additive manufacturing. Recent advances in additive manufacturing technologies for direct production of implants avoids limitations related to the constraints in shape, size and internal structure. On the other hand, from a material point of view, medical-grade PEEK filaments represent the future trends in cranioplasty, and carbon fibre reinforced PEEK filaments provide further mechanical enhancement.

References

Aitasalo, K.M., Piitulainen, J.M., Rekola, J. and Vallittu, P.K. 2014. Craniofacial bone reconstruction with bioactive fiber-reinforced composite implant. Head. Neck. 36(5): 722–728.

Arif, M.F., Kumar, S., Varadarajan, K.M. and Cantwell, W.J. 2018. Performance of biocompatible PEEK processed by fused deposition additive manufacturing. Mater. Des. 146: 249–259.

Arridge, S., Moss, J.P., Linney, A.D. and James, D.R. 1985. Three dimensional digitization of the face and skull. J. Maxillofac. Surg. 13: 136–143.

Bekeny, J.R., Swaney, P.J., Webster III, R.J., Russell, P.T. and Weaver, K.D. 2013. Forces applied at the skull base during transnasal endoscopic transsphenoidal pituitary tumor excision. Neurol. Surg. B Skull Base 74(06): 337–341.

Bloch, O. and McDermott, M.W. 2011. *In situ* cranioplasty for hyperostosing meningiomas of the cranial vault. Can. J. Neurol. Sci. 38(1): 59–64.

Bock, N., Riminucci, A., Dionigi, C., Russo, A., Tampieri, A., Landi, E. et al. 2010. A novel route in bone tissue engineering: magnetic biomimetic scaffolds. Acta Biomater. 6: 786–796.

Caro-Osorio, E., De la Garza-Ramos, R., Martínez-Sánchez, S.R. and Olazarán-Salinas, F. 2013. Cranioplasty with polymethylmethacrylate prostheses fabricated by hand using original bone flaps: Technical note and surgical outcomes. Surg. Neurol. Int. 8: 136.

Causa, F., Battista, E., Della Moglie, R., Guarnieri, D., Iannone, M. and Netti, P.A. 2010. Surface investigation on biomimetic materials to control cell adhesion: the case of RGD conjugation on PCL. Langmuir 26: 9875–9884.

Chao, M.T., Jiang, S., Smith, D., DeCesare, G.E., Cooper, G.M., Pollack, I.F. et al. 2009. Demineralized bone matrix and resorbable mesh bilaminate cranioplasty: a novel method for reconstruction of large-scale defects in the pediatric calvaria. Plast. Reconstr. Surg. 123(3): 976–982.

Cheng, C.H., Chuang, H.Y., Lin, H.L., Liu, C.L. and Yao, C.H. 2018. Surgical results of cranioplasty using three-dimensional printing technology. Clin. Neurol. Neurosurg. 168: 118–123.

Chim, H. and Schantz, J.T. 2005. New frontiers in calvarial reconstruction: integrating computer-assisted design and tissue engineering in cranioplasty. Plast. Reconstr. Surg. 116(6): 1726–1741.

Cicala, G., Ognibene, G., Portuesi, S., Blanco, I., Rapisarda, M., Pergolizzi, E. et al. 2018. Comparison of Ultem 9085 used in fused deposition modelling (FDM) with polytherimide blends. Materials 11(2): 285.

Cohen, A.J., Dickerman, R.D. and Schneider, S.J. 2004. New method of pediatric cranioplasty for skull defect utilizing polylactic acid absorbable plates and carbonated apatite bone cement. J. Craniofac. Surg. 15(3): 469–472.

Costantino, P.D., Friedman, C.D. and Lane, A. 1993. Synthetic biomaterials in facial plastic and reconstructive surgery. Facial. Plast. Surg. 9(01): 1–5.

Da Silveira, A.C., Daw, J.L., Kusnoto, B., Evans, C. and Cohen, M. 2003. Craniofacial applications of three-dimensional laser surface scanning. J. Craniofac. Surg. 14(4): 449–56.

De Santis, R., Ambrosio, L. and Nicolais, L. 2000. Polymer-based composite hip prostheses. J. Inorg. Biochem. 79(1-4): 97–102.

De Santis, R., Mollica, F., Ambrosio, L., Nicolais, L. and Ronca, D. 2003. Dynamic mechanical behavior of PMMA based bone cements in wet environment. J. Mater. Sci. Mater. Med. 14(7): 583–594.

De Santis, R., Mollica, F., Prisco, D., Rengo, S., Ambrosio, L. and Nicolais, L. 2005. 3D analysis of mechanically stressed dentin–adhesive–composite interfaces using X-ray micro-CT. Biomaterials 26(3): 257–270.

De Santis, R., Ambrogi, V., Carfagna, C., Ambrosio, L. and Nicolais, L. 2006. Effect of microencapsulated phase change materials on the thermo-mechanical properties of poly(methyl-methacrylate) based biomaterials. J. Mater. Sci. Mater. Med. 17(12): 1219–1226.

De Santis, R., Gloria, A. and Ambrosio, L. 2010. Materials and technologies for craniofacial tissue repair and regeneration. Top. Med. 16: 1–6.

De Santis, R., Gloria, A., Russo, T., D'Amora, U., Zeppetelli, S., Tampieri, A. et al. 2011. A route toward the development of 3D magnetic scaffolds with tailored mechanical and morphological properties for hard tissue regeneration: Preliminary study: A basic approach toward the design of 3D rapid prototyped magnetic scaffolds for hard-tissue regeneration is presented and validated in this paper. Virtual Phys. Prototyp. 6(4): 189–195.

De Santis, R., Russo, A., Gloria, A., D'Amora, U., Russo, T., Panseri, S. et al. 2015a. Towards the design of 3D fiber-deposited poly(ε-caprolactone)/iron-doped hydroxyapatite nanocomposite magnetic scaffolds for bone regeneration. J. Biomed. Nanotechnol. 11(7): 1236–1246.

De Santis, R., D'Amora, U., Russo, T., Ronca, A., Gloria, A. and Ambrosio, L. 2015b. 3D fibre deposition and stereolithography techniques for the design of multifunctional nanocomposite magnetic scaffolds. J. Mater. Sci. Mater. Med. 26(10): 250.

De Santis, R., Gloria, A., Viglione, S., Maietta, S., Nappi, F., Ambrosio, L. et al. 2018. 3D laser scanning in conjunction with surface texturing to evaluate shift and reduction of the tibiofemoral contact area after meniscectomy. J. Mech. Behav. Biomed. Mater. 88: 41–47.

De Santis, R., Guarino, V. and Ambrosio, L. 2019. Composite biomaterials for bone repair. pp. 273–299. *In*: Planell, J.A. (ed.). Bone Repair Biomaterials. Woodhead Publishing, Sawston, Cambridge, UK.

Dean, D., Min, K.J. and Bond, A. 2003. Computer aided design of large-format prefabricated cranial plates. J. Craniofac. Surg. 14(6): 819–832.

Díez-Pascual, A.M. and Díez-Vicente, A.L. 2015. Nano-TiO$_2$ reinforced PEEK/PEI blends as biomaterials for load-bearing implant applications. ACS Appl. Mater. Interfaces 7(9): 5561–5573.

Eltorai, A.E., Nguyen, E. and Daniels, A.H. 2015. Three-dimensional printing in orthopedic surgery. Orthopedics 38(11): 684–687.

Espalin, D., Arcaute, K., Rodriguez, D., Medina, F., Posner, M. and Wicker, R. 2010. Fused deposition modeling of patient-specific polymethylmethacrylate implants. Rapid Prototyp. J. 16(3): 164–173.

Fedore, C.W., Tse, L.Y.L., Nam, H.K., Barton, K.L. and Hatch, N.E. 2017. Analysis of polycaprolactone scaffolds fabricated via precision extrusion deposition for control of craniofacial tissue mineralization. Orthod. Craniofac. Res. 20: 12–17.

Gephart, M.G., Woodard, J.I., Arrigo, R.T., Lorenz, H.P., Schendel, S.A., Edwards, M.S. et al. 2013. Using bioabsorbable fixation systems in the treatment of pediatric skull deformities leads to good outcomes and low morbidity. Childs Nerv. Syst. 29(2): 297–301.

Gloria, A., Manto, L., De Santis, R. and Ambrosio, L. 2008. Biomechanical behavior of a novel composite intervertebral body fusion device. J. Appl. Biomater. Biomech. 6(3): 163–169.

Gloria, A., De Santis, R. and Ambrosio, L. 2010. Polymer-based composite scaffolds for tissue engineering. J. Appl. Biomater. Biomec. 8(2): 57–67.

Gloria, A., Causa, F., Russo, T., Battista, E., Della Moglie, R., Zeppetelli, S. et al. 2012. Three-dimensional poly(ε-caprolactone) bioactive scaffolds with controlled structural and surface properties. Biomacromolecules 13: 3510−3521.

Gloria, A., Russo, T., D'Amora, U., Zeppetelli, S., D'Alessandro, T., Sandri, M. et al. 2013. Magnetic poly(ε-caprolactone)/iron-doped hydroxyapatite nanocomposite substrates for advanced bone tissue engineering. J. Royal Soc. Interface 10(80): 20120833.

Han, X., Yang, D., Yang, C., Spintzyk, S., Scheideler, L., Li, P. et al. 2019. Carbon fiber reinforced PEEK composites based on 3D-printing technology for orthopedic and dental applications. J. Clin. Med. 8(2): E240.

Han, X., Sharma, N., Xu, Z., Scheideler, L., Geis-Gerstorfer, J., Rupp, F. et al. 2019. An *in vitro* study of osteoblast response on fused-filament fabrication 3D printed PEEK for dental and cranio-maxillofacial implants. J. Clin. Med. 8(6): 771.

Harris, D.A., Fong, A.J., Buchanan, E.P., Monson, L., Khechoyan, D. and Lam, S. 2014. History of synthetic materials in alloplastic cranioplasty. Neurosurg. Focus 36(4): E20.

Hassan, M.N., Yassin, M.A., Suliman, S., Lie, S.A., Gjengedal, H. and Mustafa, K. 2019. The bone regeneration capacity of 3D-printed templates in calvarial defect models: A systematic review and meta-analysis. Acta Biomater. 91: 1–23.

Honigmann, P., Sharma, N., Okolo, B., Popp, U., Msallem, B. and Thieringer, F.M. 2018. Patient-specific surgical implants made of 3D printed PEEK: material, technology, and scope of surgical application. Biomed. Res. Int. 2018: 4520636.

Hutmacher, D.W. 2000. Scaffolds in tissue engineering bone and cartilage. Biomaterials 21: 2529–2543.

Itokawa, H., Hiraide, T., Moriya, M., Fujimoto, M., Nagashima, G., Suzuki, R. et al. 2007. A 12 month *in vivo* study on the response of bone to a hydroxyapatite–polymethylmethacrylate cranioplasty composite. Biomaterials 28(33): 4922–4927.

Jaberi, J., Gambrell, K., Tiwana, P., Madden, C. and Finn, R. 2013. Long-term clinical outcome analysis of poly-methyl-methacrylate cranioplasty for large skull defects. J. Oral Maxillofac. Surg. 71(2): e81–88.

Johnson, R.O. and Burlhis, H.S. 1983. Polyetherimide: A new high-performance thermoplastic resin. J. Polym. Sci. Symp. 70(1): 129–143.

Jones, J.R. 2013. Review of bioactive glass: From hench to hybrids. Acta Biomater. 9(1): 4457–4486.

Kanczler, J.M., Sura, H.S., Magnay, J., Attridge, K., Green, D., Oreffo, R.O.C. et al. 2010. Controlled differentiation of human bone marrow stromal cells using magnetic nanoparticle technology. Tissue Eng. 16: 3241–3250.

Kyriakidou, K., Lucarini, G. Zizzi, A., Salvolini, E., Mattioli Belmonte, M. Mollica, F. et al. 2008. Dynamic co-seeding of osteoblast and endothelial cells on 3D polycaprolactone scaffolds for enhanced bone tissue engineering. J. Bioact. Compat. Polym. 23: 227–243.

Lee, J.Y., An, J. and Chua, C.K. 2017. Fundamentals and applications of 3D printing for novel materials. Appl. Mater. Today 7: 120–133.

Lee, S.C., Wu, C.T., Lee, S.T. and Chen, P.J. 2009. Cranioplasty using polymethyl methacrylate prostheses. J. Clin. Neurosci. 16(1): 56–63.

Lethaus, B., ter Laak, M.P., Laeven, P., Beerens, M., Koper, D., Poukens, J. et al. 2011. A treatment algorithm for patients with large skull bone defects and first results. J. Craniomaxillofac. Surg. 39(6): 435–440.

Lethaus, B., Safi, Y., ter Laak-Poort, M., Kloss-Brandstätter, A., Banki, F., Robbenmenke, C. et al. 2012. Cranioplasty with customized titanium and PEEK implants in a mechanical stress model. J. Neurotrauma. 29(6): 1077–1083.

Maricevich, J.P., Cezar-Junior, A.B., de Oliveira-Junior, E.X., Silva, J.A., da Silva, J.V., Nunes, A.A. et al. 2019. Functional and aesthetic evaluation after cranial reconstruction with polymethyl methacrylate prostheses using low-cost 3D printing templates in patients with cranial defects secondary to decompressive craniectomies: A prospective study. Surg. Neurol. Int. 117: 443–452.

McElhaney, J.H., Fogle, J.L., Melvin, J.W., Haynes, R.R., Roberts, V.L. and Alem, N.M. 1970. Mechanical properties on cranial bone. J. Biomech. 3(5): 495–511.

Merolli, A., Perrone, V., Leali, P.T., Ambrosio, L., De Santis, R., Nicolais, L. et al. 1999. Response to polyetherimide based composite materials implanted in muscle and in bone. J. Mater. Sci. Mater. Med. 10(5): 265–268.

Mertens, C., Wessel, E., Berger, M., Ristow, O., Hoffmann, J., Kansy, K. et al. 2017. The value of three-dimensional photogrammetry in isolated sagittal synostosis: Impact of age and surgical technique on intracranial volume and cephalic index—a retrospective cohort study. J. Craniomaxillofac. Surg. 45(12): 2010–2016.

Mi, H.Y., Salick, M.R., Jing, X., Jacques, B.R., Crone, W.C., Peng, X.F. et al. 2013. Characterization of thermoplastic polyurethane/polylactic acid (TPU/PLA) tissue engineering scaffolds fabricated by microcellular injection molding. Materials Science and Engineering C 33(8): 4767–4776.

Mills, D.K., Jammalamadaka, U., Tappa, K. and Weisman, J. 2018. Studies on the cytocompatibility, mechanical and antimicrobial properties of 3D printed poly(methyl methacrylate) beads. Bioact. Mater. 3(2): 157–166.

Minelli, E.B., Della Bora, T. and Benini, A. 2011. Different microbial biofilm formation on polymethylmethacrylate (PMMA) bone cement loaded with gentamicin and vancomycin. Anaerobe 17(6): 380–383.

Moreira-Gonzalez, A., Jackson, I.T., Miyawaki, T., Barakat, K. and DiNick, V. 2003. Clinical outcome in cranioplasty: critical review in long-term follow-up. J. Craniofac. Surg. 14(2): 144–153.

Morrison, R.J., Hollister, S.J., Niedner, M.F., Mahani, M.G., Park, A.H., Mehta, D.K. et al. 2015. Mitigation of tracheobronchomalacia with 3D-printed personalized medical devices in pediatric patients. Sci. Transl. Med. 7: 285ra64.

Msallem, B., Beiglboeck, F., Honigmann, P., Jaquiéry, C. and Thieringer, F. 2017. Craniofacial reconstruction by a cost-efficient template-based process using 3d printing. Plast. Reconstr. Surg. Glob. Open 5(11): e1582.

Ng, Z.Y. and Nawaz, I. 2014. Computer-designed PEEK implants: a peek into the future of cranioplasty? Journal of Craniofacial Surgery 25(1): e55–58.

Oei, J.D., Zhao, W.W., Chu, L., DeSilva, M.N., Ghimire, A., Rawls, H.R. et al. 2012. Antimicrobial acrylic materials with *in situ* generated silver nanoparticles. J. Biomed. Mater. Res. B 100(2): 409–415.

Orienti, I., Bigucci, F., Gentilomi, G. and Zecchi, V. 2001. Self-assembling poly(vinyl alcohol) derivatives, interactions with drugs and control of release. J. Pharm. Sci. 90(9): 1435–1444.

Parthasarathy, J. 2015. Medical applications of additive manufacturing. pp. 389–402. *In*: Srivatsan, T.S. and Sudarshan, T.S. (eds.). Additive Manufacturing: Innovations, Advances, and Applications. CRC Press, Taylor & Francis Group. Boca Raton (IPA), USA.

Peltola, M.J., Vallittu, P.K., Vuorinen, V., Aho, A.A., Puntala, A. and Aitasalo, K.M. 2012. Novel composite implant in craniofacial bone reconstruction. Eur. Arch. Otorhinolaryngol. 269(2): 623–628.

Piitulainen, J.M., Posti, J.P., Aitasalo, K.M., Vuorinen, V., Vallittu, P.K. and Serlo, W. 2015. Paediatric cranial defect reconstruction using bioactive fibre-reinforced composite implant: Early outcomes. Acta Neurochir. 157(4): 681–687.

Piitulainen, J.M., Kauko, T., Aitasalo, K.M., Vuorinen, V., Vallittu, P.K. and Posti, J.P. 2015. Outcomes of cranioplasty with synthetic materials and autologous bone grafts. World Neurosurg. 83(5): 708–714.

Piitulainen, J.M., Mattila, R., Moritz, N. and Vallittu, P.K. 2017. Load-bearing capacity and fracture behavior of glass fiber-reinforced composite cranioplasty implants. J. Appl. Biomater. Funct. Mater. 15(4): e356–361.

Posti, J.P., Piitulainen, J.M., Hupa, L., Fagerlund, S., Frantzén, J., Aitasalo, K.M. et al. 2016. A glass fiber-reinforced composite–bioactive glass cranioplasty implant: a case study of an early development stage implant removed due to a late infection. J. Mech. Behav. Biomed. Mater. 55: 191–200.

Prabha, R.D., Kraft, D.C.E., Harkness, L., Melsen, B., Varma, H., Nair, P.D. et al. 2018. Bioactive nano-fibrous scaffold for vascularized craniofacial bone regeneration. J. Tissue Eng. Regen. Med. 12(3): e1537–e1548.

Richardson Jr, R.R., Miller, J.A. and Reichert, W.M. 1993. Polyimides as biomaterials: Preliminary biocompatibility testing. Biomaterials 14(8): 627–635.

Ronca, D., Gloria, A., De Santis, R., Russo, T., D'Amora, U., Chierchia, M. et al. 2014. Critical analysis on dynamic-mechanical performance of spongy bone: the effect of an acrylic cement. Hard Tissue 3(1): 9–16.

Russo, L., Gloria, A., Russo, T., D'Amora, U., Taraballi, F., De Santis, R. et al. 2013. Glucosamine grafting on poly(ε-caprolactone): a novel glycated polyester as a substrate for tissue engineering. RSC Adv. 3(18): 6286–6289.

Russo, T., Gloria, A., D'Antò, V., D'Amora, U., Ametrano, G., Bollino, F. et al. 2010. Poly(ε-caprolactone) reinforced with sol-gel synthesized organic-inorganic hybrid fillers as composite substrates for tissue engineering. J. Appl. Biomater. Biomech. 8(3): 146–152.

Russo, T., Gloria, A., De Santis, R., D'Amora, U., Balato, G., Vollaro, A. et al. 2017. Preliminary focus on the mechanical and antibacterial activity of a PMMA-based bone cement loaded with gold nanoparticles. Bioact. Mater. 2(3): 156–161.

Santiago, L.Y., Nowak, R.W., Rubin, J.P. and Marra, K.G. 2006. Peptide-surface modification of poly(caprolactone) with laminin-derived sequences for adipose-derived stem cell applications. Biomaterials 27: 2962–2969.

Saringer, W., Nöbauer-Huhmann, I. and Knosp, E. 2002. Cranioplasty with individual carbon fibre reinforced polymere (CFRP) medical grade implants based on CAD/CAM technique. Acta Neurochir. 144(11): 1193–1203.

Schantz, J.T., Hutmacher, D.W., Lam, C.X., Brinkmann, M., Wong, K.M., Lim, T.C. et al. 2003. Repair of calvarial defects with customised tissue-engineered bone grafts II. Evaluation of cellular efficiency and efficacy *in vivo*. Tissue Eng. 9(4, Supplement 1): 127–139.

Schantz, J.T., Teoh, S.H., Lim, T.C., Endres, M., Lam, C.X. and Hutmacher, D.W. 2003. Repair of calvarial defects with customized tissue-engineered bone grafts I. Evaluation of osteogenesis in a three-dimensional culture system. Tissue Eng. 9(4, Supplement 1): 113–126.

Schantz, J.T., Lim, T.C., Ning, C., Teoh, S.H., Tan, K.C., Wang, S.C. et al. 2006. Cranioplasty after trephination using a novel biodegradable burr hole cover: technical case report. Oper. Neurosurg. 58(suppl._1): ONS-E176.

Shah, A.M., Jung, H. and Skirboll, S. 2014. Materials used in cranioplasty: a history and analysis. Neurosurg. Focus. 36(4): E19.

Sinikovic, B., Schumann, P., Winkler, M., Kuestermeyer, J., Tavassol, F., von See, C. et al. 2011. Calvaria bone chamber—a new model for intravital assessment of osseous angiogenesis. J. Biomed. Mater. Res. A 99(2): 151–157.

Spence, W.T. 1954. Form-fitting plastic cranioplasty. J. Neurosurg. 11(3): 219–225.

Thesleff, T., Lehtimäki, K., Niskakangas, T., Huovinen, S., Mannerström, B., Miettinen, S. et al. 2017. Cranioplasty with adipose-derived stem cells, beta-tricalcium phosphate granules and supporting mesh: Six-year clinical follow-up results. Stem Cells Transl. Med. 6(7): 1576–1582.

Tuusa, S.M., Peltola, M.J., Tirri, T., Puska, M.A., Röyttä, M., Aho, H. et al. 2008. Reconstruction of critical size calvarial bone defects in rabbits with glass–fiber reinforced composite with bioactive glass granule coating. J. Biomed. Mater. Res. B 84(2): 510–519.

Unterhofer, C., Wipplinger, C., Verius, M., Recheis, W., Thomé, C. and Ortler, M. 2017. Reconstruction of large cranial defects with poly-methyl-methacrylate (PMMA) using a rapid prototyping model and a new technique for intraoperative implant modeling. Neurol. Neurochir. Pol. 51(3): 214–220.

van Putten Jr, M.C. and Yamada, S. 1992. Alloplastic cranial implants made from computed tomographic scan-generated casts. J. Prosthet. Dent. 68(1): 103–108.

Wang, S., Zhao, Z., Yang, Y., Mikos, A.G., Qiu, Z., Song, T. et al. 2018. A high-strength mineralized collagen bone scaffold for large-sized cranial bone defect repair in sheep. Regen. Biomater. 5(5): 283–292.

Wink, J.D., Gerety, P.A., Sherif, R.D., Lim, Y., Clarke, N.A., Rajapakse, C.S. et al. 2014. Sustained delivery of rhBMP-2 via PLGA microspheres: cranial bone regeneration without heterotopic ossification or craniosynostosis. Plast. Reconstr. Surg. 134(1): 51–59.

Worm, P.V., do Nascimento, T.L., do Couto Nicola, F., Sanches, E.F., dos Santos Moreira, C.F. and Dos Reis, M. 2016. Polymethylmethacrylate imbedded with antibiotics cranioplasty: An infection solution for moderate and large defects reconstruction? Surg. Neurol. Int. 7(Suppl. 28): S746–S751.

Wurm, G., Tomancok, B., Holl, K. and Trenkler, J. 2004. Prospective study on cranioplasty with individual carbon fiber reinforced polymere (CFRP) implants produced by means of stereolithography. Surg. Neurol. 62(6): 510–521.

Yang, S., Leong, K.F., Du, Z. and Chua, C.K. 2001. The design of scaffolds for use in tissue engineering. Part I. Traditional factors. Tissue Engineering 7(6): 679–689.

Yang, W.F., Choi, W.S., Leung, Y.Y., Curtin, J.P., Du, R., Zhang, C.Y. et al. 2018. Three-dimensional printing of patient-specific surgical plates in head and neck reconstruction: A prospective pilot study. Oral Oncology 78: 31–36.

Zebarjad, S.M., Sajjadi, S.A., Sdrabadi, T.E., Yaghmaei, A. and Naderi, B. 2011. A study on mechanical properties of PMMA/hydroxyapatite nanocomposite. Engineering 3(8): 720–726.

8

Biomimetic Approaches for the Design and Development of Multifunctional Bioresorbable Layered Scaffolds for Dental Regeneration

Campodoni Elisabetta, Dozio Samuele Maria, Mulazzi Manuela, Montanari Margherita, Montesi Monica, Panseri Silvia, Sprio Simone, Tampieri Anna and *Sandri Monica**

Introduction

The inability of most tissues and organs in adult humans to regenerate after damage, has been a serious problem that affects doctors, dentists and, of course, patients. Among them, the tooth is a complex organ made of highly mineralized tissues (alveolar bone, cement, dentin, enamel) and non-mineralized periodontal ligament (PDL), which connects the alveolar bone to the cement and guarantees the functionality and stability of the tooth. Dental diseases are extremely widespread all over the world, such as periodontitis, pulpitis and others that require a reconstruction of the dental apparatus. Biocompatible prosthetic devices have provided options in many cases helping millions of people, but the body's reaction to these devices is far from ideal. Complications including thrombosis, infections, ongoing inflammatory reactions, excessive fibrosis, impaired functions, mobilization and extrusion, are problematic for patients and expensive for the healthcare system. In dentistry, titanium and bioceramics implants or materials such as stainless steel, silicone rubber and poly methacrylate are widely used for the reconstruction of dental apparatus. However, the failure of these devices due to healing problems, infection and overload problems is well recognized. Nowadays, the accepted

Institute of Science and Technology for Ceramics-National Research Council (ISTEC-CNR), Via Granarolo 64, 48018 Faenza, Italy.

Emails: elisabetta.campodoni@istec.cnr.it; samuele.dozio@istec.cnr.it; manuela.mulazzi@istec.cnr.it; margherita.montanari@istec.cnr.it; monica.montesi@istec.cnr.it; silvia.panseri@istec.cnr.it; simone.sprio@istec.cnr.it; anna.tampieri@istec.cnr.it

* Corresponding author: monica.sandri@istec.cnr.it

standard for defining success in dental implants, has become a phenomenon called osseointegration, associated to a concept of years of reliable function rather than device life. The challenge for osseointegration is the stimulation of the right response in the surrounding tissue, which is strictly related to the material and the implant chemistry and features.

In the last 50 years, innovative technologies and materials inspired by nature have been designed, among them biomaterials, that play an exciting role in the field of regenerative medicine where biomimetic has now become a driving concept. Biomimetic in this field means examining the nature, its models, systems, processes and elements to take inspiration and emulate them to solve technological problems for human health. In fact, the close reproduction of the physical-chemical, morphological and mechanical characteristics of the targeted tissues provides biomaterials with the ability to exchange information with cells and trigger the tissue regeneration cascade (Preti et al. 2019).

In particular, new synthetic methods that allow the controlled growth of crystals and the multi-scale organized structures are attracting increasing attention. In this way, nature is studied not only to develop new and biomimetic materials, but also to imitate the natural processes and design innovative methods of syntheses. An example of a highly biomimetic process investigated to develop bone-like biomaterials is biomineralization, a natural assembly process that has been successfully reproduced in the laboratory to induce a heterogeneous nucleation of inorganic nanocrystals on an organic matrix, to produce hybrid scaffolds with compositional, morphological and structural characteristics similar to natural mineralized tissues. This process is the basis of bones, shells and exoskeletons generation and allows the design of bio-hybrid materials with unique properties that cannot be obtained from conventional approaches (Campodoni et al. 2018).

As in bone, the event active in the formation of the different dental tissues involves collagen and proteoglycans as macromolecular assembled matrix, mineralized with variable ratio of hydroxyapatite (HA) nanocrystals and organized into complex fibrous constructs with graded morphologies and structures. Being able to develop novel biomimetic scaffolds providing suitable chemical and mechanical cues allows to promote the exchanges of effective information with the endogenous cells and trigger tissue regeneration, avoiding the use of inert and non-biomimetic prostheses often associated with problems related to inflammations and infections reactions (Sprio et al. 2018).

Biomimesis and Biomineralization: From Nature to the Laboratory

The word *biomimetics* was first used in the late 1950s when it was coined by the biophysics Otto Schmitt who used it to describe the transfer of ideas and analogues from biology to technology. More precisely, he studied and described the formation, the structure and the function of biological substances and materials and the biological mechanisms (as protein synthesis or photosynthesis) especially for the purpose of synthesizing similar products by artificial processes which mimic natural ones (Ramalingam et al. 2013; Hwang et al. 2015).

Some effective examples of biomimetics are, actually, very well-known everyday usage items like, for example, Velcro®. Velcro® is considered the first practical application of a biomimetic principle. In the early 1940s, the Swiss inventor George de Mestral went for a walk with his dog and on returning home, he noticed that his dog coat and his pants were covered with cockleburs. His curiosity led him to study the burrs under a microscope, discovering their natural hook-like shape. This was to become the basis for a unique, two-sided fastener, one side with stiff 'hooks' like the burrs and the other side with the soft 'loops' like the fabric of his pants. The result was Velcro® (1959) brand hook and loop fasteners, named for the French words 'velour' and 'crochet' (Saunders 2015).

Biomimetics means, to look towards nature for ideas that may be adapted and adopted for solving problems or create new solutions. One of the natural processes to which biomimetic takes inspiration is without any doubt biomineralization, a very peculiar, natural, as well as transversal phenomenon, of interest to many different living organisms from diatoms to humans, passed by plants. It refers to the processes by which organisms form minerals starting from simple compounds and are defined biogenic minerals or biominerals. It is, therefore, by definition, a true multidisciplinary field that spans from both the inorganic and the organic world. But, why should living organisms form minerals? The main functions of mineralized tissues can be many, among them protection, motion, cutting, grinding, optical, magnetic and gravity sensing as well as storage (Talham 2002; Weiner 2003).

Among biogenic minerals approximately 80% are crystalline and 20% are amorphous. Calcium minerals account for about 50% of biogenic minerals and the most widespread are carbonates, phosphates, silicates and iron oxides. About 60% of the known biogenic minerals types contain hydroxyl groups and/or bound water molecules and the abundance of hydrated biominerals is not accidental. Hydrated phases are favoured over anhydrous counterparts by significantly lowering energetic barriers to nucleation and growth from aqueous solution (Rao and Cölfen 2016; Crichton 2019). Organisms are metabolic misers, thus, they use the Ostwald-Lussac rule to their advantage by favouring the precipitation of the lowest energy phases (Coombs 1984).

Biomineralized materials do not contain just mineral components but both inorganic and organic ones finely organized and chemically interacting thus to create a new hybrid material endowed with new smart properties and are used to build exo- as well as endoskeletons. Most invertebrates use chitin based structures, that are largely organic while vertebrates, in their bones and teeth, use biomineral constructions made of both organic and inorganic components consisting of calcium phosphate nano-crystals growth on collagen. The inorganic component confers mechanical resistance without which large land-living animals could not exist, but it also brings brittleness, therefore, an organic matrix is required to provide elasticity and tensile strength to the final hybrid material. Moreover, the organic matrix is also responsible for the control of nucleation and growth of crystals of the inorganic component. Biogenic minerals are mainly characterized by mild conditions of temperature and pressure synthesis, unusual morphology and unusual mechanical properties. Typically, the organic matrix is laid down first and the inorganic reinforcing phase grows within this organic matrix template that confers important and unique feature to the crystals (Weiner 2003; Chen et al. 2019).

One of the major characteristics of biomineralization strategies is, the use of the bottom up approaches, in which the material is built starting at the atomic and molecular scale, leading to the formation of nanostructured building blocks, which in turn organize into complex hierarchical structures. This take the name of biomineral tectonics, an integrated process of construction that extends across many length scales and results in higher ordered structures. Although these tissues are made of relatively weak components at ambient conditions, their hierarchical structural organization and intimate interactions between different elements lead to superior mechanical properties (Talham 2002; Beniash 2011).

Biomineralization is one of the natural processes that has been successfully reproduced in the laboratory, with the induction of the inorganic phase heterogeneous nucleation onto the organic matrix through fine mechanisms, controlled and driven by the organic matrix itself (Fig. 8.1). The chemical interaction between the elected organic matrix and the grown-into-it inorganic phases allow to obtain hybrid materials with unique features that are more than just a pro sum of the single parts, exactly as it happens in nature. Moreover, a proper doping with inorganic ions [e.g., Mg^{2+}, Sr^{2+}, CO_3^{2-}, $Fe^{2+/3+}$] makes it possible to adapt the composite material properties, like biodegradation, cell adhesion and growth, rather than magnetism, for the final purpose (Campodoni et al. 2016; Sprio et al. 2018; Tampieri et al. 2008).

In the early 2000s a nature-inspired fabrication process to obtain 3D scaffolds strongly mimicking hard human tissues was successfully drawn up. The developed process takes advantage of the complex chemical-physical, topological and ultrastructural information stored into the collagen molecule to induce hierarchical self-assembly and biological-like low-crystalline hydroxyapatite nucleation, with the help of foreign ions incorporation into the apatite lattice.

In details, to in-laboratory reproducing the biomineralization process performing a neutralization reaction where the acid solution containing the PO_4^{3-} was mixed with collagen gel and then dropped into a Ca^{2+} ions-containing alkaline solution. The pH of the suspension was then decreased until it

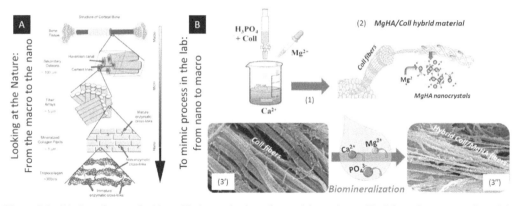

Figure 8.1. (A) Overview on the hierarchical organization of natural bone tissue. (B) Schematic representation of the biomineralization process developed in lab-scale through a neutralization acid-base reaction (1). Mg-doped hydroxyapatite nanocrystals (MgHA) were grown on collagen fibers (Coll) by to obtain a hybrid biomaterial (MgHA/Coll) mimetic of the bone mineralized matrix (2). In detail the collagen fibers, investigated with ESEM, before (3') and after (3'') the biomineralization process.

reached the neutral value to trigger two different mechanisms at a time: meeting the isoelectric point of collagen so that it starts to assemble in fibers and then into a 3D network and starting the mineral nucleation in correspondence of the carboxylic groups exposed by collagen. The presence in the basic suspension of foreign ions that can be incorporated in the apatite lattice during its formation helps in obtaining a more amorphous and natural-like mineral phase and therefore a highly biomimetic hybrid material. The possibility for the apatite lattice to host many different isovalent and heterovalent ions allow to synthesize tailor-substituted apatite for different applications, an example is the MgHA/Coll biomimetic and fully bioresorbable hybrid material described by Tampieri et al. for the regeneration of the subchondral bone in a multi-layer scaffold designed for osteochondral regeneration (Fig. 8.1) (Tampieri et al. 2008).

Besides collagen, many other polymers have been exploited as possible matrices for biomineralization; among them gelatin, chitosan, alginate and nano-cellulose (Campodoni et al. 2016; Panseri et al. 2016). Thanks to their chemical structure, all of the above-mentioned biopolymers are able to undertake the biomineralization process with interesting different final results in terms of mechanical and physical properties, biological activity and biodegradation kinetic that can be exploited to achieve disparate regenerative goals.

The possibility to in-laboratory the polymer assembly directly and the inorganic phase nucleation and growing on it, paves the way for obtaining better customized hybrid mineralized gels possible to be further processed in 3D scaffolds or carriers depending on the therapeutic need (Campodoni et al. 2018).

Biomaterials for Dental Regeneration

After several years of research, the evolution of biomaterials for tissue regeneration has seen a quick upgrade of fundamental concepts which biomaterials have to respect. At the beginning of the biomaterials research, the first requirement was non-toxicity, which meant avoiding to stimulate undesired reactions and the death of the surrounding tissue, in that period bio-inert products were developed.

Nowadays, new emerging concepts have been taking hold in this field such as bioactivity, bioresorbability and biomimicry that allow to obtain materials capable of stimulating specific responses from the biological system (Tampieri et al. 2003; Bakopoulou et al. 2016). Firstly, a bioresorbable material is designed to stimulate regenerative reactions inducing the formation of a new tissue and to gradually degrade in safe by-products during the new tissue formation until complete material substitution. Secondly, bioactivity allows to stimulate a specific biological response producing a bond between the material and tissue; two different responses can be stimulated depending on the material which can be osteoconductive or osteoinductive (Kokubo et al. 2003; Panseri et al. 2014). An osteoconductive material is able to be colonized by the host tissue cells and promote the bone tissue formation thanks to the permissive morphology, e.g., pore dimensions and interconnection however, it is not able to induce a cell differentiation. This capability is instead typical of an osteoinductive material, which is able to recruit undifferentiated cells and stimulate them to evolve in osteoblastic cells promoting bone formation. This family of materials is the ideal solution only if they are endowed with the required chemical stability, degradation kinetics and mechanical strength. Together with features described above, the new generation of materials are generally defined as 'smart' because they can intrinsically and actively promote the capability of tissue to heal and self-repair by responding to internal and external environment stimuli, thus taking part in the regeneration of the damaged tissue (Tampieri et al. 2005; Iafisco et al. 2014; Mohammed and Gomaa 2016). To achieve this interaction, a smart biomaterial may also have biomimetic abilities, meaning the ability to mimic the function of extracellular matrix to stimulate cellular invasion, attachment and proliferation. This means that well-designed biomaterials are able to recruit resident stem cells and orchestrate their behaviors and functions to promote tissue regeneration, exploiting the intrinsic regenerative potential of endogenous tissues.

Biomimetic properties have been obtained by technological adaptation of the macro, micro and nanostructure, by trying to replicate the hierarchical organization of natural tissues. Taking inspiration from nature it is possible to develop better materials in terms of chemical composition, 3D organization and mechanical performances, which lead to a better interaction with the cell and to improve their responses, for the achievement of vascularization and tissue ingrowth (Panseri et al. 2012; Sandri et al. 2016; Krishnakumar et al. 2018).

The concept mimicking the natural tissues to develop a bioresorbable material is largely applied in several areas such as orthopaedic and dental fields. In the orthopaedic field, several previous studies carried out by Tampieri et al. showed substantial mimicry of the composition and multi-scale structure of target native tissues enhancing regenerative ability to develop different kind of materials such as porous 3D scaffolds, injectable self-hardening bone cements, 3D hybrid scaffold for osteochondral and bone regenerations, biomimetic scaffolds obtained by biomorphic transformation for the regeneration of load-bearing segmental bones (Tampieri et al. 2008; 2019; Dapporto et al. 2016; Sprio et al. 2016). The same approaches were translated to the dental field, for which several biomaterials mimicking structure and composition of various tissues have been developed for the regeneration of different parts of the tooth; the next paragraphs will better describe examples of biomaterials for periodontium and dentin regeneration (Panseri et al. 2016; Sprio et al. 2018).

Mimicry of the Natural Gradient of the Periodontal Region

The tooth is a complex organ combining hard and soft tissues such as enamel, dentin, cementum and vascularized dental pulp; the latter is the central part of the tooth filled with soft connective tissue in close connection with dentin. It contains blood vessels and nerves that enter the tooth from a hole at the apex of the root and along the border between the dentin and the pulp are odontoblasts, which initiate the formation of dentin. Periodontium includes tissues surrounding and supporting the tooth such as cementum, periodontal ligament (PDL) and the alveolar bone (Fig. 8.2B) (Maeda et al. 2014). It is well known that stem cells, scaffolds and signal molecules are requisites for tissue engineering, especially, with regard to scaffolds, in addition to the chemistry of the material, it is important to mimic the 'scale' of these tissues. The periodontal ligament, in fact, has a thickness of about 20 μm, and the cementum of about 10 μm, each layer has a unique morphology and chemistry fundamental for its role and recognized by cells. A good mimesis of natural tissue reproducing the biophysical properties such as surface tomography, internal microstructure, scale and highly interconnected

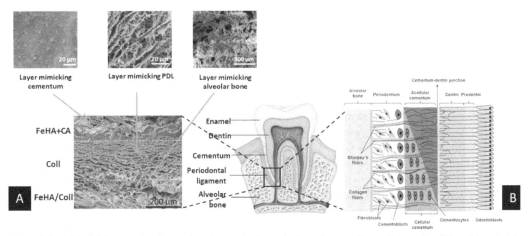

Figure 8.2. (A) SEM images showing at the micro scale the multi-layer feature of a scaffold made of three biomaterials mimetic of cementum, periodontal ligament and alveolar bone. (B) Details of the complex structure of natural periodontal tissue that inspired the fabrication of the three-layer biomimetic periodontal scaffold.

porous structure together with a good elasticity are key points to promote cell adhesion, migration, proliferation and differentiation (Lee et al. 2005; 2010; Dangaria et al. 2011).

Several materials are used for tissue regeneration classifiable in synthetic, natural, bioceramic and composite materials. Among synthetic materials available for periodontal regeneration are largely investigated poly-lactic acid (PLA), poly-glicolic acid (PLGA) polycaprolactone (PCL), all these materials have been developed with a suitable porous structure ideal for cell engraftment and proliferation (Galler et al. 2011; Kawase et al. 2011; Ba Linh et al. 2013; Bölgen et al. 2014). Galler et al. developed a polyethylene glycol blended with fibrin and dental stem cell from the human PDL and pulp displaying a good *in vivo* cell growth and differentiation due to a suitable 3D fabrication (Galler et al. 2011). Among natural materials, several of those are exploited for dental application such as gelatin, chitosan, collagen, fibrin and hyaluronic acid due to their favorable properties such as biodegradability, biocompatibility and bioresorbability. These features together to the possibility to generate 3D porous structures through well-known fabrication technologies allow to promote cell adhesion and proliferation, important for a suitable tissue regeneration (Panos et al. 2008; Wolf et al. 2013; Jimbo et al. 2014; Sugawara and Sato 2014). Furthermore, bioceramic materials for the fabrication of cement and bone replacement devices were also studied, including hydroxyapatite (HA) and β-tricalcium phosphate (β-TCP), both calcium phosphates are widely used in the field of dentistry, orthopedic and plastic surgery (Xia et al. 2011; Kasaj et al. 2012; Lee et al. 2012; Suto et al. 2013).

Finally, composites of bioactive ceramic and polymers, in particular biomimetic biodegradable scaffolds are currently being developed by taking advantage of each component to increase mechanical stability and improve tissue interaction (Takechi et al. 2012; Tsuzuki et al. 2012; Maeda et al. 2014). The combination of HA and collagen has frequently been used to create biomimetic scaffolds in regenerative medicine because they greatly resemble natural tissues especially if they are synthesized by biomineralization process (Tampieri et al. 2003; Krishnakumar et al. 2018).

A novel strategy to develop scaffolds for periodontal regeneration mimicking the natural periodontal region well is to create a multi-layer construct with different region-specific pore/channel sizes and material composition (Lee et al. 2014; Sprio et al. 2018). In particular, Sprio et al. developed a device that mimics the entire periodontium through three different layers, each of which mimics a specific tissue and is responsible for a well-defined function: PDL, cementum, alveolar bone. This construct has a biomimetic composition and a gradient structure able to instruct specialized cellular components of the periodontium to participate in the regenerative process (Fig. 8.2A) (Sprio et al. 2018). Alveolar bone-like layer is composed of mineralized collagen with Fe-doped HA (FeHA/Coll) highlighting chemic-physical features of natural tissue together with superparamagnetic properties able to stimulate the cell to reproduce and differentiate under weak magnetic fields (Silvia Panseri et al. 2012; Tampieri et al. 2012; 2014). The presence of Fe-ions, responsible for the generation of the magnetic properties, do not affect the biocompatibility and the degradability of the hybrid material, instead their effectiveness in accelerating tissue regeneration was reported (Glazer et al. 1997; Bañobre-López et al. 2011; Assiotis et al. 2012; Panseri et al. 2012). PDL-like layer is composed of a highly porous collagen membrane (Coll) in order to allow a high vascularization of cementum metabolically relating to nutrients diffused by PDL. Collagen is an excellent candidate because it is highly biomimetic and biocompatible, but its degradability is very fast, for this reason biocompatible chemical cross-linking agents, such as 1,4-butanediol diglycidyl ether (BDDGE), are selected to improve its permanence *in vivo* (Zeeman et al. 2000). Finally, a cementum-like layer is a very thin layer processed by electrospinning a composite blend of nano-sized FeHA with cellulose acetate (FeHA+CA). The technique chosen allows to obtain a mat composed of non-woven micrometric fibers very close to natural cementum, furthermore, the presence of nano-sized hydroxyapatite confers to the layer a high biomimicry creating a structure very similar to bone, but with a disordered and less porous structure, suitable for anchoring the tooth to the alveolus. The introduction of FeHA confers to the biomaterial a magnetic property, normally not present in nature, but is assessed to be useful for improving the regeneration capability of the biomaterial by the activation of cell processes in hard tissue regeneration.

Layers were assembled all together before freeze-drying because the presence of water improves the layers' integration minimizing their delamination. In conclusion a three-layers device was developed by changing for each layer, the composition, degree of mineralization, morphology and fabrication technology in order to mimic the different tissues of periodontal region and obtaining a biocompatible and bioactive construct with instructive abilities for cells due to its multi-scale structural features (Fig. 8.2A) (Sprio et al. 2018).

Mimicry of the Natural Aligned Morphology of the Dentin

As described earlier, the development of a suitable biomaterial for dentin regeneration starts from the study of natural tissue. Teeth are hard, calcified structures found in the jaws (or mouths) with complex structures consisting of enamel and dentin supported by connective tissues (cementum, PDL and alveolar bone). As bone tissue, the dentin's structure is the most voluminous mineralized tooth's tissue formed by microscopic longitudinal dentinal tubules and is made of mineral phase, organic and water (75, 20 and 5%) (Burwell et al. 2012; Besinis et al. 2015). Several causes such as periodontal diseases, profound caries and trauma lead to partial or full edentulism decreasing the individual's quality of life. Current solutions are still based on fixed prosthesis and full/partial dentures, however, due to their great limitation, finding a better solution is still a challenge. Material science intends to explore new approaches based on biomimesis of the damage tissue, designing biomaterials conceived as ideal matrices for cells homing and stimulation in the regenerative direction.

To do this the choice of the biomaterial plays a key role due to specific dentin characteristics, which has a mineralized structure and a 3D architecture arranged in aligned channels (Panseri et al. 2016), furthermore, to guarantee a suitable odontointegration, it should be resistant to chemical and physical abrasion providing a suitable mechanical strength and esthetic (Hakki et al. 2016) (Fig. 8.3).

Cell-instructive materials are chosen for the regeneration of dentin due to their affinity with the cell and the ability to recruit them; among these materials, bioceramics refers to a group of bioactive glasses and calcium phosphate ceramics, displaying several advantages such as bioactivity, biocompatibility, biodegradability and a suitable porous structure that allows them to be widely used in reconstructive, orthopedic, maxillofacial and craniofacial application (Yuan et al. 2011). Moreover, due to a highly interconnected porous structure, these materials are able to support odontogenic differentiation and biomineralization. For instance, Zn-doped and Mg-doped bioceramic scaffold obtained through a foam replica technique are synthesized and combined with human Treated-Dentin Matrices (hTDMs growth

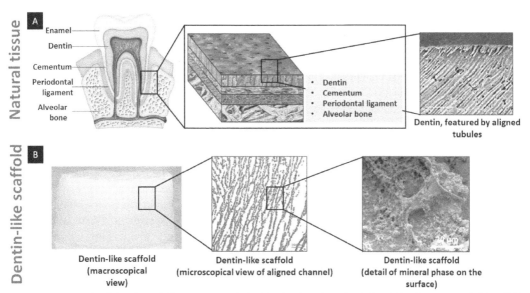

Figure 8.3. (A) In detail the channel-like structure typical of the dentin and (B) the designed biomimetic 3D hybrid scaffold indicating a porous aligned structure. On the right, in evidence, the mineral apatitic phase exposed on the surface of its channels.

factors); scaffolds display tailored bio-interactivity to allow a constant release of low concentrations of Mg^{2+}, Ca^{2+}, Zn^{2+}, and Si^{4+} ions and highly interconnected porosity supporting cell attachment. Results showed an abundant source of dentin morphogens and a targeted odontogenic differentiation due to the combination of growth factor and biomaterial (Bakopoulou et al. 2016).

Several scaffolds containing both mineral and organic phase were developed; for tooth reconstruction scaffold should meet specific requirements such as ease of handling, adequate porosity, biodegradability, bioactivity, good physical and mechanical strengths and the ability to support vascularity (Yuan et al. 2011; Sharma et al. 2014). Beside this, the scaffold degradation should match with the rate of new tissue formation, by exploiting biocompatible cross-liking agents it is possible to modulate the degradation rate without leaving any toxic by-products (Shankar et al. 2017; Campodoni et al. 2019). In this kind of synthetic hybrid composites, each phase provides proper advantages while minimizing disadvantages such as, increased stiffness without being brittle. For instance, biomineralized hybrid materials, such as natural bone tissue, present nanocrystals of HA reinforcing the polymer structure, while the collagen allowing the control of HA growth which remain nanometric and bioavailable. Blaker et al. indicated improved cell adhesion, spreading and viability of cells grown on polymer-bioglass-composites and also confirmed the high bioactivity and biocompatibility of the material for hard tissue repair (Blaker et al. 2003), moreover, scaffolds composed of PLGA/TCP display good performances for tooth regeneration (Zheng et al. 2011).

Exploring a biomimetic approach, Panseri et al. have designed a bio-hybrid scaffold exploiting two key points: the biomineralization process to synthesize hydroxyapatite, as the mineral phase, nucleated on biopolymer matrix conferring low crystallinity and features very close to natural tissue and the development of a 3D porous structure capable to promote cell colonization and differentiation (Panseri et al. 2016).

Among biopolymers used as an organic phase, gelatin (Gel) is a biodegradable polymer which is obtained by the collagen's denaturation, this process allows to break the triple-helix structure into random coil maintaining the typical collagen properties such as biocompatibility, biodegradability and the ability to be mineralized; on the other hand, Gel has poor mechanical properties and stability that limit its application (Dash et al. 2013; Hossan et al. 2014). Alginate (Alg) is a natural polymer extracted by brown algae largely used for the fabrication of biomaterials and nanobeads well known for their biodegradability, biocompatibility and low production costs (Drury and Mooney 2003; Tampieri et al. 2005). Alg is capable of gelation with multivalent cations, undergoing through an aqueous sol-gel transformation from water-soluble alginate salt to water-insoluble salt. Calcium is the most frequently used divalent ions for Alg gelation, as it forms an ideal matrix for hydrogel, film, beads, nanoparticles thanks to its ability to link M and G units, getting a typical structure called 'egg-box' (De and Robinson 2003; Campodoni et al. 2016).

Biomineralization in-laboratory process allows to develop low crystalline Mg-doped hydroxyapatite nanoparticles nucleated on gelatin molecules as organic template, the features of the mineral phase obtained through this process have been widely investigated by Tampieri et al. (Tampieri et al. 2003; Tampieri et al. 2008) highlighting that the developed compounds are bestowed with high biomimetism, low crystallinity and cell affinity very close to natural tissue and so, very useful for hard tissue regeneration. In particular, for dentin regeneration, the combination of the biomineralized gelatin with alginate, allow to generate 3D scaffold with channel-like structures replicating that of the dentin. The high affinity of alginate with Ca generates a good interaction with the hydroxyapatite determining a homogeneous compound and scaffold where the biomineralized gelatin is well-spread on its surface. Moreover, the addition of calcium chloride as alginate cross-linker on the dried scaffold allows in stabilizing the polymer structure and modulating the scaffold degradation rate. To shape the 3D device with an anisotropic and channel-like structure a freeze-drying process was performed. Due to the control and optimization of process parameters, it was possible to produce thin and lamellar pores disposed into aligned microtubules favoring a cell colonization feature required for dentin regeneration (Fig. 8.3B) (Panseri et al. 2016).

In general, the mechanical performance of composite materials with biopolymers is lower than those with synthetic polymers and, of course, of natural dentin. These types of materials were conceived in a context of a biomimetic approach in which the biomimesis of the developed matrix and thus the signal is able to deliver to the surrounding cells, assume a greater importance than their mechanical performance. However, they are sufficient to preserve their structural integrity and load-bearing properties also in wet conditions and to sustain cell colonization and proliferation during new tissue formation and scaffold degradation.

Cell-Biomaterial Interaction

The tooth is a small yet complex organ, composed by hard calcified tissues which constitute the dentin, enamel and cementum, and a connective tissue represented by the periodontal ligaments and dental pulp. Its peculiar structure and cell organization define it as one of the resilient tissues in the human body (Chai et al. 2009). Unfortunately, it comes with a drawback: the tooth has no self-regeneration capacity, excluding enamel reconstruction helped by a conscientious diet and oral care, and the slightest fracture or cavity reaching the dentin will become permanent. Moreover, tooth diseases tend to be also irreversible and are generally widespread. For instance, periodontitis is listed second among the most common chronic diseases, with an incidence rate like that of diabetes and cardiovascular diseases (Williams et al. 2008; Haumschild and Haumschild 2009). Therefore it is no surprise that tooth regeneration, especially when compared to the current conservative approaches and resective surgeries, is considered to be a challenging but appealing scientific goal (Du et al. 2006; Duraccio et al. 2015). To achieve this goal, many issues must be overcome. One of the most relevant is how to drive the correct development of the new dental tissues (Sloan and Lynch 2012).

The events constituting tooth formation are complex, they require the active presence of all cellular components of the periodontium: fibroblasts for soft connective tissues such as PDL, cementoblasts for cementogenesis, osteoblasts for bone and endothelial cells for angiogenesis (Bartold et al. 2000; Maeda et al. 2014). All these cell lineages must correctly interact with each other, as well as with a variety of molecules of the extracellular matrix (ECM) (Thesleff and Hurmerinta 1981).

Starting from this complex pattern of signals and actors, researchers must rely on smarter approaches involving a certain degree of simplicity and reproducibility, while increasing the similarities with the *in vivo* condition, to reach their goal. These approaches involve the use of 3D scaffolds which are designed to chemically and physically mimic a portion of the tooth ECM for cell-material interaction evaluation and exploitation, trying to narrow the gap between the *in vitro* and *in vivo* findings (Hollister 2005). The reasons behind the necessity of developing a 3D environment for tooth regeneration is that standard 2D cell cultures, although have proven to be pillars for the discovery and understanding of several biological pathways in the past, lack the variety and complexity of signals physiologically present *in vivo*. As a matter of fact, it is well known in literature that data obtained by 2D cultures could lead to misleading findings when trying to recreate more complex biological events (Baharvand et al. 2006; Imamura et al. 2015).

Focusing on 3D biomaterials design, there are two main features by which they can interact with cultured cells directing their fate. One is their 3D structure, ranging from nano to macroscale hierarchy. A well-designed 3D structure is much more instructive than its chemical composition. Starting from the nanometric scale of a biomaterial hierarchy, its degree of surface rugosity impact on cell attachment ability, fundamental for avoiding anoikis, a form of apoptosis cell-surface anchoring dependent, and essential for biomaterial colonization after implantation (Valentijn et al. 2004). Increasing the micrometric and macrometric levels, the overall biomaterial porosity and resulting 3D structure play an important role in cell-colonization capacity and cell development (Loh and Choong 2013). If hierarchically organized, the pores could resemble the dentin aligned channels serving as a way for deep cell colonization and spatial orientation (Fig. 8.4G–J) (Panseri et al. 2016). Or it could be shaped in a multilayer biomaterial with differential porosity which, combined with differential chemical compositions, mimic the different tissues composing the periodontium. By adding different instructor layers, the biomaterial could direct the cell fate in a refined fashion, also diminishing external infection occurrence through a protective layer with reduced porosity composed by FeHA plus cellulose acetate, fundamental for dental applications (Fig. 8.4A–F) (Li et al. 2003; Sprio et al. 2018).

The chemical composition of a biomaterial for dental applications is also extremely relevant. It is well known that calcium phosphate (Ca/P) like β-tricalcium phosphate (β-TCP) or hydroxyapatite (HA) can constitute the inorganic phase of a biomaterial, likewise the mineral phase of the bone, fostering resorption, biocompatibility, low immunogenicity and osteoconductivity (Li et al. 2003; Sharma et al. 2014). Conveniently, the inorganic phase of the dentin is constituted by hydroxyapatite as well, and there are studies proving that Ca/P granules can provide a suitable substrate for dental pulp stem cells growth and differentiation toward odontoblasts (Nam et al. 2011). Therefore, the realization of biomaterials for dental applications with an appropriate Ca/P content can greatly increase their instructive capabilities, directing cell differentiation (Phadke et al. 2012; Panseri et al. 2016). Besides by doping the mineral phase of the biomaterials with different ions such as Mg^{2+}, CO_3^{2-} and Sr^{2+}, it is possible to further enhance the biomaterial bioactivity (Sprio et al. 2018). In particular, a smart approach for cell differentiation enhancement is based on the substitution of Fe^{2+} and Fe^{3+} ions into Ca/P crystals, allowing the Ca/P based biomaterials to also gain superparamagnetic properties (Tampieri et al. 2012; Iannotti et al. 2017). This property, normally not present in nature, is convenient for hard tissue regeneration as magnetic signalling proves to foster an activation effect on osteoblast cells controlling bone growth, improving the regeneration capability of the biomaterial (Panseri et al. 2012; Panseri et al. 2013). This feature could also be applied for dental regeneration biomaterials as both cell types involved in dentin and bone growth have partially overlapping pathways (Yang et al. 2013).

A further addition to the inorganic phase of biomaterial, natural polymers like collagen, gelatin, alginate, chitosan, hyaluronic acid and others, may be used to synthesize an organic phase, adding mechanical features and increasing the overall biocompatibility and biodegradation (Yuan et al. 2011). Several of these polymers have already been extensively tested for regenerative medicine applied to dental applications (Inuyama et al. 2010). Moreover, the synergy between the organic and inorganic

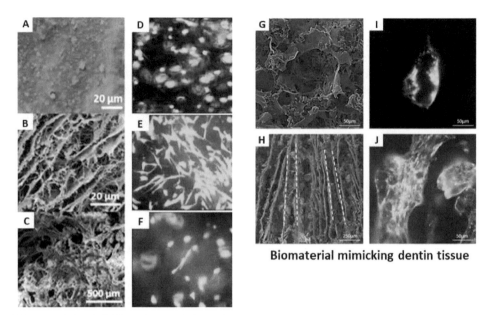

Biomaterial mimicking dentin tissue

Tri-layer biomaterial mimicking the periodontal apparatus

Figure 8.4. (A, B, C) SEM images of the three different layers obtained through different synthesis approaches and chemical compositions; from top to bottom, FeHA+CA mimicking the cementum, Coll mimicking the periodontal ligament and FeHA/Coll mimicking the alveolar bone. (D, E, F) Top to bottom cell viability of cell-seeded FeHA+CA, Coll and FeHA/Coll scaffolds through the Live/Dead assay. (G, H) SEM images of a transversal section and longitudinal section of a freeze-dried hybrid scaffold of MgHA Cell/Alg constituted by an aligned porosity channel-like, resembling the same channeled structure of the dentin. (I, J) Fluorescent staining of the cellular cytoskeleton of a population of mesenchymal stem cells seeded on the hybrid scaffold, growing inside and alongside the channel-like porosity.

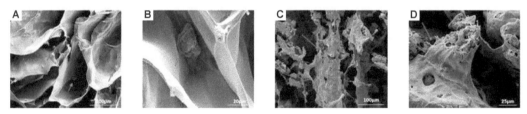

Figure 8.5. SEM analyses of MSCs seeded on hybrid materials composed by (A, B) Col/Alg were cells that do not tolerate a slippery surface and do not spread their cytoskeleton, remaining instead in a globular shape, typical of a stressful condition; and (C, D) MgHA-Cell/Alg increased surface rugosity of the biomaterial allows the cells to grow and extend their cytoskeleton, indicating a non-stressful environment.

phases, through the biomineralization of the latter, could transform a cheap natural polymer like alginate, which is normally not ideal for cell attachment and growth, to become suitable via increased superficial rugosity (Fig. 8.5) (Panseri et al. 2016; Zan et al. 2016).

Conclusions and Future Perspective

Teeth have no self-regeneration capacity and, to date, treatment strategies rely on the insertion of bioinert materials such as metal alloys, ceramics, cements and composite resins to restore dental defects and rely on the intrinsic cellular competence to form surrounding tissues.

The emerging concept of biomimetics, applied to material science, provides an innovative and more effective response to extremely widespread and life-threatening diseases such as, periodontitis and edentulism, by designing smart biomimetic hybrid composites that have been identified as the perfect solution able to respect and fit the complexity of the human body. To induce tissue regeneration, it is essential to dispose the novel biomimetic scaffolds capable of providing chemical and mechanical cues to promote multiple specific interactions and orchestrate processes such as cell adhesion, migration, differentiation, matrix synthesis, mineralization and/or vasculogenesis. For the creation of new smart biomaterials, the indepth study and understanding of the tricks nature uses to create them in their original environment is fundamental to mimic and sometimes even improve them.

Starting from a wide and deep study and characterization of the target site, and through a biomimetic approach, it was possible to obtain new materials based on biopolymeric matrix and a biomimetic inorganic phase presenting the potential to act as a source of biochemical and topographical signals for the various cell lineages present in dental environment. This means that well-conceived biomimetic hybrid materials are able to recruit endogenous cells and promote tissue regeneration, exploiting the intrinsic regenerative potential of tissues.

Nevertheless, the development of biomaterials mimicking structure, composition and mechanical properties of biological tissues still presents important limitations. In fact, oral tissues engineering, as a multidisciplinary approach to build complex structures such as bone, teeth or soft dental tissues, remains a challenge that will require further significant development of materials chemistry, biochemistry and biology for the formulation of a multifunctional system. This means that, by interdisciplinary approaches and close communication between material scientists, biologists and clinicians, future dental treatment will be able to combine biomimetic scaffolds, stem cells, growth factors and antibacterial or anti-inflammatory molecules, to support the whole tissue regeneration instead of its replacement.

References

Assiotis, A., Sachinis, N.P. and Chalidis, B.E. 2012. Pulsed electromagnetic fields for the treatment of tibial delayed unions and nonunions. A prospective clinical study and review of the literature. J. Orthop. Surg. Res. 7(1): 24. DOI: 10.1186/1749-799X-7-24.

Ba Linh, N.T., Lee, K. and Lee, B. 2013. Functional nanofiber mat of polyvinyl alcohol/gelatin containing nanoparticles of biphasic calcium phosphate for bone regeneration in rat calvaria defects. J. Biomed. Mater. Res. Part A. Wiley Online Library 101(8): 2412–2423.

Baharvand, H., Hashemi, S.M., Ashtiani, S.K. and Farrokhi, A. 2006. Differentiation of human embryonic stem cells into hepatocytes in 2D and 3D culture systems *in vitro*. Int. J. Dev. Biol. 50(7): 645–652. DOI: 10.1387/ijdb.052072hb.

Bakopoulou, A., Papachristou, E., Bousnaki, M., Hadjichristou, C., Kontonasaki, E., Theocharidou, A. et al. 2016. Human treated dentin matrices combined with Zn-doped, Mg-based bioceramic scaffolds and human dental pulp stem cells towards targeted dentin regeneration. Dent. Mater. DOI: 10.1016/j.dental.2016.05.013.

Bañobre-López, M., Piñeiro-Redondo, Y., De Santis, R., Gloria, A., Ambrosio, L., Tampieri, A. et al. 2011. Poly(caprolactone) based magnetic scaffolds for bone tissue engineering. J. Appl. Phys. DOI: 10.1063/1.3561149.

Bartold, P.M., Mcculloch, C.A.G., Narayanan, A.S. and Pitaru, S. 2000. Tissue engineering: a new paradigm for periodontal regeneration based on molecular and cell biology. Periodontol. John Wiley & Sons, Ltd. (10.1111) 24(1): 253–269. DOI: 10.1034/j.1600-0757.2000.2240113.x.

Beniash, E. 2011. Biominerals—hierarchical nanocomposites: the example of bone. Wiley Interdiscip. Rev.: Nanomedicine and Nanobiotechnology. John Wiley & Sons, Inc. 3(1): 47–69. DOI: 10.1002/wnan.105.

Besinis, A., Peralta, T., De Tredwin, C.J., Handy, R.D., Sciences, B., Circus, D. et al. 2015. Review of Nanomaterials in Dentistry: Interactions with the Oral. (3): 2255–2289. DOI: 10.1021/nn505015e.

Blaker, J.J., Gough, J.E., Maquet, V., Notingher, I. and Boccaccini, A.R. 2003. *In vitro* evaluation of novel bioactive composites based on Bioglass®-filled polylactide foams for bone tissue engineering scaffolds. J. Biomed. Mater. Res. - Part A.

Bölgen, N., Korkusuz, P., Vargel, İ., Kılıç, E., Güzel, E., Çavuşoğlu, T. et al. 2014. Stem cell suspension injected HEMA-lactate-dextran cryogels for regeneration of critical sized bone defects. Artif. Cells, Nanomedicine, Biotechnol. Taylor & Francis 42(1): 70–77.

Burwell, A.K., Thula-Mata, T., Gower, L.B., Habeliz, S., Kurylo, M., Ho, S.P. et al. 2012. Functional remineralization of dentin lesions using polymer-induced liquid-precursor process. PLoS ONE 7(6). DOI: 10.1371/journal.pone.0038852.

Campodoni, E., Adamiano, A., Dozio, S.M., Panseri, S., Montesi, M., Sprio, S. et al. 2016. Development of innovative hybrid and intrinsically magnetic nanobeads as a drug delivery system. Nanomedicine (Lond.) 11(16): 2119–2130. DOI: 10.2217/nnm-2016-0101.

Campodoni, Elisabetta, Patricio, T., Montesi, M., Tampieri, A., Sandri, M. and Sprio, S. 2018. Biomineralization process generating hybrid nano- and micro-carriers. Core-Shell Nanostructures for Drug Delivery and Theranostics. Elsevier Ltd. DOI: 10.1016/B978-0-08-102198-9.00003-X.

Campodoni, Elisabetta, Heggset, E.B., Rashad, A., Ramírez-Rodríguez, G.B., Mustafa, K., Syverud, K. et al. 2019. Polymeric 3D scaffolds for tissue regeneration: Evaluation of biopolymer nanocomposite reinforced with cellulose nanofibrils. Mater. Sci. Eng. C. Elsevier B.V. 94: 867–878. DOI: 10.1016/j.msec.2018.10.026.

Chai, H., Lee, J.J.-W., Constantino, P.J., Lucas, P.W. and Lawn, B.R. 2009. Remarkable resilience of teeth. Proc. Natl. Acad. Sci. 106(18): 7289 LP – 7293. DOI: 10.1073/pnas.0902466106.

Chen, Y., Feng, Y., Deveaux, J.G., Masoud, M.A., Chandra, F.S., Chen, H. et al. 2019. Biomineralization forming process and bio-inspired nanomaterials for biomedical application: A review. Minerals. DOI: 10.3390/min9020068.

Coombs, T.L. 1984. Biomineralization and biological metal accumulation: Biological and geological perspectives. Trends Biochem. Sci. DOI: 10.1016/0968-0004(84)90190-7.

Crichton, R. 2019. Chapter 19 Biomineralization. *In*: pp. 517–544. DOI: 10.1016/b978-0-12-811741-5.00019-9.

Dangaria, S.J., Ito, Y., Yin, L., Valdré, G., Luan, X. and Diekwisch, T.G.H. 2011. Apatite microtopographies instruct signaling tapestries for progenitor-driven new attachment of teeth. Tissue Eng. Part A. 2010/10/08. Mary Ann. Liebert, Inc. 17(3-4): 279–290. DOI: 10.1089/ten.TEA.2010.0264.

Dapporto, M., Sprio, S., Fabbi, C., Figallo, E. and Tampieri, A. 2016. A novel route for the synthesis of macroporous bioceramics for bone regeneration. J. Eur. Ceram. Soc. DOI: 10.1016/j.jeurceramsoc.2015.10.020.

Dash, R., Foston, M. and Ragauskas, A.J. 2013. Improving the mechanical and thermal properties of gelatin hydrogels cross-linked by cellulose nanowhiskers. Carbohydr. Polym. Elsevier Ltd. 91(2): 638–645. DOI: 10.1016/j.carbpol.2012.08.080.

De, S. and Robinson, D. 2003. Polymer relationships during preparation of chitosan-alginate and poly-l-lysine-alginate nanospheres. J. Control. Release. Elsevier 89(1): 101–112. DOI: 10.1016/S0168-3659(03)00098-1.

Drury, J.L. and Mooney, D.J. 2003. Hydrogels for tissue engineering: Scaffold design variables and applications. Biomaterials. Elsevier, pp. 4337–4351. DOI: 10.1016/S0142-9612(03)00340-5.

Du, N., Xiang, Y.L., Narayanan, J., Li, L., Lim, M.L.M. and Li, D. 2006. Design of superior spider silk: From nanostructure to mechanical properties. Biophys. J. Elsevier 91(12): 4528–4535. DOI: 10.1529/biophysj.106.089144.

Duraccio, D., Mussano, F. and Faga, M.G. 2015. Biomaterials for dental implants: Current and future trends. J. Mat. Sci. DOI: 10.1007/s10853-015-9056-3.

Galler, K.M., Cavender, A.C., Koeklue, U., Suggs, L.J., Schmalz, G. and D'Souza, R.N. 2011. Bioengineering of dental stem cells in a PEGylated fibrin gel. Regen. Med. Future Medicine 6(2): 191–200.

Glazer, P.A., Heilmann, M.R., Lotz, J.C. and Bradford, D.S. 1997. Use of electromagnetic fields in a spinal fusion: A rabbit model. Spine. DOI: 10.1097/00007632-199710150-00007.

Hakki, S.S., Karaoz, E., Hakki, S.S. and Karaoz, E. 2016. Dental Stem Cells: Possibility for Generation of a Bio-tooth. Stem Cell Biol. Regen. Med. Springer International Publishing Switzerland, 167. DOI: 10.1007/978-3-319-28947-2_9.

Haumschild, M.S. and Haumschild, R.J. 2009. The importance of oral health in long-term care. J. Am. Med. Dir. Assoc. Elsevier 10(9): 667–671. DOI: 10.1016/J.JAMDA.2009.01.002.

Hollister, S.J. 2005. Porous scaffold design for tissue engineering. Nat. Mater. 4(7): 518–524. DOI: 10.1038/nmat1421.

Hossan, J., Gafur, M.A., Kadir, M.R. and Mainul, M. 2014. Preparation and characterization of gelatin-hydroxyapatite composite for bone tissue engineering. Int. J. Eng. Technol. 57(01): 113–122.

Hwang, J., Jeong, Y., Park, J.M., Lee, K.H., Hong, J.W. and Choi, J. 2015. Biomimetics: Forecasting the future of science, engineering, and medicine. Inter. J. Nanomed. Dove Medical Press Ltd., pp. 5701–5713. doi: 10.2147/IJN.S83642.

Iafisco, M., Ruffini, A., Adamiano, A., Sprio, S. and Tampieri, A. 2014. Biomimetic magnesium-carbonate-apatite nanocrystals endowed with strontium ions as anti-osteoporotic trigger. Mater. Sci. Eng. C 35(1): 212–219. DOI: 10.1016/j.msec.2013.11.009.

Iannotti, V., Adamiano, A., Ausanio, G., Lanotte, L., Aquilanti, G., Coey, J.M.D. 2017. Fe-doping-induced magnetism in nano-hydroxyapatites. Inorg. Chem. 56(8): 4446–4458. DOI: 10.1021/acs.inorgchem.6b03143.

Imamura, Y., Mukohara, T., Shimono, Y., Funakoshi, Y., Chayahara, N., Toyoda, M. et al. 2015. Comparison of 2D- and 3D-culture models as drug-testing platforms in breast cancer. Oncol. Rep. DOI: 10.3892/or.2015.3767.

Inuyama, Y., Kitamura, C., Nishihara, T., Morotomi, T., Nagayoshi, M., Tabata, Y. et al. 2010. Effects of hyaluronic acid sponge as a scaffold on odontoblastic cell line and amputated dental pulp. J. Biomed. Mater. Res. Part B Appl. Biomater. John Wiley & Sons, Ltd. 92B(1): 120–128. DOI: 10.1002/jbm.b.31497.

Jimbo, R., Tovar, N., Janal, M.N., Mousa, R., Marin, C., Yoo, D. et al. 2014. The effect of brain-derived neurotrophic factor on periodontal furcation defects. PloS One. Public Library of Science 9(1): e84845.

Kasaj, A., Willershausen, B., Junker, R., Stratul, S.-I. and Schmidt, M. 2012. Human periodontal ligament fibroblasts stimulated by nanocrystalline hydroxyapatite paste or enamel matrix derivative. An *in vitro* assessment of PDL attachment, migration, and proliferation. Clin. Oral Investig. Springer 16(3): 745–754.

Kawase, T., Tanaka, T., Nishimoto, T., Okuda, K., Nagata, M., Burns, D.M. et al. 2011. Improved adhesion of human cultured periosteal sheets to a porous poly (l-lactic acid) membrane scaffold without the aid of exogenous adhesion biomolecules. J. Biomed. Mater. Res. Part A. Wiley Online Library 98(1): 100–113.

Kokubo, T., Kim, H.M. and Kawashita, M. 2003. Novel bioactive materials with different mechanical properties. Biomaterials. Elsevier 24(13): 2161–2175. DOI: 10.1016/S0142-9612(03)00044-9.

Krishnakumar, G.S., Gostynska, N., Dapporto, M., Campodoni, E., Montesi, M., Panseri, S. et al. 2018. Evaluation of different crosslinking agents on hybrid biomimetic collagen-hydroxyapatite composites for regenerative medicine. Int. J. Biol. Macromol. Elsevier B.V. 106: 739–748. DOI: 10.1016/j.ijbiomac.2017.08.076.

Lee, Chang Hun, Shin, H.J., Cho, I.H., Kang, Y.-M., Kim, I.A., Park, K.-D. et al. 2005. Nanofiber alignment and direction of mechanical strain affect the ECM production of human ACL fibroblast. Biomaterials. Elsevier 26(11): 1261–1270.

Lee, Chang H., Cook, J.L., Mendelson, A., Moioli, E.K., Yao, H. and Mao, J.J. 2010. Regeneration of the articular surface of the rabbit synovial joint by cell homing: a proof of concept study. The Lancet. Elsevier 376(9739): 440–448.

Lee, Chang H., Hajibandeh, J., Suzuki, T., Fan, A., Shang, P. and Mao, J.J. 2014. Three-dimensional printed multiphase scaffolds for regeneration of periodontium complex. Tissue Eng. Part A. Mary Ann. Liebert, Inc. 140 Huguenot Street, 3rd Floor New Rochelle, NY 10801 USA 20(7-8): 1342–1351.

Lee, J.-S., Park, W.-Y., Cha, J.-K., Jung, U.-W., Kim, C.-S., Lee, Y.-K. et al. 2012. Periodontal tissue reaction to customized nano-hydroxyapatite block scaffold in one-wall intrabony defect: a histologic study in dogs. J. Periodontal Implant Sci. 42(2): 50–58.

Li, S., De Wijn, J.R., Li, J., Layrolle, P. and De Groot, K. 2003. Macroporous biphasic calcium phosphate scaffold with high permeability/porosity ratio. Tissue Eng. DOI: 10.1089/107632703322066714.

Loh, Q.L. and Choong, C. 2013. Three-dimensional scaffolds for tissue engineering applications: role of porosity and pore size. Tissue Eng. Part B Rev. Mary Ann Liebert, Inc., Publishers 19(6): 485–502. DOI: 10.1089/ten.teb.2012.0437.

Maeda, H., Tomokiyo, A., Wada, N., Koori, K., Kawachi, G. and Akamine, A. 2014. Regeneration of the periodontium for preservation of the damaged tooth. Histol. Histopathol. 29(10): 1249–1262.

Mohammed, L., Ragab, D. and Gomaa, H. 2016. Bioactivity of hybrid polymeric magnetic nanoparticles and their applications in drug delivery. Current Pharmaceutical Design, pp. 3332–3352. DOI: http://dx.doi.org/10.2174/1381612822666160208143237.

Nam, S., Won, J.E., Kim, C.H. and Kim, H.W. 2011. Odontogenic differentiation of human dental pulp stem cells stimulated by the calcium phosphate porous granules. J. Tissue Eng. DOI: 10.4061/2011/812547.

Panos, I., Acosta, N. and Heras, A. 2008. New drug delivery systems based on chitosan. Curr. Drug Discov. Technol. Bentham Science Publishers 5(4): 333–341.

Panseri, S., Russo, A., Giavaresi, G., Sartori, M., Veronesi, F., Fini, M. et al. 2012. Innovative magnetic scaffolds for orthopedic tissue engineering. J. Biomed. Mater. Res. Part A. Wiley Subscription Services, Inc., A Wiley Company 100A(9): n/a–n/a. DOI: 10.1002/jbm.a.34167.

Panseri, S., Cunha, C., D'Alessandro, T., Sandri, M., Giavaresi, G., Marcacci, M. et al. 2012. Intrinsically superparamagnetic Fe-hydroxyapatite nanoparticles positively influence osteoblast-like cell behaviour. J. Nanobiotechnology 10(1): 32. DOI: 10.1186/1477-3155-10-32.

Panseri, S., Russo, A., Cunha, C., Bondi, A., Di Martino, A., Patella, S. et al. 2012. Osteochondral tissue engineering approaches for articular cartilage and subchondral bone regeneration. Knee Surgery, Sport. Traumatol. Arthrosc. 20(6): 1182–1191. DOI: 10.1007/s00167-011-1655-1.

Panseri, S., Russo, A., Sartori, M., Giavaresi, G., Sandri, M., Fini, M. et al. 2013. Modifying bone scaffold architecture *in vivo* with permanent magnets to facilitate fixation of magnetic scaffolds. Bone. Elsevier 56(2): 432–439. DOI: 10.1016/j.bone.2013.07.015.

Panseri, S., Russo, L., Montesi, M., Taraballi, F., Cunha, C., Marcacci, M. et al. 2014. Bioactivity of surface tethered osteogenic growth peptide motifs. MedChemComm. 5(7): 899–903. DOI: 10.1039/c4md00112e.

Panseri, S., Montesi, M., Dozio, S.M., Savini, E., Tampieri, A., and Sandri, M. 2016. Biomimetic scaffold with aligned microporosity designed for dentin regeneration. Front. Bioeng. Biotechnol. 4: 48. DOI: 10.3389/fbioe.2016.00048.

Phadke, A., Shih, Y.R.V. and Varghese, S. 2012. Mineralized synthetic matrices as an instructive microenvironment for osteogenic differentiation of human mesenchymal stem cells. Macromol. Biosci. DOI: 10.1002/mabi.201100289.

Preti, L., Lambiase, B., Campodoni, E., Sandri, M., Ruffini, A., Pugno, N. et al. 2019. Nature-inspired processes and structures: new paradigms to develop highly bioactive devices for hard tissue regeneration. In Bio-Inspired Technology [Working Title]. DOI: 10.5772/intechopen.82740.

Ramalingam, M., Wang, X., Chen, G.P., Ma, P. and Cui, F.Z. 2013. Advancing Nanobiomaterials and Tissue Engineering. Biomed. Sci. Eng. Technol., John Wiley & Sons Ltd. Wiley Online Library.

Rao, A. and Cölfen, H. 2016. Morphology control and molecular templates in biomineralization. pp. 51–93. *In*: Biomineralization and Biomaterials. Elsevier.

Sandri, M., Filardo, G., Kon, E., Panseri, S., Montesi, M., Iafisco, M. et al. 2016. Fabrication and pilot *in vivo* study of a collagen-BDDGE-elastin core-shell scaffold for tendon regeneration. Front. Bioeng. Biotechnol. 4(June): 1–14. DOI: 10.3389/fbioe.2016.00052.

Saunders, B.E. 2015. A biomimetic study of natural attachment mechanisms—Arctium minus part 1. Robot. Biomimetics. DOI: 10.1186/s40638-015-0028-5.

Shankar, K.G., Gostynska, N., Montesi, M., Panseri, S., Sprio, S., Kon, E. et al. 2017. Investigation of different cross-linking approaches on 3D gelatin scaffolds for tissue engineering application: A comparative analysis. Int. J. Biol. Macromol. Elsevier B.V. 95: 1199–1209. DOI: 10.1016/j.ijbiomac.2016.11.010.

Sharma, S., Srivastava, D., Grover, S. and Sharma, V. 2014. Biomaterials in tooth tissue engineering: a review. J. Clin. Diagnostic Res.: JCDR. JCDR Research & Publications Private Limited 8(1): 309.

Sloan, A.J. and Lynch, C.D. 2012. Dental tissue repair: novel models for tissue regeneration strategies. Open Dent. J. 2012/12/28. Bentham Open. 6: 214–219. DOI: 10.2174/1874210601206010214.

Sprio, S., Dapporto, M., Montesi, M., Panseri, S., Lattanzi, W., Pola, E. et al. 2016. Novel osteointegrative sr-substituted apatitic cements enriched with Alginate. Materials. DOI: 10.3390/ma9090763.

Sprio, S., Campodoni, E., Sandri, M., Preti, L., Keppler, T., Müller, F.A. et al. 2018. A graded multifunctional hybrid scaffold with superparamagnetic ability for periodontal regeneration. Int. J. Mol. Sci. DOI: 10.3390/ijms19113604.

Sugawara, A. and Sato, S. 2014. Application of dedifferentiated fat cells for periodontal tissue regeneration. Hum. Cell. Springer 27(1): 12–21.

Suto, M., Nemoto, E., Kanaya, S., Suzuki, R., Tsuchiya, M. and Shimauchi, H. 2013. Nanohydroxyapatite increases BMP-2 expression via a p38 MAP kinase dependent pathway in periodontal ligament cells. Arch. Oral Biol. Elsevier 58(8): 1021–1028.

Takechi, M., Ohta, K., Ninomiya, Y., Tada, M., Minami, M., Takamoto, M. et al. 2012. 3-dimensional composite scaffolds consisting of apatite-PLGA-atelocollagen for bone tissue engineering. Dent. Mater. J. The Japanese Society for Dental Materials and Devices, pp. 2011–2182.

Talham, D.R. 2002. Biomineralization: Principles and Concepts in Bioinorganic Materials Chemistry Stephen Mann. Oxford University Press, New York, 2001. Cryst. Growth Des. doi: 10.1021/cg020033l.

Tampieri, A., Celotti, G., Landi, E., Sandri, M., Roveri, N. and Falini, G. 2003. Biologically inspired synthesis of bone-like composite: Self-assembled collagen fibers/hydroxyapatite nanocrystals. J. Biomed. Mater. Res.—Part A 67(2): 618–625. DOI: 10.1002/jbm.a.10039.

Tampieri, A., Sandri, M., Landi, E., Celotti, G., Roveri, N., Mattioli-Belmonte, M. et al. 2005. HA/alginate hybrid composites prepared through bio-inspired nucleation. Acta Biomater. Elsevier 1(3): 343–351. DOI: 10.1016/j.actbio.2005.01.001.

Tampieri, A., Celotti, G. and Landi, E. 2005. From biomimetic apatites to biologically inspired composites. Anal. Bioanal. Chem. 381(3): 568–576. DOI: 10.1007/s00216-004-2943-0.

Tampieri, A., Sandri, M., Landi, E., Pressato, D., Francioli, S., Quarto, R. et al. 2008. Design of graded biomimetic osteochondral composite scaffolds. Biomaterials. Elsevier 29(26): 3539–3546. DOI: 10.1016/j.biomaterials.2008.05.008.

Tampieri, A., D'Alessandro, T., Sandri, M., Sprio, S., Landi, E., Bertinetti, L. et al. 2012. Intrinsic magnetism and hyperthermia in bioactive Fe-doped hydroxyapatite. Acta Biomater. Acta Materialia Inc. 8(2): 843–851. DOI: 10.1016/j.actbio.2011.09.032.

Tampieri, A., Iafisco, M., Sandri, M., Panseri, S., Cunha, C., Sprio, S. et al. 2014. Magnetic bioinspired hybrid nanostructured collagen-hydroxyapatite scaffolds supporting cell proliferation and tuning regenerative process. ACS Appl. Mater. Interfaces 6(18): 15697–15707. DOI: 10.1021/am5050967.

Tampieri, A., Ruffini, A., Ballardini, A., Montesi, M., Panseri, S., Salamanna, F. et al. 2019. Heterogeneous chemistry in the 3-D state: An original approach to generate bioactive, mechanically-competent bone scaffolds. Biomater. Sci. DOI: 10.1039/c8bm01145a.

Thesleff, I. and Hurmerinta, K. 1981. Tissue Interactions in Tooth Development. Differentiation. John Wiley & Sons, Ltd. (10.1111), 18(1-3): 75–88. DOI: 10.1111/j.1432-0436.1981.tb01107.x.

Tsuzuki, N., Otsuka, K., Seo, J., Yamada, K., Haneda, S., Furuoka, H. et al. 2012. *In vivo* osteoinductivity of gelatin β-tri-calcium phosphate sponge and bone morphogenetic protein-2 on an equine third metacarpal bone defect. Res. Vet. Sci. Elsevier 93(2): 1021–1025.

Valentijn, A.J., Zouq, N. and Gilmore, A.P. 2004. Anoikis. Biochem. Soc. Trans. 32(3): 421–425. DOI: 10.1042/bst0320421.

Weiner, S. 2003. An overview of biomineralization processes and the problem of the vital effect. Rev. Mineral. Geochemistry. DOI: 10.2113/0540001.

Williams, R.C., Barnett, A.H., Claffey, N., Davis, M., Gadsby, R., Kellett, M. et al. 2008. The potential impact of periodontal disease on general health: A consensus view. Curr. Med. Res. Opin. DOI: 10.1185/03007990802131215.

Wolf, M., Lossdoerfer, S., Abuduwali, N., Meyer, R., Kebir, S., Götz, W. et al. 2013. *In vivo* differentiation of human periodontal ligament cells leads to formation of dental hard tissue. J. Orofac. Orthop. Der Kieferorthopädie. Springer 74(6): 494–505.

Xia, L., Zhang, Z., Chen, L., Zhang, W., Zeng, D., Zhang, X. et al. 2011. Proliferation and osteogenic differentiation of human periodontal ligament cells on akermanite and β-TCP bioceramics. Eur. Cell Mater. 22(68): e82.

Yang, H.-Y., Kwon, J., Kook, M.-S., Kang, S.S., Kim, S.E., Sohn, S. et al. 2013. Proteomic analysis of gingival tissue and alveolar bone during alveolar bone healing. Mol. Cell. Proteomics: MCP. 2013/07/03. The American Society for Biochemistry and Molecular Biology 12(10): 2674–2688. DOI: 10.1074/mcp.M112.026740.

Yuan, Z., Nie, H., Wang, S., Lee, C.H., Li, A., Fu, S.Y. et al. 2011. Biomaterial selection for tooth regeneration. Tissue Eng. Part B Rev. Mary Ann. Liebert, Inc. 140 Huguenot Street, 3rd Floor New Rochelle, NY 10801 USA 17(5): 373–388.

Zan, X., Sitasuwan, P., Feng, S. and Wang, Q. 2016. Effect of roughness on *in situ* biomineralized CaP-collagen coating on the osteogenesis of mesenchymal stem cells. Langmuir. American Chemical Society 32(7): 1808–1817. DOI: 10.1021/acs.langmuir.5b04245.

Zeeman, R., Dijkstra, P.J., Van Wachem, P.B., Van Luyn, M.J.A., Hendriks, M., Cahalan, P.T. et al. 2000. The kinetics of 1,4-butanediol diglycidyl ether crosslinking of dermal sheep collagen. J. Biomed. Mater. Res. 51(4): 541–548.

Zheng, L., Yang, F., Shen, H., Hu, X., Mochizuki, C., Sato, M. et al. 2011. The effect of composition of calcium phosphate composite scaffolds on the formation of tooth tissue from human dental pulp stem cells. Biomaterials. Elsevier 32(29): 7053–7059.

9

Craniofacial Regeneration—Bone

Laura Guadalupe Hernández Tapia,[1] *Lucia Pérez Sánchez,*[1]
Rafael Hernández González[2] and *Janeth Serrano-Bello*[1,*]

Introduction

Bone tissue engineering is a multidisciplinary field that includes biology, medicine and engineering areas, of which the main objective is bone regeneration. The craniofacial defects are challenges because of their complicated structure and function, the various etiological factors are trauma, tumor or cyst resection, infectious diseases, and also the congenital and developmental condition.

Therefore, it is important to know the anatomical and physiological structures of the cranio-maxillofacial bone for further understanding of the complex bone regeneration and repair of normal bone tissue, in order to find new alternatives to improve the treatment used to date in the clinic (autografts, allografts, xenografts and synthetic substitutes), such as 3D printing or scaffolding manufacturing that meet the characteristics of biocompatibility, osteogenicity, osteoconductivity, osteoinduction and osseointegration. The ideal bone substitute should replicate the essential traits of the native bone and evaluate the metabolism and microarchitecture through molecular imaging technology.

Generalities of the Craniofacial Structure

Anatomy

Craniofacial bone tissue (CF) is a complex physiological structure consisting of bone and soft tissue (Gaihre et al. 2017; Datta et al. 2017). The bone tissue provides essential structural support and projection to overlying soft tissue structures such as tendons, ligaments, muscles, facial skin, nerves, blood vessels and sensory organs (Kawecki et al. 2018; Visscher et al. 2017).

The craniofacial bone can be divided according to their functions and structures in the neurocranium, face skeleton and oral apparatus (Fig. 9.1).

[1] Tissue Bioengineering Laboratory, Division of Graduate Studies and Research, Faculty of Dentistry, National Autonomous University of Mexico (UNAM), Circuito Exterior s/n. Col. Copilco el Alto, Alcaldía de Coyoacán, C.P. 04510, CDMX, México.
[2] Unit Bioterium, Medical School, National Autonomous University of Mexico (UNAM), Circuito Exterior s/n. Col. Copilco el Alto, Alcaldía de Coyoacán, C.P. 04510, CDMX, México.
* Corresponding author: janserbe@comunidad.unam.mx, janserbe@hotmail.com

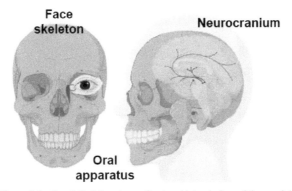

Figure 9.1. Craniofacial anatomy, front and lateral view of the cranial.

The neurocranium encloses the brain, the 12 cranial nerves, and the vascular supply within the brain. Its primary function is to protect the brain and is composed by eight bones: the frontal, ethmoid, sphenoid, two parietal, the occipital and the temporal bones (Chu et al. 2014). The face skeleton is defined by 14 bones: two palatine bones, two lacrimal bones, the maxilla bones, the mandible, the vomer, the zygomatic, two inferior nasal turbinates, as well as two nasal bones (Gaihre et al. 2017; Kawecki et al. 2018).

The oral apparatus includes both soft and hard tissues. The soft tissues are the hard tissues comprising the alveolar bone, the alveolar process and the teeth, which include three structures: the enamel, cementum and dentine. The soft-tissue component is the pulp (Gaihre et al. 2017; Chu et al. 2014). All the above-mentioned bones have specific shapes, different volumes and provide a frame on which the soft tissues of the face can act to facilitate facial expression, eating, breathing and speech. These key characteristics will have to be considered when selecting the ideal graft for craniofacial reconstruction (Kawecki et al. 2018).

Neovascularity

Bone is a richly vascularized connective tissue, the process of neovascularization plays a significant role in the process of bone development (endochondral and intramembranous ossification), regeneration and remodeling, which involves both angiogenesis and vasculogenesis (Fishero et al. 2014; Filipowska et al. 2017).

Blood vessels of the bone develop through the process of angiogenesis, which involves the proliferation of local endothelial cells to produce new blood vessels from pre-existing vessels in a remodeling process. The vasculogenesis is the formation of a vascular network, from a progenitor cell, angioblast or hemangioblast. The blood vessels supply the bone system with nutrients and oxygen, excrete waste biological materials, remove metabolites from the bone, provide the bone with specific hormones, growth factors and neurotransmitters secreted by other tissues, maintaining the bone cells survival and stimulating their activity. The craniofacial bones develop by two processes: intramembranous ossification and endochondral ossification. Intramembranous ossification is the main mechanism leading to a development of flats bones (e.g., maxillae, palatal bones, nasal bones, zygomatic bones) this process is related to a direct differentiation of mesenchymal stem cells into osteoblasts, initially with a fibrous membrane and finally replaced by a spongy bone, whereas the endochondral ossification is typical of long bones and the cranial base, this process has an intermediate stage with cartilage (Chu et al. 2014; Fishero et al. 2014; Filipowska et al. 2017). The development and maintenance of the endochondral and intramembranous bone formation are dependent on the bone vascular network (Filipowska et al. 2017; Prisby 2017).

The bone is a dynamic tissue which is in a continuous process of remodeling. Bone remodeling is performed by groups of cells called Bone Multicellular Units (BMU) (Fig. 9.2), the main characters

Cell	Molecules
aOCY	HMGB-1 M-CSF
OCY	Sclerostin TGF-B
OBA	RANKL OPG PDGF IGF FGF
OBL	BMPs
OCL	RANK

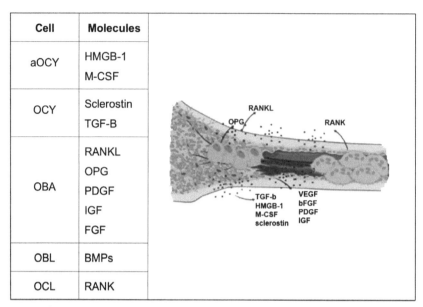

Figure 9.2. BMU: Main molecules induced in the process of remodeled bone. aOCY: apoptotic osteocytes, OCY: Osteocytes, OBA: activated osteoblasts, OBL: osteoblasts, OCL: osteoclasts (modified from Arias et al. 2018).

of this process are osteoclasts (bone-resorbing) recruited when needed from their cell precursors, osteoblasts (bone-forming) biosynthesized of/by new bone to replace the former is carried out by osteoblasts and osteocytes that are mechanosensors which monitor mechanical stress within bone tissues (Visscher et al. 2017; Prisby 2017; Arias et al. 2018).

Remodeling has basically four stages, the activation phase; in which initiation with the osteocyte cell death is caused by bone microdamage, traumatic bone fractures, the decreased sclerostin levels active in BMU, in this phase the osteoclasts precursor cells are recruited to the altered bone surface, alert signals are produced for recruit immune cells and the mediator's inflammatory process such as Vascular Endothelial Growth Factor (VEGF), angiopoietins, HMGB-1 and cytokines such as IGF-1, growth factors such as Transforming Growth Factor-β (TGF-β) induce osteoblast differentiation, basic Fibroblast Growth Factor (bFGF) and PDGF activate osteoblast and inhibition osteoclast action, Platelet-Derived Growth Factor-BB (PDGF-BB), and Insulin-like Growth Factor-I (IGF-I) contribute to the induction of callus formation, released from bone matrix, and then activating osteoblast differentiation. Next is the resorption phase, when the mature osteoclast degrade the mineralized matrix, the main signals in this step are Macrophage Colony Stimulating Factor M-CSF and ligand Receptor Activator of Nuclear Factor kappa-B RANKL they promote the differentiation of osteoclasts precursors; following the reversal phase, activated osteoblasts advance in the wake of bone-destroying cutting cones to replenish the cavity left behind by the latter, the osteoblasts are recruited and the osteoclasts undergo apoptosis; later during the formation phase, where the osteoblasts lay down new organic bone matrix that subsequently mineralizes, some of the active osteoblasts become trapped in the matrix that they secrete and subsequently differentiate into mature osteocytes. All these phases together contribute to the formation of a complete remodeled bone which is both structurally and functionally similar (Langdahl et al. 2016; Arias et al. 2018).

Etiologies of the Craniofacial Bone Defects

The craniofacial bone defects are challenges for tissues engineering because of the presence of complex physiological structures, including cartilage, facial skeletal features, muscles, skin, ligaments, blood vessels and nerves.

The bones of the craniofacial skeleton provide an essential role in supporting the adjacent soft tissues, providing anchorage for dental structures, maintaining structural stability for many physical functions and composing the aesthetics of the human body.

There are various conditions or etiologies for craniofacial defects such as congenital malformation, cleft palate defects, trauma, infections and tumor or cysts resection, often leading to large psychomedical burdens as well as difficult reconstructive (Wan et al. 2006; Ward et al. 2010; Si et al. 2015; Datta et al. 2017). In this chapter, we will briefly describe the most important etiologies of craniofacial defects.

Congenital Anomalies and Disorders

The most common birth congenital anomalies, include orofacial clefts, craniosynostoses, the mandibulofacial dysostoses and craniofacial macrosomia. Congenital anomalies occur in approximately 3 to 5% of all live births. Furthermore, many of these conditions have a genetic etiology such as chromosomal, single-gene disorders or epigenetic mutation or may be caused by teratogens (Saal 2016).

Whitaker proposed a classification of craniofacial anomalies according to a major structure involving (the lip and palate, eyes, nose, mandibular and maxillary abnormalities) on etiology and treatment principles (Pashayan and Reichman 1980) see Table 9.1.

Most recognizable craniofacial syndromes are monogenic Mendelian disorders, different mutations in the same gene. However, it is not a rule because, more than 7000 single-gene disorders have been identified in craniofacial disorders such as Stickler syndrome and the craniosynostosis syndromes involving different FGFR genes (FGFR1, FGFR2, FGFR3).

In genetic medicine, there are a growing number of clinically indistinguishable or overlapping phenotypes of craniofacial disorders that may be caused by mutations in different genes (locus heterogeneity). These include rasopathies, cohesinopathies, mandibulofacial dysostoses and Stickler syndrome, whereby differential diagnosis is crucial for treatment (Sanchez-Lara 2015).

Table 9.1. Whitaker classification of craniofacial anomalies.

Type		Definition
I	Clefts	Centric (surrounding structures are displaced laterally, require reposition) • Facial clefts • Cranial extensions Acentric • Facial clefts • Cranial clefts
II	Synostoses	Symmetric • Metopic • Coronal • Sagittal Asymmetric • Unilateral coronal closure • Lambdoid closure
III	Atrophy (hypoplasia)	Atrophy of skin, subcutaneous tissue, muscle, bone (e.g., Romberg syndrome, coup the sabre)
IV	Neoplasia (hyperplasia)	Lymphangioma, hemangioma, fibrous Dysplasya
V	Unclassified	Multiorgan involvement Single-organ involvement

Modified from Buchanan, P.E., Xue, S.A. and Hollier Jr, H.L. 2014. Craniofacial syndromes. Plastic and Reconstructive Surgery 134: 128–153.

Cleft lip with or without cleft palate (CLP) is among the most common birth defects caused by genetic and nongenetic factors. Several genes could be implicated in CLP; IFR6, TGF-A, TGF-B3 and MSX1 (Howard et al. 2008). Nongenetic factors may include, fetal environment, teratogenic exposures, placental factors and the health of the mother.

Maternal illnesses could be because of congenital craniofacial disorder, the greatest risks associated with type 1 diabetes mellitus are a cleft lip, Cleft Palate (CP), and Pierre Robin Sequence (PRS). On the other hand, illness relating to craniofacial anomalies are phenylketonuria affecting women who do not follow a phenylalanine-restricted diet. The elevated levels of the metabolites of phenylalanine can cause multiple anomalies, including microcephaly, ear anomalies, congenital heart defects and CP.

The craniosynostosis have been associated with maternal hyperthyroidism and serious diseases (Howard et al. 2008).

Other etiologies cause craniofacial congenital anomalies are teratogens which are exogenous substances or physical agents, there are many teratogens such as tobacco, medications (nitrofurantoin and warfarin used during pregnancy), infectious agents, physical agents, radiation, alcohol, toluene and cocaine. The main anomalies relation with teratogens are microcephaly, brain anomalies, holoprosencephaly, limb anomalies, short stature, and behavior disorders, also including CLP, CP, and PRS (Saal 2016; Nagy et al. 2014).

Trauma

One common cause of craniofacial defects is trauma including acute trauma, falls, assaults, sport injuries and vehicle accidents (Zhang and Yelick 2019).

According to Detroit Medical Center (DMC) records 30,260 adult facial fractures were identified, these included nasal (30.1%), mandible (22.7%), malar-maxillary (15.4%), orbital floor (15.7%), and other (16.1%) fractures.

Facial injuries are more common in men than women (68% compared with 32%), furthermore half of all the patients were between 15 and 45-years-old (Walker et al. 2011). Different authors mention other causes of craniofacial defects such as intoxication with alcohol, illegal drugs, Motor Vehicle Accidents (MVAs) in adults. However, children account for approximately 14% of all facial fractures, three trauma mechanisms of pediatric facial fractures: MVA (43%), Intentional Trauma (IT) (17%) and falls (11%) the incidence of inpatient facial fractures due to MVA and IT increases with age. Nevertheless, African American and Hispanic patients accounted for most patients in the IT group and came from the poorest neighborhoods (Streubel and Mirsky 2016).

Treatments for Craniofacial Bone Defects

Current surgical treatment for craniofacial defects has been improved during the last year. Nowadays various techniques have been used (autogenous grafting, allogeneic grafting, and prosthetic materials) for bone reconstruction. However, tissues engineering has played a crucial role in bone regeneration for craniofacial defects, as they are common critical size defects and their complicated structure and function. However, despite the use of grafting, none of these modalities have yet to prove a consummate tool for craniofacial bone reconstruction, therefore the emergence of tissue engineering shows great potential as a future treatment for craniofacial defects as tissue engineering must consider three main factors to achieve bone regeneration such as cells, scaffold and growth factors, which influence cellular activity. All of them have proposed one which mimics the extracellular matrix of tissues.

Here the principal treatments for bone regeneration of craniofacial defects, including grafting, scaffolds, cells, growth factor, etc., are described.

Grafting

Bone grafting is one of the most commonly used surgical methods to augment bone regeneration, and the second most frequent tissue transplantation just after blood transfusion. Bone grafts differ in terms of their properties of osteoconduction, osteoinduction, osteogenesis and structural support. As a result, in order to identify the ideal graft, surgeons should understand the requirements of the clinical situation and of the specific properties of the different types of bone graft. Among available grafts, autografts and allografts are considered the best approach. However, these strategies are associated with their own disadvantages, including limited availability in case of autografts and potential immunogenic rejection when it comes to the utilization of allografts (Gaihre et al. 2017; Wang and Yeung 2017; Fillingham and Jacobs 2016).

Other forms of bone grafts, such as xenografts and synthetic grafts, eliminate the need for secondary procedures and obviate donor site complications, which make Bone Grafts and Substitutes (BGS) among the most promising in the orthopedic industry. We also describe the most common bone graftings (Shibuya and Jupiter 2015; Wang and Yeung 2017).

Autografts

Autologous bone is still considered as the gold standard since all the necessary properties required in bone regeneration are present, it holds viable cells that can form new bone tissue (osteogenic), it provides a scaffold for the ingrowth of cells necessary for bone regeneration (osteoconductive); and promotes the proliferation of stem cells and their differentiation into osteogenic cells (osteoinductive). It is ideal in many situations because it is harvested from the patient himself or herself, thus less likely to be rejected and more likely to be incorporated. However, the use of an autograft has limitations, including donor site morbidity, limited availability of tissue, an additional operation and prolonged healing time (Henkel et al. 2013). The major and minor complication rates of autogenous bone graft harvest have been reported at 8.6 and 20.6%, respectively (Fillingham and Jacobs 2016).

Other forms of bone grafts, such as allografts, xenografts and synthetic grafts, eliminate the need for secondary procedures and obviate donor site complications. However, rejection and slower incorporation can be disadvantages to the use of these grafts. In well-vascularized bones, such as the calcaneus, it has been documented that there is no difference in the complication rate at the osteotomy site between autograft and allografts. Nevertheless, in less vascularized areas, incorporation can be difficult.

Allografts

Allografted bone is harvested from human donors, mainly cadavers. Its ready availability in various shapes and sizes, avoidance of the need to sacrifice a host structure and no donor-site morbidity are some of the advantages. The disadvantages include infection such as transmission of Human Immunodeficiency Virus (HIV), Hepatitis C Virus (HCV), Human T-Lymphocytic Virus (HTLV), unspecified hepatitis, tuberculosis and other bacteria has been documented for allografts (mainly from those containing viable cells) and immune resistance (Henkel et al. 2013).

Allografts are procured from humans and undergo vigorous sterilization processes before they are ready for surgeons to use. They can be prepared using a combination of different processing procedures. Cadaveric allograft bone is available in either cancellous or cortical forms, or as a Demineralized Bone Matrix (DBM). Allografts are primarily osteoconductive, while DBM is processed in such a way as to retain osteoinductive properties (Fillingham and Jacobs 2016).

In general, allogenic bone grafts can be classified into fresh, fresh-frozen, freeze-dried and demineralized types, depending on the preparation process. Although a more vigorous sterilization process can eliminate the chances of disease transmission and infection, it can also reduce osteogenic and osteoinductive properties. In general, fresher grafts are more expensive and less readily available than other grafts that have a longer shelf life.

Xenografts

A xenograft comes from a nonhuman species. They are materials with their organic components totally removed, so concern about immunological reactions becomes nonexistent. Therefore, antigenicity is significantly greater than that of allografts. The remaining inorganic structure provides a natural architectural matrix as well as an excellent source of calcium. The inorganic material also maintains the physical dimension of the augmentation during the remodeling phases (Hoexter 2002). Naturally, it requires more sterile processing, which can result in reduced osteoinductive properties. However, owing to the abundance of donors, these grafts may be less expensive and more readily available. Also because of the extensive sterilization processes, the shelf life is generally long. The most common xenogenous bone graft used in orthopedic surgery is porous natural bone hydroxyapatite from animal bones (bovine, equine, porcine, etc.), are also part of this group. Phytogenic materials such as bone-analog calcium phosphate originally obtained from marine algae or coral derived materials, also fall into this category (Wang and Yeung 2017; Shibuya and Jupiter 2015; Henkel et al. 2013).

Bone Graft Substitute Materials (BSM)

Bone substitutes can be defined as "a synthetic, inorganic or biologically organic combination—biomaterial—which can be inserted for the treatment of a bone defect instead of autogenous or allogenous bone" (Henkel et al. 2013), an ideal bone substitute material should offer an osteoinductive three-dimensional structure, contain osteogenic cells and osteoinductive factors, have sufficient mechanical properties and promote vascularization.

BSM is classified according to their origin (Diagram 9.1).

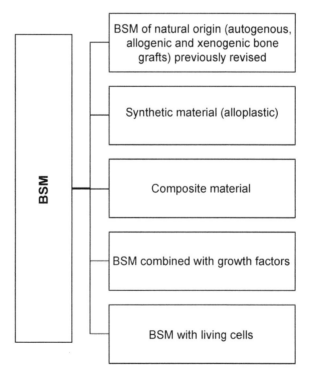

Diagram 9.1. Bone grafts substitute materials (BSM).

Synthetic Materials (Alloplastic)

Artificial bone can be created from ceramics such as calcium phosphates, bioglass and calcium sulfate that are biologically active depending on solubility in the physiological environment. The varying nature of this graft material (porosity, geometries, differing solubilities and densities) will determine the resorption of calcium phosphate-based graft materials, bioceramics are neither osteogenic nor osteoinductive, but work by creating an osteoconductive scaffold to promote osteosynthesis. Today there are four main types of bioceramics available: calcium sulfate, calcium phosphate, tricalcium phosphate and coralline hydroxyapatite; composite bioceramics use a combination of these types to provide materials with improved properties (Kumar et al. 2013; Martin and Bettencourt 2018; Fillingham and Jacobs 2016).

Titanium metal also belongs to this group. Moreover, polymers including polymethylmethacrylate (PMMA), polylactides/polyglycolide and copolymers as well as polycaprolactone (PCL) are also part of this group (Henkel et al. 2013).

Composite Materials

BSM combine different materials such as ceramics and polymers with the objective of a modifier to the more favorable characteristic of merging materials with different structural and biochemical properties (Henkel et al. 2013).

BSM Combined with Growth Factors

Most bone graft substitutes, especially synthetic ceramics and cement do not possess any osteoinductive property. Growth factors present at the site of bone injury during bone healing is critically important because of their osteogenesis is induced and this is the reason for using growth factor and grafts.

The factors and proteins that exist in bone are responsible for regulating cellular activity. Growth factors bind to receptors on cell surfaces and stimulate intracellular environment to act. Generally, this activity translates to a protein kinase that induces a series of events resulting in transcription of messenger ribonucleic acid (mRNA) and ultimately into the formation of a protein to be used intracellularly or extracellularly (Kumar et al. 2013; Wang and Yeung 2017).

Growth factors enhanced grafts are produced using recombinant DNA technology. They consist of either human growth factors or morphogens (BMPs in conjunction with a carrier medium, such as collagen) such as natural or recombinant growth factors like a Bone Morphogenic Protein (BMP), Platelet-Derived Growth Factor (PDGF), Transforming Growth Factor-ß (TGF-ß), Insulin-like Growth Factor (IGF-1), Vascular Endothelial Growth Factor (VEGF) and fibroblast growth factor (Henkel et al. 2013; Kumar et al. 2013).

Bone Morphogenetic Proteins (BMPs), work to promote osteoinduction by binding to specific transmembrane receptors on mesenchymal stem cells, osteoblasts and mature chondrocytes are possibly the most extensively investigated growth factors in treating skeletal defects, mainly BMP-2 and BMP-7, are members of the Transforming Growth Factor beta (TGF-b) superfamily.

BMP-2 is able to induce osteoblastic differentiation from mesenchymal stem cells, and BMP-7 can directly promote angiogenesis, both have an important role in osteoinductive properties (Wang and Yeung 2017).

Platelet-Rich Plasma (PRP)

A typical blood specimen comprises 93% red blood cells, 6% platelets, and 1% white blood cells. Platelets are responsible for hemostasis, construction of new connective tissue and revascularization because, following an injury that causes bleeding, platelets are activated and aggregate together to release their granules containing growth factors that stimulate the inflammatory cascade and healing process. Degranulation causes the granules to fuse with the cell membrane of the platelets, where some of the secretory proteins (e.g., PDGF and TGF-) enter the active state when they are added to the histones and side chains of carbohydrates. Thus, the proteins are secreted, allowing them to bind to the receptors of the target cells (e.g., mesenchymal stem cells, osteoblasts, fibroblasts, endothelial cells or epidermal cells) (Rodríguez Flores et al. 2012).

PRP may be defined as a portion of plasma fraction of autologous blood with the platelet count above the baseline, these concentrated platelets are suspended in a small volume of the plasma, PRP is more than just a platelet concentrate; it also contains the proteins in blood known to act as cell adhesion molecules for osteoconduction and as a matrix for bone, connective tissue and epithelial migration. These cell adhesion molecules are fibrin itself, thrombospondin (TSP)-1, fibronectin and vitronectin.

The other notable components of PRP include *transforming growth factor (TGF)-β*, whose main function stimulates undifferentiated mesenchymal cell proliferation, regulates endothelial, fibroblastic and osteoblastic mitogenesis and regulates collagen synthesis and collagenase secretion, *platelet-derived growth factors (PDGF-AB and PDGF-BB)* their functions are mitogenic for mesenchymal cells and osteoblasts, stimulates chemotaxis and mitogenesis in fibroblast, glial or smooth muscle cells, *insulin-like growth factor (IGF)* enhancing bone formation and chemotactic for fibroblasts and stimulating protein synthesis *vascular endothelial growth factors (VEGFs)* stimulating mitogenesis for endothelial cells, and increasing angiogenesis and vessel permeability, *epidermal growth factor (EGFs)* stimulating endothelial chemotaxis or angiogenesis and regulating collagenase secretion, *fibroblast growth factor (FGF)-2* promoting growth and differentiation of chondrocytes and osteoblasts, *connective tissue growth factor (CTGF)* promoting angiogenesis and cartilage regeneration, *platelet factor 4 (PF-4)* stimulating the initial influx of neutrophils into wounds and chemo-attractant for fibroblasts, *interleukin 8 (IL-8)* pro-inflammatory mediator and recruitment of inflammatory cells, *keratinocyte growth factor (KGF)* promoting endothelial cell growth, migration, adhesion and survival (Dhillon et al. 2012).

Their principal clinical applications for platelet-rich plasma are being used in various applications, including orthopedics, cardiovascular surgery, cosmetics, facio-maxillary surgery and urology (Lansdown and Fortier 2017).

In field bone engineering PRP uses bone formation because it stimulates the osteoblastic differentiation of myoblasts and osteoblastic cells in three-dimensional cultures in the presence of bone morphogenetic protein BMP-2, BMP-4, BMP-6 or BMP-7. These results suggest that platelets contain not only growth factors for proliferation but also novel potentiator(s) for BMP-dependent osteoblastic differentiation. This is the reason why PRP could be osteoinductive when used in grafts (Dhillon et al. 2012).

BSM with Living Cells

Mesenchymal stem cells, bone marrow stromal cells, periosteal cells, osteoblasts and embryonic as well as adult stem cells have been used in bone tissue engineering, as they have the potential to augment the performance of current bone graft substitutes and are the focus of a great deal of ongoing research (Henkel et al. 2013).

The most likely candidate for such therapies remains the Mesenchymal Stem Cell (MSC) for their characters such as potent immunomodulatory and anti-inflammatory properties MSCs can thus regulate the intensity of immune response by inducing T-cell apoptosis, which could have great therapeutic potential when biomaterials are used for tissue engineering applications.

The goal of the scaffold with the cell is to achieve cell adherence, proliferation, differentiate and form the required tissues. Another strategy is implantation of acellular scaffold into the defect, the body can live inside the scaffold with cells (Abou Neel et al. 2014).

Various biomaterials can be chosen for combination with the cells, depending on the goal (mechanical strength or filling) and approach (percutaneous or surgical). The most widely used biomaterials for bone regeneration are calcium-phosphate ceramics, which usually combine hydroxyapatite and β-tricalcium phosphate as granules or other inorganic molecules. However, natural and synthetic polymers can also be used. All of them are examples of biomaterials uses such as scaffolds.

The scaffold is a framework and mechanical support for cell migration, proliferation and differentiation where the main goal is mimicking extracellular matrix (ECM).

The main property is porosity because it facilitates cell ingrowth and vascularization and the biodegradation process. Other important characteristics are biocompatible, surface chemistry, mechanical properties and topography which could help interactions between cells and material (Padial-Molina et al. 2015).

Studies have shown that scaffold architecture modifies the response of cells and subsequent tissue formation and nano to microscale topography and has been demonstrated to affect cell behavior by modification of cytoskeleton arrangements (Howard et al. 2008).

Natural polymers, synthetic polymers and ceramics have been successfully investigated to be applied in developing scaffolds. The most common synthetic polymers are poly (α-hydroxy ester) such as Poly (Glycolic Acid) (PGA), Poly (Lactic Acid) (PLA), poly (ε-caprolactone) (PCL) and their copolymers. These polymers are decomposed by hydrolysis of the ester bonds, their degradation products are in some cases resorbed through the metabolic pathways and their structures can be tailored by altering their degradation rates. In the case of natural polymers, some examples are alginate, chitosan, starch, cellulose, collagen, silk fibroin and albumin. These polymers are a frequently used option utilized for tissue regeneration applications because they can more closely mimic the ECM and may improve the biological recognition in the growing new tissue.

Bioceramics have limited their potential use in the clinical side, because of insufficient biocompatibility and biodegradability. However, they react with physiological fluids and create tough interactions with hard and soft tissue via cellular activity. The main bioceramics are hydroxyapatite (HA), tricalcium phosphate (TCP) and some compositions of silicate and phosphate glasses and glassceramics. It is important mention that by combining degradable polymers and inorganic bioactive particles, it is feasible to produce scaffolds with engineered physical, biological and mechanical features (Raeisdasteh Hokmabad et al. 2017).

There are various scaffold fabrication methods, such as solvent casting, particle leaching, gas foaming, phase separation and electro-spinning. However, they have many limitations, in comparison with 3D printing technology which has emerged as a valuable tool, it can design a scaffold to closely match the geometry of a defect site that allow in predicting greater integration because of the improving host-scaffold interaction (Wang and Yeung 2017; Zhou et al. 2016).

Alternative Strategies on the Craniofacial Regeneration

Animal Models

Animal models are strategies used in tissue engineering to evaluate the process of bone regeneration, this process is induced by bone graft substitutes, scaffolds and cellular therapy in bone defects, with the main goal of obtaining clues as to how the disease develops in the body (Prasad 2016; McGovern et al. 2018), understanding key processes (Mardas et al. 2014), these provide information for systematically assessing the efficacy and risks of recently created biomaterials, medical devices, drugs and provide important reference data for clinical trials (Li et al. 2017; Otterloo et al. 2017).

Animal studies for investigation are considered a long-standing practice and proof-of-concept, which are based on biological responses, degradation time and dose response of the biomaterials (Morais et al. 2018; Otterloo et al. 2017).

Bone injuries are a significant clinical challenge and animal models have been used to evaluate the concepts of bone regeneration. This represents an important phase between *in vitro* studies and clinical applications. However, the use of animal models is controversial because the procedure is considered as highly invasive and not always successful. The main reasons for the use of animal models are:

1. Similarity. There exists a physiological and anatomical similarity between humans and animals, principally mammals (Morais et al. 2018).
2. Feasibility. The management of animal models is relatively easy since different factors can be controlled from the composition of food intake to genetic background and environmental conditions such as temperature and lighting. Hence they represent good models since they can be studied over their entire life cycle.
3. Medicaments tests. Preclinical toxicity testing, pharmacokinetics and pharmacodynamics, before using in humans (Morais et al. 2018; Otterloo et al. 2017).

Several and small large laboratory animal models have been developed over the last half-century to study bone regeneration. Specific to craniofacial regeneration, small animal species including mice, rats, guinea pigs and rabbits have been used. These models are used to investigate the healing capabilities of various scaffolds, growth factors and stem cells (Mardas et al. 2014; Murphy et al. 2017).

Small animal models have some benefits:

a) Are easy to handle and maintain
b) Have well-defined genetic backgrounds
c) Large groups of animals can be used
d) Are economically suited to most type of research.

Among large animal models some of the species included are dog, pig, sheep, goat, monkey and horse. However, these models are expensive and there are ethical concerns regarding their use as laboratory species for experimental research (Li et al. 2017; Ning et al. 2017; Mardas et al. 2014; Liu et al. 2019).

With the characteristics described above, the most frequently used animal models for repairing defect are small animal models, rats and rabbits are mostly used (Ning et al. 2017; Li et al. 2017).

The critical size defect has been extensively used to simulate a greater degree of bone loss and as experimental control defects for testing the efficacy of different biomaterials to promote bone regeneration (Ning et al. 2017; Mardas et al. 2014).

Critical-size bone defect refers to the range of critical bone defects that are unable to spontaneously re-ossify at a specific bone trauma site and require complex reconstructive approaches, due to the lack of muscle tissue and the absence of blood. In craniofacial regeneration, the main types of defects are the calvarial bone defects, which are usually performed on the frontal, parietal and mandibular bones. Rats and mice are commonly used as animal models for studying critical-size defects. In rats, a critical-size cranial defect is about to 8 mm and for mice 4–5 mm. Overall the defect is performed by removing the pericranium with a trephine. This is used to create a circular defect in the skull, after taking care to avoid damage to the dura mater (Ning et al. 2017; McGovern et al. 2018).

In our research group, we worked with males, 7–8 weeks of age, *Wistar* rats of approximately 250–300 grams. The rats were obtained from the animal facility of National Autonomous University of Mexico, School of Medicine. Calvarial bone defects were evaluated by PLA scaffolds designed from microtomographic images, the defect size was 9 mm placed in the frontal bone subsequently, microtomography studies were carried out at different times to assess the regeneration of the area (Fig. 9.3).

The strategies of bone-tissue engineering for craniofacial defects are multistep and multicomponent process approaches involved in improving models that are established in combination with biocompatible scaffolds, cells and growth factors.

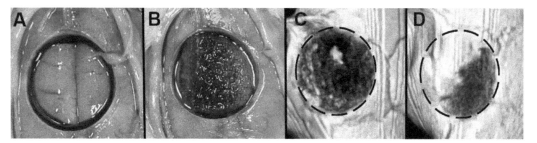

Figure 9.3. Calvaria bone defects. (a) critical size defect, (b) scaffolding of PLA printed in 3d in the area of the defect implanted, (c) μCT of bone defect after 7 days post-surgery, (d) μCT of bone defect after 3 weeks post-surgery.

3D printing

3D printing is a technology of additive manufacturing, takes care of the fabricating of scaffolds of constructs with complex structure with the purpose of mimic the extracellular matrix of native tissue (Thrivikraman et al. 2017).

The most important advantages are:

- Fabrication of scaffolds precise for patient-specific customization (Lee and Dai 2016).
- Ability to control the internal and external 3D structure of scaffolds.
- Obtain rapid and fabricating of complex structures.
- Shape fidelity and precise.
- Fabrication of scaffolds with multiple material and multiscale.
- Ability of controlling cell behavior and mechanical response (Derakhshanfar et al. 2018).
- Accessibility costs (Li et al. 2017).

Nowadays during the fabrication process of this technique, printing of cells, cell aggregates or growth factors into the bio-ink, deposited layer by layer, and subsequently crosslinked, having an effective control over scaffold fabrication and cell distribution, which is named as 3D bioprinting (Derakhshanfar et al. 2018; Sigaux et al. 2017).

There are several methods used for printing such as; extrusion printing, inkjet printing, laser assisted printing and lithography printing (Thrivikraman et al. 2017; Visscher et al. 2017).

Extrusion 3D printer/bioprinter: Is the modality most commonly used compared with other printers, and is based on fused deposition modeling. This method uses a syringe and piston-driven or pneumatic-driven to dispense material through nozzles under constant pressure, extruding continuous strands. The technique allows a rapid fabrication, better structure integrity, printing capability of highly-viscous material, flexibility with materials and curing options and also using inexpensive equipment. The major advantage is the printing of cell spheroids or multicellular cell-spheroids, however their disadvantages are the high shear trigger that kills the cells during or after the printing process and limited material selection due to rapid cell encapsulation. Basically, the materials used are hydrogels such as alginate, collagen, fibrin, chitosan, agarose and synthetic polymers such as Pluronic F-127, polycaprolactone (PCL), PLGA and polyethylene glycol (PEG) and ceramics such as hydroxyapatite (Munaz et al. 2016; Bishop et al. 2017; Derakhshanfar et al. 2018).

Laser-assisted printing allows the deposition of solid material or liquids photoactive through laser energy absorption, this technology is nozzle free which means that it is a non-contacting method, therefore does not have problems of nozzle clogging with cells or materials (Derakhshanfar et al. 2018). This develops very high-resolution patterning and precision, which is generally compatible with a wide range of biomaterial with a high viscosity (1–300 mPa/s) (Lee and Dai 2016). Also, this has made it possible to bioprint a very high density of cells, about of 10^8 cells/ml, as well as peptides and DNA, however because of the long printing time a heat generated from the laser leads to a higher rate of damaged cells. Currently, the materials most used by this technique are the hydrogels such as sodium alginate, collagen, gelatin, fibrin and metal such as titanium and gold (Bishop et al. 2017).

Inkjet printer is a technique that deposit the material or cells by means of droplets, this can be used as a system piezoelectric, thermal or electromagnetic, basically consisting of an ink chamber with a number of nozzles, depending on the system, expelling successive droplets out of the nozzle, the drop size is about 1 to 300 picoliter and drop deposition rate of 1–10,000 droplets/sec, an inkjet printer offers many advantages such as availability, versatility, high speed, can generate relatively high-resolution structures (20–100 lm), can be controlled electronically, is relatively of low cost and with minimal contamination of the cells (Mir and Nakamura 2017), however high temperatures in the nozzle, mechanical stress and vibration could affect the cell viability, another disadvantage includes the requirement for low viscosity bio ink, for avoiding frequent nozzle clogging, cellular distortion and lack of precision with regards to droplet size in comparison to other methods. Some materials used are PEG, polyethylene glycol dimethacrylate (PEGDMA), bioglass, polyethylene glycol-gelatin methacrylate (PEG-GelMA), Hydroxyapatite (HA) (Visscher et al. 2017; Derakhshanfar et al. 2018).

Stereolithography is another modality of printing, this is based on a laser-controlled system, allowing the polymerization of light-sensitive polymers such as Ultraviolet (UV) light or laser, which fabricate 3D structures layer by layer as a hydrogel precursor solution. This is considered a technique of versatility, controllability, with high printing quality and precision, speed and complex fabrication with micrometer-scale resolution. However, it has been reported that UV exposure usually causes damage to cells, DNA and, even causes skin cancer, but in different studies it has shown relatively high cell viability. The material most used in this technique are synthetic hydrogels as Polylactic acid (PLA), Polyglycolic acid (PLG), Polycaprolactone (PCL), biphasic calcium phosphate, tricalcium phosphate, and combinations (Lee and Dai 2016; Hikita et al. 2017).

The ability to repair and regenerate bone depends on several factors, such as the biological environment, the composition of scaffold, fabrication method, macroscopic and microscopic characteristics, mechanical properties, modifications with cells and growth factors, because of this, the 3D printing technique needs the multidisciplinary fabrication to create functional biological constructs and to optimize bone healing and vascularization (Visscher et al. 2017). In our laboratory, we fabricated scaffolds by an extrusion printer, using a gamma of materials such as hydrogels, synthetic polymer, ceramics as well as the combination of them, and bio-ink with cells, for bone regeneration. Some examples are shown in Fig. 9.4.

Methods to Evaluate Bone Regeneration by Means of Images

Nuclear medicine is a medical speciality that uses radiopharmaceuticals (radiotracers) to evaluate metabolic functions as well as to diagnose and treat diseases. The radiotracers are formed by carrier molecules tightly bound to a radioactive atom. These carrier molecules vary enormously depending on the purpose of the scan. Some tracers use molecules that interact with a specific protein or sugar in the body (Salvatori et al. 2019; Phelps 2000; Melédez-Alafort et al. 2019).

Approved tracers are called radiopharmaceuticals because they must comply with the strict, safety and appropriate performance standards of the FDA for clinical use.

Single Photon Emission Computed Tomography (SPECT) and Positron Emission Tomography (PET) are the two most common modalities in nuclear medicine.

In the case of images through SPECT tomography, they provide three-dimensional (tomographic) images of the distribution of radioactive tracer molecules that have been introduced into the patient's body. 3D images are generated by a computer from a large number of body projection images, recorded at different angles. Scanners for SPECT have gamma camera detectors that can detect gamma ray emissions from tracers that have been injected into the patient. Gamma rays are a form of light that moves at a wavelength different from visible light. The cameras are mounted on a rotating gantry that allows the detectors to move in a closed circle around a patient lying on a platform without moving (Phelps 2000).

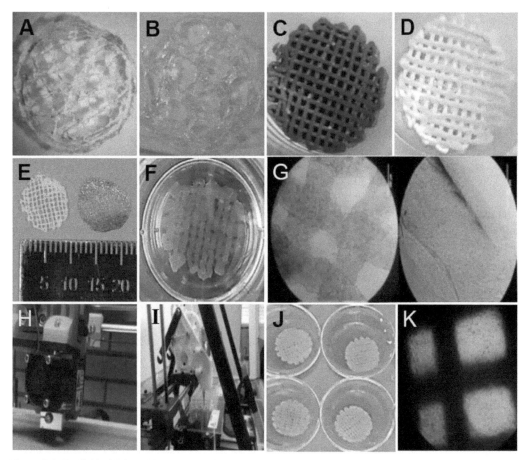

Figure 9.4. Scaffolds fabricated by extrusion 3D printing. (a) rectilinear hydrogel scaffold, (b) honeycomb hydrogel scaffold, (c) rectilinear TiO$_2$ scaffold, (d) rectilinear β-TCP scaffold (e) rectilinear polymer scaffolds with different porosity, (f) hydrogel scaffold printer with bone cell, (g) optical images of the bone cells 1×10^6 cells/mL into scaffold in different dimensions 10x and 40x, (h) 3D printer for thermoplastic polymers, (i) 3D printer for hydrogels and pastes, (j) β-TCP scaffold encapsulated into bio ink, (k) optical image of β-TCP scaffold and cells to 10x (the black spot are the cells).

PET scans also use radiopharmaceuticals to create three-dimensional images. The main difference between SPECT and PET scans is the type of radiotracers used. While SPECT scans measure gamma rays, the decomposition of radiotracers used with PET scans produces small particles called positrons. A positron is a particle with approximately the same mass as an electron, but with an opposite charge. These react with the electrons in the body and when these two particles combine they annihilate each other. This annihilation produces a small amount of energy in the form of two photons that fire in opposite directions. The PET scanner detectors measure these photons and use this information to create images of the internal organs. With this non-invasive technique, quantitative information on biological processes at the molecular level is obtained through tomographic images that reflect the concentration of activity of a radiopharmaceutical administered to a patient. The information obtained depends on the metabolic pathway, the targeted target receptor, the biodistribution and the rates of accumulation and elimination of the radiopharmaceutical. This makes it possible to perform an early detection of pathological processes with PET (Phelps 2000).

Both the SPECT and PET images provide information on what is happening in a metabolic way, today hybrid images are performed such as single photon emission computed tomography/computed tomography (SPECT/CT) and positron emission tomography/computed tomography (PET/CT) showed a rapid diffusion in recent years because of their high sensitivity, specificity and accuracy. However, CT component (Salvatori et al. 2019), complements the information because, in an anatomical way, the area of interest can be seen anatomically.

In hybrid imaging, the total radiation dose to patients is the sum of the dose due to the radiopharmaceutical used for SPECT/PET imaging and the dose derived from the CT component of the study. The combined acquisition of functional and anatomical images can lead to increased radiation exposure to patients (Salvatori et al. 2019), for this reason the amount of radiopharmaceutical injected into a patient must be sufficient to obtain quality images and, at the same time, be low as possible so as not to affect the metabolic process to be studied.

In preclinical research, the same fundamentals of PET and SPECT are used to carry out investigations in small experimental animals, mainly rats and mice, with these studies, microPET, microSPECT (μPET/μSPECT), new radiopharmaceuticals can be evaluated, metabolic pathways and biodistribution times can be explored, the specificity, the stability and, if it exists, the negative or counterproductive effects of radiopharmaceuticals. Likewise, these trials allow to carry out research projects related to the monitoring of diseases, development of innovative treatments that allow non-invasive monitoring in animal models.

In addition, other studies use radiopharmaceuticals that cannot be used in human subjects, but that are synthesized to answer a specific biological question in animal models (Funk and Hasegawa 2004).

Although research in SPECT and PET is mainly for diagnosis and treatment in cancer, today this technology is applied to evaluate bone metabolism in regeneration processes in critical size defects or fractures, since radiopharmaceuticals can accumulate in areas of skeletal metabolic activity, which means the technology holds great potential in visualizing and quantifying bone tissue regeneration and remodeling with high sensitivity and high resolution (Mathavan et al. 2019), as well as of the quantification of osteoblastic activity (Yamane 2018). Lara (2004), used the images of micro-SPECT to study the effectiveness of the filling of surgically created defects in rat tibiae with freeze-dried bovine bone. The objective was to evaluate the effectiveness of ^{99m}Tc-MDP uptake as a reliable marker of bone remodeling (Lara 2004).

These compounds accumulate rapidly in bone, and by 2–6 hours after the injection, about 50% of the injected dose is in the skeletal system. Lienemann et al. 2015 mentioned that the bisphosphonate tracers were shown to adsorb to hydroxyapatite and visualize bone and joint disorders using SPECT. At 1 week of treatment, injection of ^{99m}Tc-MDP led to its subsequent accumulation in many metabolically active anatomical areas, such as joints, teeth, vertebrae and also to the created bone defects. Although the ^{99m}Tc-MDP signals as well as the increase in bone volume reached a maximum after 4 weeks of treatment, significantly high SPECT activities remained up to week 12 (Lienemann et al. 2015).

The uptake mechanisms of diphosphonates have not been completely elucidated. Presumably they are adsorbed to the mineral phase of bone, with relatively little binding to the organic phase. The degree of radiotracer uptake depends primarily on two factors: blood flow and, perhaps more importantly, the rate of new bone formation (Love et al. 2003).

In the case of PET images Mathavan et al. mention that positron-emitting isotopes which accumulate in areas of skeletal metabolic activity, means the technology holds great potential in visualizing and quantifying bone tissue regeneration and remodeling with highly sensitive and high resolution (Mathavan et al. 2019).

Mathavan et al. also speculated that ^{18}F-fluoride PET characterizes cellular activity of bone forming cells in the early stages of the regenerative process. The exact uptake mechanisms by which ^{18}F-fluoride is incorporated into bone remains controversial, on intravenous administration, ^{18}F-fluoride rapidly diffuses through bone capillaries and ionic exchange transpires at the bone surface where the fluoride ion replaces the hydroxyl group in hydroxyapatite to form flour-apatite (Mathavan et al. 2019).

As previously mentioned, using microPET/microSPECT images (μPET/μSPECT) to evaluate bone regeneration is a very efficient alternative to perform hybrid images with microcomputed

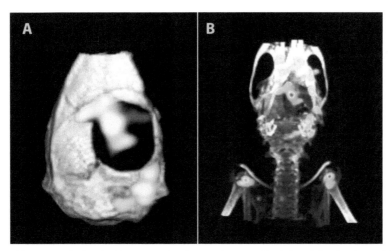

Figure 9.5. Hybrid images of μSPECT/CT. (a) Fusion image where a critical size defect is observed in rat calvaria. (b) Another image alternative in the same area. In both images the uptake of ^{99m}Tc-MDP is observed in green.

tomography (microCT or μCT), which is a non-destructive imaging tool for the production of high-resolution three-dimensional (3D) images of two-dimensional (2D) trans-axial projections, or "slices", of a target specimen (Boerckel et al. 2014). Figure 9.5 shows a hybrid μSPECT/μCT image of a rat to which a critical size defect was made in calvaria and an alginate/hydroxyapatite scaffold was placed to evaluate the regeneration of the area.

By performing hybrid images μPET/μCT and μSPECT/μCT in addition to giving us information on bone metabolism in the regeneration process (μPET/μSPECT), can also give information about the quality of the tissue that is being formed, that is, we can quantify the bone mineral density as well as parameters of the microarchitecture such as the Bone Volume fraction (BV/TV), trabecular number (Tb.N.), trabecular thickness (Tb.Th.), trabecular separation (Tb.Sp.) (Bouxsein et al. 2010), among others.

Imaging techniques, provide valuable information in the regeneration of mainly bone tissues and reduce the number of experimental animals. Imaging techniques can provide data over the time of the same experimental unit and hence reduce the biological variability among animals, since the same group can be followed up at different periods of time, in addition to decreasing in this manner.

Conclusion

It is essential to be able to give a better quality of life to patients who, due to some pathology or trauma, have lost or damaged some part of their body, for example, the craniofacial area. Conventional treatments use in the clinic so far have been limited in terms of regeneration of critical-size defects. Therefore, tissue bioengineering seeks new treatment options, such as the manufacture of scaffolds that mimic the extracellular matrix of the tissue for regeneration. For example, 3D designed and printed scaffolds which include some growth factors, molecules, as well as the use of mesenchymal stem cells that promote bone regeneration.

The main objective is to have promising and efficient treatments that achieve regeneration anatomically and physiologically of the craniofacial area, thus promoting personalized therapies, which can be custom designed using tomographic images, which can subsequently be evaluated with tomography or by positron emission tomography, monitoring the evaluation and predicting the efficiency of the new treatments.

Acknowledgements

Authors would like to thank DGAPAUNAM: PAPIIT IA205818 for the financial support and Elein Hernandez for reviewing this chapter.

References

Abou Neel, Ensanya Ali, Wojciech Chrzanowski, Vehid M. Salih, Hae Won Kim and Jonathan C. Knowles. 2014. Tissue engineering in dentistry. J. Dent. 42(8): 915–28.

Arias, Clemente F., Miguel A. Herrero, Luis F. Echeverri, Gerardo E. Oleaga and José M. López. 2018. Bone remodeling: A tissue-level process emerging from cell-level molecular algorithms. PLoS One 13(9): 1–19.

Bishop, Elliot S., Sami Mostafa, Mikhail Pakvasa, Hue H. Luu, Michael J. Lee, Jennifer Moriatis Wolf et al. 2017. 3-D bioprinting technologies in tissue engineering and regenerative medicine: Current and future trends. Genes and Diseases 4(4): 185–95.

Boerckel, Joel D., Devon E. Mason, Anna M. McDermott and Eben Alsberg. 2014. Microcomputed tomography: Approaches and applications in bioengineering. Stem Cell Res. Ther. 5(6): 1–12. https://doi.org/10.1186/scrt534.

Bouxsein, Mary L., Stephen K. Boyd, Blaine A. Christiansen, Robert E. Guldberg, Karl J. Jepsen and Ralph Müller. 2010. Guidelines for assessment of bone microstructure in rodents using micro-computed tomography. JBMR 25(7): 1468–86.

Chu, Tien-Min Gabriel, Sean Shih-YAo Liu and Wilian J. Babler. 2014. Craniofacial biology, orthodontics, and implants. In Basic and Applied Bone Biology, 225–42.

Datta, Pallab, Veli Ozbolat, Bugra Ayan, Aman Dhawan and Ibrahim T. Ozbolat. 2017. Bone tissue bioprinting for craniofacial reconstruction. Biotechnol. Bioeng. 114(11): 2424–31. https://doi.org/10.1002/bit.26349.

Derakhshanfar, Soroosh, Rene Mbeleck, Kaige Xu, Xingying Zhang, Wen Zhong and Malcolm Xing. 2018. 3D bioprinting for biomedical devices and tissue engineering: A review of recent trends and advances. Bioact. Mater. 3(2): 144–56.

Dhillon, Robinder S., Edward M. Schwarz and Michael D. Maloney. 2012. Platelet-rich plasma therapy—future or trend? Dhillon et al. Arthritis Res. Ther. 14(219): 1–10.

Filipowska, Joanna, Krzysztof A. Tomaszewski, Łukasz Niedźwiedzki, Jerzy A. Walocha and Tadeusz Niedźwiedzki. 2017. The role of vasculature in bone development, regeneration and proper systemic functioning. Angiogenesis 20(3): 291–302.

Fillingham, Y. and Jacobs, J. 2016. Bone grafts and their substitutes. Bone Joint J. 6–9.

Fishero, Brian Alan, Nikita Kohli, Anusuya Das, John Jared Christophel and Quanjun Cui. 2014. Current concepts of bone tissue engineering for craniofacial bone defect repair. Craniomaxillofac. Trauma Reconstr. 08: 23–30.

Funk, Tobias and Bruce H. Hasegawa. 2004. Radiation dose estimate in small animal SPECT and PET. Med. Phys. 2680–86.

Gaihre, Bipin, Suren Uswatta and Ambalangodage C. Jayasuriya. 2017. Reconstruction of craniomaxillofacial bone defects using tissue-engineering strategies with injectable and non-injectable scaffolds. J. Funct. Biomater. 8(49): 1–19. https://doi.org/10.3390/jfb8040049.

Henkel, Jan, Maria A. Woodruff, Devakara R. Epari, Roland Steck, Vaida Glatt, Ian C. Dickinson et al. 2013. Bone regeneration based on tissue engineering conceptions—a 21st century perspective. Bone Res. Nature Publishing Group 1(3): 216–48. https://doi.org/10.4248/BR201303002.

Hikita, Atsuhiko, Ung-il Chung, Kazuto Hoshi and Tsuyoshi Takato. 2017. Bone regenerative medicine in oral and maxillofacial region using a three-dimensional printer. Tissue Eng. Part A 23(11-12): 515–21. https://doi.org/10.1089/ten.tea.2016.0543.

Hoexter, David L. 2002. Bone regeneration graft materials. J. Oral Implantol. 3–7.

Howard, Daniel, Lee D. Buttery, Kevin M. Shakesheff and Scott J. Roberts. 2008. Tissue engineering: Strategies, stem cells and scaffolds. J. Anat. 213(1): 66–72.

Kawecki, Fabien, William P. Clafshenkel, Michel Fortin, François A. Auger and Julie Fradette. 2018. Biomimetic tissue-engineered bone substitutes for maxillofacial and craniofacial repair: The potential of cell sheet technologies. Adv. Healthc. Mate. 7(6): 1–16.

Kumar, Prasanna, Belliappa Vinitha and Ghousia Fathima. 2013. Bone grafts in dentistry. J. Pharm. Bioallied Sci. 5: 125–28.

Langdahl, Bente, Serge Ferrari and David W. Dempster. 2016. Bone modeling and remodeling: Potential as therapeutic targets for the treatment of osteoporosis. Ther. Adv. Musculoskelet. Dis. 8(6): 225–35.

Lansdown, Drew A. and Lisa A. Fortier. 2017. Platelet-rich plasma: formulations, preparations, constituents, and their effects. Oper. Tech. Sports 25(1): 7–12.

Lara, Pedro Fernandes. 2004. Radioisotopic Evaluation of Bone Repair After Experimental 12(1): 78–83.

Lienemann, Philipp S., Stéphanie Metzger, Anna Sofia Kiveliö, Alain Blanc, Panagiota Papageorgiou, Alberto Astolfo et al. 2015. Longitudinal *in vivo* evaluation of bone regeneration by combined measurement of multi-pinhole SPECT and micro-CT for tissue engineering. Sci. Rep. 5(May).

Lee, vivian k. and Guohao Dai. 2016. Printing of three-dimensional tissue analogs for regenerative medicine. Ann. Biomed. Eng. 45(1): 115–31.

Li, G., Zhou, T., Lin, S., Shi, S. and Lin, Y. 2017. Nanomaterials for craniofacial and dental tissue engineering. J. Dent. Res. 96(7): 725–32.

Liu, Guanqi, Yuanlong Guo, Linjun Zhang, Xiaoshuang Wang, Runheng Liu, Peina Huang et al. 2019. A standardized rat burr hole defect model to study maxillofacial bone regeneration. Acta Biomater. 86: 450–64.

Liu, Mengying and Yonggang Lv. 2018. Reconstructing bone with natural bone graft: a review of *in vivo* studies in bone defect animal model. Nanomaterials 8: 1–19.

Love, C., Anabella, S., Tomas, M. and Palestro, J. 2003. Radionuclide bone imaging: and illustrative review. RG 23: 341–358.

Mardas, Nikos, Xanthippi Dereka, Nikolaos Donos and Michel Dard. 2014. Experimental model for bone regeneration in oral and cranio-maxillo-facial surgery. J. Invest. Surg. 27(1): 32–49.

Martin, Victor and Ana Bettencourt. 2018. Bone regeneration: Biomaterials as local delivery systems with improved osteoinductive properties. Mater. Sci. Eng. C Mater. Biol. Appl. 82: 363–71. https://doi.org/10.1016/j.msec.2017.04.038.

Mathavan, Neashan, Janine Koopman, Deepak Bushan Raina, Magnus Tägil and Hanna Isaksson. 2019. F-Fluoride as a Prognostic Indicator of Bone Regeneration (20160599), the Faculty of Medicine at Lund University and the Foundations of Greta And. Acta Biomater.

McGovern, Jacqui Anne, Michelle Griffin and Dietmar Werner Hutmacher. 2018. Animal models for bone tissue engineering and modelling disease. Dis. Model Mech. 11(4): 1–14.

Mélendez-Alafort, Ferro-Flores, De Nardo, Paiusco, Negri, Zorz, Uzunov, Esposito and Rosato. 2019. Internal radiation dose assessment of radiopharmaceuticals prepared with cyclotron-produced ^{99m}Tc. Med. Phys. 46(3): 1437–1446.

Mir, Tanveer Ahmad and Makoto Nakamura. 2017. Three-dimensional bioprinting: toward the era of manufacturing human organs as spare parts for healthcare and medicine. Tissue Eng. Part B Rev. 23(3): 245–56.

Morais, Alain da Silva, Miguel J. Oliveira and Rui L. Reis. 2018. Small animal models. *In*: Miguel J. Oliveira (ed.). Osteochondral Tissue Engineering, Advances in Experimental Medicine and Biology, 1059: 423–39.

Munaz, Ahmed, Raja K. Vadivelu, James St. John, Matthew Barton, Harshad Kamble and Nam-Trung Nguyen. 2016. Three-dimensional printing of biological matters. J. Sci.: Adv. Mat. & Dev. 1: 1–17.

Murphy, Matthew P., Natalina Quarto, Michael T. Longaker and Derrick C. Wan. 2017. Calvarial defects: Cell-based reconstructive strategies in the murine model. Tissue Eng. Part C Methods 23(12): 971–81. https://doi.org/10.1089/ten.tec.2017.0230.

Nagy, Laszlo and Joshua C. Demke. 2014. Craniofacial anomalies. Facial Plast. Surg. Clin. North Am. 22(4): 523–48. https://doi.org/10.1016/j.fsc.2014.08.002.

Ning, Bin, Yunpeng Zhao, John A. Buza, Wei Li, Wenzhao Wang and Tanghong Jia. 2017. Surgically-induced mouse models in the study of bone regeneration: Current models and future directions (review). Mol. Med. Rep. 15: 1017–23.

Otterloo, Eric Van, Trevor Williams and Kristin B. Artinger. 2017. The old and new face of craniofacial research: How animla models inform human craniofacial genetic and clinical data. Dev. Biol. 415(2): 171–87.

Padial-Molina, Miguel, Francisco O'Valle, Alejandro Lanis, Francisco Mesa, David M. Dohan Ehrenfest, Hom Lay Wang et al. 2015. Clinical application of mesenchymal stem cells and novel supportive therapies for oral bone regeneration. Biomed. Res. Int.

Pashayan, Hermine and Joseph Reichman. 1980. A Review of Some of the More Widely Used Classification Systems Is given below. Anoma-Lies of the Face Have Most Often Been Classified According to a Major Structure Involved, i.e., Lip and Palate, Eye, Nose, Mandibular, and Maxillary Abnormaliti, no. 1976.

Phelps, Michael E. 2000. Positron emission tomography provides molecular imaging biological processes. Proc. Natl. Acad. Sci. 97(16).

Prasad, Bhanu. 2016. A review on drug testing in animals. Transl. Biomedicine 7(4): 1–4.

Prisby, Rhonda D. 2017. Mechanical, hormonal and metabolic influences on blood vessels, blood flow and bone. J. Endocrinol. 235(3): 1–40.

Raeisdasteh Hokmabad, Vahideh, Soodabeh Davaran, Ali Ramazani and Roya Salehi. 2017. Design and fabrication of porous biodegradable scaffolds: A strategy for tissue engineering. J. Mat. Sci., Polym. Ed. 28(16): 1797–1825.

Rodríguez Flores, Jordi, María Angustias Palomar Gallego and Jesús Torres García-Denche. 2012. Plasma Rico En Plaquetas: Fundamentos Biológicos y Aplicaciones En Cirugía Maxilofacial y Estética Facial. Rev. Esp. Cirug. Oral y Maxilofac. 34(1): 8–17.

Saal, Howard M. 2016. Genetic evaluation for craniofacial conditions. Facial Plast. Surg. Clin. N Am. 24(4): 405–25.

Salvatori, Massimo, Alessio Rizzo, Guido Rovera, Luca Indovina and Orazio Schillaci. 2019. Radiation dose in nuclear medicine: the hybrid imaging. La Radiologia Medica, no. 0123456789.

Sanchez-Lara, Pedro A. 2015. Clinical and genomic approaches for the diagnosis of craniofacial disorders. Current Topics in Developmental Biology. 1st ed. Vol. 115. Elsevier Inc.

Shibuya, Naohiro and Daniel C. Jupiter. 2015. Bone graft substitute: allograft and xenograft. Clin. Pod. Med. & Surg. 32(1): 21–34.

Si, Jia-wen, Xu-dong Wang, Steve G.F. Shen, Jia-wen Si, Xu-dong Wang and Steve G.F. Shen. 2015. Perinatal stem cells: a promising cell resource for tissue engineering of craniofacial bone. World J. Stem Cells 7(1): 149–59.

Sigaux, Nicolas, Léa Pourchet, Marion Albouy, Amélie Thépot and Christophe Marquette. 2017. Is 3D bioprinting the future of reconstructive surgery? Plast. Reconstr. Surg. Glob. Open 5: 1–2.

Streubel, Sven Olrik and David M. Mirsky. 2016. Craniomaxillofacial trauma. Facial Plast. Surg. Clin. N. Am. 24(4): 605–17. https://doi.org/10.1016/j.fsc.2016.06.014.

Thrivikraman, Greeshma, Avathamsa Athirasala, Chelsea Twohig, Sunil Kumar Boda and Luiz E. Bertassoni. 2017. Biomaterials for craniofacial bone regeneration. Dent. Clin. North Am. 61: 835–56.

Visscher, Dafydd O., Elisabet Farré-Guasch, Marco N. Helder, Susan Gibs and Jan Wolff. 2017. Advances in bioprinting technologies for craniofacial reconstruction. Trends Biotech. 34(9): 700–710.

Walker, T., Byrne, Donnellan and McArdle. 2011. West of Ireland facial injury study. Part 1. J. Oral Maxilofac. Surg. 50: 631–635.

Wan, Derrick, C., Randall P. Nacamuli and Michael T. Longaker. 2006. Craniofacial bone tissue engineering. Dent. Clin. N. Am. 50: 175–90.

Wang, Wenhao and Kelvin W.K. Yeung. 2017. Bone grafts and biomaterials substitutes for bone defect repair: A review. Bioact. Mater. 2(4): 224–47.

Ward, B., Brown, S. and Krebsbach, P.H. 2010. Bioengineering strategies for regeneration of craniofacial bone: A review of emerging technologies. Oral Dis. 16(8): 709–16.

Yamane, Tomohiko. 2018. Quantification of osteoblastic activity in epiphyseal growth plates by quantitative bone SPECT/CT. Skeletal Radiol. 47(6): 805–810.

Zhang, Weibo and Pamela Crotty Yelick. 2019. Craniofacial tissue engineering. Cold Spring Harb. Perspect. Med. 8(1): a025775.

Zhou, Xuan, Nathan J. Castro, Wei Zhu, Haitao Cui, Mitra Aliabouzar and Kausik Sarkar. 2016. Improved human bone marrow mesenchymal stem cell osteogenesis in 3D bioprinted tissue scaffolds with low intensity pulsed ultrasound stimulation. Scientific report. Nature Publishing Group.

10

Gingiva and Periodontal Tissue Regeneration

Avita Rath,[1,*] *Preena Sidhu,*[1]
Priyadarshini Hesarghatta Ramamurthy,[1]
Bennete Aloysius Fernandes,[1] *Swapnil Shankargouda*[2] and
Sultan Omer Sheriff[3]

Introduction

The fact that periodontal tissues possess the capacity, albeit minimally, to repair/regenerate, has led to substantial efforts focused on understanding the biological basis for this activity.

In the past few decades, various materials that can accelerate the reconstruction of periodontal tissues have been use. Matrix-based scaffolds, stem cells and growth factors have shown high potential for regenerative therapies. In dentistry, such approaches have been explored to promote wound healing, restore periodontal ligament attachment, provide a broader zone of attached gingiva and cover the exposed root surfaces. Esposito et al. conducted a systematic review of soft tissue management for dental implants. They concluded that there is not sufficient evidence to provide recommendations for the optimal soft tissue augmentation technique. (1)

In the current concept, naturally occurring tissue regeneration can be reproduced *in vitro*. A tissue-engineering approach to utilize the regenerative capacity of stem cells residing within the periodontium (or other related tissues) involves the isolation of these cells and their subsequent proliferation/differentiation within a three-dimensional (3D) framework that could be surgically inserted in a given defect.

[1] Jalan Teknologi, Kota Damansara, 47810 Petaling Jaya, Selangor.
  Emails: reachddocpriya@gmail.com; drben17@yahoo.com
[2] KLE Academy of Higher Education and Research's Vishwanath Katti Institute of Dental Sciences, Nehru Nagar, Belgaum, Karnataka 590010, India.
  Email: swapnilshankargouda@gmail.com
[3] International Medical University 126, Jalan Jalil Perkasa 19, Bukit Jalil, 57000, Kuala Lumpur, Federal Territory of Kuala Lumpur.
  Emails: sidhu.preena@gmail.com; dromersheriff@gmail.com
* Corresponding author: drrathavita@yahoo.com

Anatomy

The tooth is attached to the jaw by a specialized supporting apparatus that consists of the alveolar bone, the PDL and the cementum, all of which are protected by the gingiva.

The mucosa immediately surrounding an erupted tooth is known as the gingiva. In functional terms, the gingiva consists of two parts: (1) the part facing the oral cavity, which is masticatory mucosa, and (2) the part facing the tooth, which is involved in attaching the gingiva to the tooth and forms part of the periodontium (Tencate 2017).

In an adult, healthy gingiva covers the alveolar bone and tooth root to a level just coronal to the cementoenamel junction. The gingiva is divided anatomically into marginal, attached and interdental areas. Although each type of gingiva exhibits considerable variation in differentiation, histology and thickness according to its functional demands, all types are specifically structured to function appropriately against mechanical and microbial damage (Carranza 2014).

The human gingiva is the oral mucosal tissue that surrounds the teeth and forms a mucoperiosteum covering the alveolar bone. Since the epithelial layer shows the capacity for continuous renewal, it is anticipated that this tissue could also be a source of stem cells.

Periodontal regeneration is a complex set of tissues and structures around the tooth; hence, an ideal biomaterial-driven approach should include a functionally-graded scaffold where the chemical composition and 3D architecture of each layer (or 'compartment') match the fine organization, biochemical structure, and mechanical properties of the tissues to regenerate (Giorgio 2019).

Periodontitis, an oral disease with a high prevalence worldwide, affects the function of teeth and constitutes one of the primary oral health burdens (Esposito et al. 2005). An epidemiological survey has suggested that more than half of all adults are affected by periodontal disease to varying degrees (Niklaus and Linde 2015; Iviglia et al. 2019) and a remarkable surge (25.4% increase) in the prevalence rates of periodontal disease was observed from 2005 to 2015 (Prichard 1957). Periodontitis can consistently disrupt tooth-investing tissues and lead to tooth loss if left untreated (Prichard 1957; Jepsen et al. 2002). Periodontitis is also closely associated with the occurrence and prognosis of various systemic diseases, including cardiovascular diseases, cancer, obesity, diabetes and chronic nephritis (Needleman et al. 2006; Trombelli and Farina 2008; Hammarström 1997; Giannobile 1996). Therefore, the exploration of effective and safe periodontal therapies that can be translated into the clinic is an urgent health need worldwide.

The ambitious purpose of periodontal therapy is to regenerate multiple periodontal tissues, including the alveolar bone, cementum and periodontal ligament (PDL) in the damaged periodontium (Whang et al. 1998). Although nonsurgical periodontal therapies (e.g., scaling and root planing) can prevent disease progression by physically removing the pathogens and necrotic tissues, only a small amount of periodontal tissue can be regenerated at the treated sites (Jepsen et al. 2002).

Overview of Past and Current Clinical Periodontal Regeneration Techniques

Past Techniques

The work of Pritchard (1957) provided early evidence that regeneration was not only theoretically possible but also clinically achievable in 'ideal' periodontal defects. This work showed that reconstruction of the periodontium in the apical aspects of three wall intrabony defects could be obtained following subgingival debridement through a surgical approach. Regeneration was not achieved in all cases, but this provided the 'proof of principle' evidence that if the conditions were optimal, then healing outcomes other than repair could be obtained (Prichard 1957). Since then, a great deal of research has focused on establishing a clinical technique to achieve these optimal conditions in a variety of clinical situations, not just three wall defects.

Guided Tissue Regeneration (GTR)

Guided tissue regeneration uses biocompatible barrier membranes to enable selective cellular recolonization of periodontal defects. The principle involves the use of a barrier membrane to exclude tissues that are unable to promote periodontal regeneration, such as the rapidly proliferative epithelium that grows along the root surface and the gingival connective tissues that fill the periodontal defect. Instead, this technique facilitates the repopulation of the defect with cells that can re-establish the periodontal attachment apparatus, namely those from the periodontal ligament and alveolar bone. The concept of GTR is supported by histological evidence of periodontal regeneration following the use of this technique (Jepsen et al. 2002). The membranes used can be either nonresorbable or resorbable, but due to decreased postoperative complications and a reduced number of surgical procedures, resorbable membranes have become more popular.

Resorbable membranes come in a variety of materials, with the most common being collagen and copolymers of polylactic/polyglycolic acid. The benefit of using GTR over the conventional use of open flap debridement has been reviewed in the literature, and the findings support the use of GTR in two clinical situations, Type II mandibular furcations (Jepsen et al. 2002; Needleman et al. 2006). However, although GTR is conceptually sound, the clinical results are unreliable and predictable regeneration is elusive in most periodontal defects, with the possible exception of a few 'ideal' situations, such as narrow three wall defects. These ideal clinical scenarios are not frequently encountered in clinical practice.

Platelet-Rich Plasma

The use of PRP, a platelet concentrate from autologous blood, is one of the strategies available for modulating and enhancing periodontal wound healing and regeneration (Iviglia et al. 2019). PRP includes a pool of GFs such as TGF-b, Vascular Endothelial Growth Factor (VEGF), PDGF, IGF-1, epidermal growth factor and FGFs. For this reason, it has been suggested that the use of PRP might increase the rate of bone deposition and bone quality in such dental treatments such as sinus lifts, placement of autogenous mandibular bone grafts, implants and periodontal surgery. Both preclinical and clinical studies demonstrate the effectiveness of PRP in bone and periodontal augmentation when used in conjunction with bone graft materials (reviewed in (Iviglia et al. 2019)). Unfortunately, the literature on the topic is contradictory and the published data are difficult to sort out and interpret. For example, in a series of clinical studies performed by Döri and co-workers on the healing of intrabony defects, the addition of PRP failed to improve the total outcomes in terms of probing depth reductions and CAL gains (Chang et al. 2009). Furthermore, PRP may not provide any additional effect when associated with GBR around dental implants (Bhatia and Ingber 2014). Although data published thus far suggest that PRP does not always exert additional effects, a systematic review of literature does give evidence for beneficial effects of PRP in the treatment of periodontal defects. Nonetheless, evidence for beneficial effects of PRP in sinus elevation appeared to be weak (Mao and Mooney 2015). While PRP has been somewhat beneficial when used in periodontal or implant regeneration, it has fallen out of favor recently because of the lack of controlled clinical trials providing strong evidence of its efficiency (Lee and Shin 2007). However, the use of a patient's own biologically active proteins, GFs and biomaterial scaffolds for therapeutic purposes has opened a new way of understanding regenerative medicine. These simple and cost-effective procedures may have a potential impact in reducing the economic costs for standard medical treatments, soon achieving a 'golden age' by the development of user-friendly platelet concentrate procedures and the definition of new and efficient concepts and clinical protocols (Iviglia et al. 2019).

Growth Factors in Periodontal Tissue Engineering

Growth Factors and Biologically Active Regenerative Materials

The rationale for the application of growth factors to the periodontal defect is based on their ability to influence critical cellular functions such as proliferation, migration and differentiation. However, their use is not widespread in periodontal regenerative medicine, due to problems with dosage, rapid metabolic clearance, appropriate delivery and carrier systems and cost (Trombelli and Farina 2008). The most widely studied biologically active material that has been in widespread clinical use is enamel matrix derivative. Enamel Matrix Derivative (EMD), Enamel Matrix Proteins (EMPs) are secreted by ameloblasts and play a role in the regulation and growth of hydroxyapatite crystals that comprise enamel (Hammarström 1997).

The incorporation of bioactive molecules into periodontal biomaterials and scaffolds and their local and controlled release over time has been proposed as a valuable approach to obtaining osteoinductive implants (Giannobile 1996). Two different procedures to embed growth factors are available, i.e., either during the preparation of the material or after the fabrication (Whang et al. 1998; Fournier and Doillon 1996). Platelet-rich growth factor (PDGF), Bone Morphogenetic Proteins (BMPs) and Enamel Matrix Derivatives (EMDs) are among the most potent biomolecules for accelerating the periodontal wound repair in preclinical and clinical studies (Giannobile 1996; Barron and Pandit 2003; Taba et al. 2005; Lyngstadaas et al. 2009). Growth factors have the potential of stimulating the interaction between Mesenchymal Stem Cells (MSCs) and epithelial stem cells during tooth formation, alongwith all the following processes such as collagen formation, mineralized matrix deposition and fibroblast proliferation (Giannobile 1996; Giannobile and Somerman 2003). The morphology of the scaffold where signaling molecules are incorporated is the basis to achieve an extended and effective release. These are released by a diffusion mechanism which strongly depends on pore interconnectivity—of the material. The degradation properties of the graft may also affect the release rate of growth factors, which might be faster because of the high solubility of the matrix. Furthermore, how the carrier degrades—either by surface or by bulk degradation, resulting in a controlled or burst release, respectively—plays a key role, too (Tabata 2003; Yamamoto et al. 2003; Woo et al. 2001). It is clear that growth factors have a significant impact on tissue engineering outputs, which deserves to be explored.

Platelet-Rich Growth Factor (PDGF)

PDGF is the primary growth factor involved in wound healing, and there are a lot of studies showing its ability to enhance the proliferation and migration of periodontal ligament cells. It is naturally made of the conjugation of polypeptides of growth factor-BB and growth factor-AA, which is encoded by two different genes. It has been demonstrated that all isoforms have an effect on cell proliferation *in vitro* (Dennison et al. 1994; Oates et al. 1993). It has been shown that PDGF has a chemotactic effect, by which it can promote collagen synthesis, and stimulates hyaluronate synthesis by gingival fibroblasts and fibroblast proliferation.

Furthermore, if added to a culture with osteoblast-like cells, PDGF can regulate the ALP and osteocalcin expression (Giannobile et al. 1997). Lynch et al. (Lynch et al. 1991) applied PDGF in combination with the Insulin-like Growth Factor-1 (IGF-1) in dogs, and the results demonstrated significant effectiveness in periodontal regeneration.

Furthermore, the results of clinical trials revealed that the synergistic effect of these two growth factors could lead to the stimulation of bone regeneration inperiodontal defects in humans, too (Nevins et al. 2003). Even if used alone, PDGF can significantly stimulate the formation of new cementum and the production of collagen (Giannobile et al. 1996). Molecular cloning and large-scale purification have permitted the production of recombinant human PDGF, which has been mixed with

TCP and has been made commercially available to clinicians (GEM 21s, Osteohealth, Shirley, NY, USA) (Izumi et al. 2011).

Platelet derived growth factor is a dimeric molecule produced by a variety of cells and tissues, including fibroblasts and osteoblasts (Antoniades et al. 1991; Hauschka et al. 1988). PDGF is mitogenic for various cells of mesenchymal origin, such as glial, smooth muscle and bone cells, as well as fibroblasts (Antoniades and Owen 1982; Ross et al. 1986). In fibroblastic systems, the primary effect of PDGF is that of a mitogen. It initiates cell division by acting as a competence factor, thus making the cell competent for division (Pledger et al. 1977). Platelet derived growth factor is a dimeric molecule produced by a variety of cells and tissues, including fibroblasts and osteoblasts (Antoniades et al. 1991; Hauschka et al. 1988). PDGF is mitogenic for various cells of mesenchymal origin, such as glial, smooth muscle and bone cells, as well as fibroblasts (Antoniades and Owen 1982; Ross et al. 1986; Stiles 1983). In fibroblastic systems, the primary effect of PDGF is that of a mitogen. It initiates cell division by acting as a competence factor, thus making the cell competent for division (Pledger et al. 1977), promoting mitogenesis of PDL cells (Bartold and Raben 1996). At concentrations of 10e20 ng/ml, *in vitro* studies suggested that PDGF-BB stimulated the proliferation of fibroblasts and osteoblasts (Hutmacher and Cool 2007; Woodruff and Hutmacher 2010), whereas a higher concentration (50 ng/ml) is required for the adhesion of PDL fibroblasts to diseased roots; concentrations of 5e20 ng/ml were not effective (Shue et al. 2012). PDGF has also been shown to be effective when combined with either IGF, as shown by significant bone fill on re-entry into the defects in animal models (Li et al. 2014) and in patients with periodontal disease (Zhang et al. 2009), or with dexamethasone (Inanc et al. 2009), the latter being a well-known osteogenic differentiation factor.

PDGF and IGF-1 had additive effects on calvarial DNA synthesis, but PDGF opposed the stimulatory effect of IGF-1 on collagen synthesis, and IGF-1 prevented the PDGF effect on collagen degradation (Hutmacher and Cool 2007). In further research, IGF-1 alone at a dose of 10 mg did not significantly alter periodontal wound healing, while PDGF-BB alone at the same dose significantly stimulated new attachment, with trends of effect on other parameters. For example, the PDGF-BB/IGF-1 combination resulted in significant increases in new attachment and osseous defect fill at both 4 and 12 weeks (Bottino et al. 2011). In humans, the use of purified rhPDGF-BB mixed with bone allograft resulted in robust periodontal regeneration in both class II furcations (Vaquette et al. 2012) and interproximal intrabony defects (Carlo-Reis et al. 2011). Based on these studies, recombinant human PDGF-BB homodimer in b-TCP is approved for the treatment of intrabony and furcation defects, as well as gingival recession in periodontal disease and is commercially available as Gem-21 (Osteohealth Co., Shirley, NY, USA). Furthermore, a large multicenter randomized controlled trial study of PDGF-BB homodimer, together with b-TCP, in the surgical treatment of a 4 mm or greater intrabony periodontal defect demonstrated significant increases in Clinical Attachment Levels (CAL) reduced gingival recession at three months post-surgery, and improved bone fill when compared with those of b-TCP alone at six months (Carlo-Reis et al. 2011). The safety and effectiveness of this product was further demonstrated recently in a randomized, multicenter clinical trial involving 54 patients with periodontal osseous defects (Park et al. 2010). Although far from ideal for meeting the needs of complex periodontal therapy, the road from basic research to clinical applications of PDGF-BB, or BMP-2 suggests a potential use of protein-based therapeutics for stimulating and accelerating periodontal tissue healing and bone regeneration (Park et al. 2013).

Current Strategies

At present, the majority of tissue-engineering strategies applied to treat periodontal diseases claim at regenerating the alveolar bone through the implantation of suitable scaffold-based constructs. Specifically, periodontal regeneration relies on four basic paradigms (Dabra et al. 2012):

i)   Implanted material (scaffold) acts as a three-dimensional (3D) template supporting new tissue growth;

ii) Cells are the primary building blocks of the reconstructive strategies of periodontal tissue because of their proliferation and differentiation;

iii) Bioactive molecules (i.e., growth factors) promote cell activity that results in improved cell proliferation and differentiation;

iv) The blood supply delivers oxygen, nutrients and essential elements to tissue, and thus promotes the growth of newly formed tissue and helps to maintain homeostasis inside the 3D scaffold.

The healing process of the periodontal tissue is traditionally achieved by tissue repair; however, the advent of tissue engineering and regenerative medicine provides new possibilities in this regard (Shimauchi et al. 2013). Based on the definition, the repair process includes the healing of a wound by the growth of tissue that, however, may not fully restore the fine microstructure and, consequently, the function of the lost tissue (Lindhe et al. 1984; Wikesjö et al. 1992).

Two main objectives are simultaneously pursued in the current periodontal tissue engineering approaches, including (i) the stimulation of the growth of the key surrounding tissues by applying barrier membranes and bonegrafting materials and (ii) the prevention of the growth and proliferation of undesired cell types like epithelial cells (Melcher 1976).

The periodontium in its native form is a result of the interaction between mesenchymal and epithelial cells in the embryo (Palumbo 2011; Nanci and Bosshardt 2006). A series of independent and subsequent events are needed to complete the periodontal regeneration, which include osteogenesis, cementogenesis and connective tissue formation (Bei 2009). It has been well documented that various physico-chemical and mechanical stimuli can cause appropriate responses by the cells during the healing process. The type of cells used in regeneration controls the quality of the healing process, since they are initially responsible for repopulation of the wound (Melcher 1976). In the case of the periodontal regeneration, four distinct cell types compete—namely, periodontal ligament cells, alveolar bone cells, cementoblasts and epithelial cells.

Among these cells, the first three types are responsible for regenerating the periodontal tissue, while the last type—i.e., epithelial cells—are responsible for soft tissue regeneration. It is worth mentioning that the higher migration rate of epithelial cells (10 times faster) in comparison to the other periodontal cell types is the reason for observing the formation of the long junctional epithelium in the periodontal therapy (Engler et al. 1966). Infiltration of epithelial cells inside the defect can promote repair by the formation of an unusual architecture with a loss of function (Caton et al. 1987). Therefore, guided tissue membranes are implanted to limit the infiltration of the epithelial cells (Nyman et al. 1987). If epithelial cells are ruled out from the wound, other cell types with regenerative potential are thus allowed to become established, and epithelial down-growth can be successfully prevented (Linde et al. 1993). A combination of bonegraft materials, promoting the migration and differentiation of osteoblast cells, and GTR is the most commonly-used approach for achieving an optimal periodontal regeneration (Frost 1989a; Frost 1989b). Two reasons have been identified behind this synergic activity, i.e., the biological effects of bone grafts and the 'Melcher hypothesis', which explains the importance of cells used for the periodontal regeneration. According to this hypothesis, the origin of cells dictates the nature of the attachment in periodontal healing and the complete periodontal regeneration may be achieved when we apply cells with an origin from the periodontal ligament and the perivascular bone cells (Aurer and Jorgić-Srdjak 2005; Koop et al. 2012).

The biological actions induced by bone grafting materials can be divided into three interrelated healing processes: osteogenesis, osteoconduction and osteoinduction (Vuong and Otto 2002). The autologous bonegraft is the 'gold standard' option to achieve osteogenesis (Albrektsson et al. 1981). In this case, the undifferentiated and pluripotent cells derived from the transplanted material differentiate into the bone-forming cell lineage, i.e., osteoblasts, which will form a new bone (Burchardt 1983). Osteoconduction refers to the ability of a biomaterial to act as a suitable surface for guiding the bone growth by pre-existing preosteoblasts/osteoblasts (Vuong and Otto 2002; Hojo et al. 2009).

Osteoinduction is usually identified as a specific property of materials regarding the induction of the differentiation of the host osteoprogenitor cells into osteoblasts (Fürst et al. 2007). In this case, the graft material excludes the undifferentiated connective cells and induces the differentiation and proliferation of the osteoprogenitor ones, thus leading to new bone formation (Vuong and Otto 2002; Quirynen et al. 2002; Price et al. 1996). The Melcher hypothesis also points out the importance of barrier materials for enabling cell migration from the connective tissue to avoid the repair process (Linde et al. 1993; Frost 1989b; Adeli and Parvizi 2012). Numerous previously performed animal studies clarified the efficacy of non-resorbable and resorbable membranes for GTR (Frost 1989b; Adeli and Parvizi 2012). Three correlated steps are involved in the healing process of periodontal tissue injuries. The first step is the epithelization of the inner face of the flap that forms the so-called long epithelium attachment. The maturation of the connective tissue is the second step in which we observe the formation of the so-called connective attachment. The final step is related to the recovery of bone architecture and the periodontal ligament at the level of the alveolar bone and at the deepest point of injury.

The morphology of the structure of newly formed tissue indicates whether repair or regeneration took place (Shimauchi et al. 2013). In the case of repair, fibroblasts mostly are involved in the healing process instead of osteoblasts. Thus the deposition of the osteoid matrix and new bone formation are inhibited (Lindhe et al. 1984; Wikesjö et al. 1992). On the other hand, regeneration involves the complete recovery of both the structure and the function of the periodontal tissue (Shimauchi et al. 2013; Koop et al. 2012). Despite the implementation of this approach, the success of periodontal regeneration depends on the capacity to control local infections, that are caused by microbial pathogens contaminating periodontal wounds (Vuong and Otto 2002; Pihlstrom et al. 2005; Zimmerli et al. 2004). The removal of necrotic tissue or a zone with acute infection are often identified as the major complications of periodontal defects and are high-risk points for bacterial growth and re-infection of new tissue (Hojo et al. 2009; Fürst et al. 2007). Hence, there is a need for implementing appropriate strategies to regenerate periodontal tissue and restrain bacterial growth (Quirynen et al. 2002). There are critical limitations for the currently-available strategies (e.g., conventional systemic antibiotic therapy) applied for reducing the risk of wound infection, including the systemic toxicity in the body which can lead to the need of hospitalization for monitoring (Price et al. 1996; Adeli and Parvizi 2012; Shahi and Parvizi 2015; Turgut et al. 2005). To overcome these problems, several research groups are working on advanced systems able to release a low dosage of antibiotics directly *in situ* (Ketonis et al. 2010; Zhao et al. 2009; Zilberman and Elsner 2008; Jose et al. 2005; Gristina 1987; Springer et al. 2004; Stigter et al. 2004). Promising approaches for avoiding bacterial adhesion and minimizing the side effects of systemic antibiotics include the coating of titanium implants with antiseptic molecules, the implant modification with functional groups exerting a bactericidal action, and the use of specific signaling molecules which selectively act as bactericidal substances.

Gene Therapy Approach in Periodontology

The critical drawbacks associated with the local delivery of growth factors include their short biological half-life *in vivo* as well as the high cost (Taba et al. 2005; Fang et al. 1996). Even more importantly, using a high dosage of bioactive molecules is needed to promote tissue regeneration, which could lead to unpredictable reactions and side effects; hence, an alternative approach to the local release of growth factors is the use of gene therapy for periodontal regeneration (Fang et al. 1996). Gene therapy involves the insertion of the genes of interest into an individual's cells to obtain the desired functions, i.e., in most cases, upregulation of the expression of a specific growth factor (Sheikh et al. 2015; Barron and Pandit 2003; Nussenbaum and Krebsbach 2006). For this purpose, two main strategies have been proposed and developed including (i) *in vivo* technique, in which the gene vector is directly inserted into the target site (Fang et al. 1996; Schek et al. 2004), and (ii) *ex vivo* technique, in which selected cells are harvested, expanded, genetically transduced and eventually re-implanted (Gansbacher 2003).

Gene therapy has been applied for the upregulation of the expression of PDGF (Anusaksathien et al. 2003; Anusaksathien et al. 2004; Jin et al. 2004) and BMPs (Jin et al. 2003). In the *in vivo* technique, the gene of interest is directly delivered in the body, thus altering the normal expression of the target cells. On the contrary, the *ex vivo* technique involves the use of an adenovirus vector to introduce the genetic material into the target cells that have been harvested by a biopsy; eventually, transfected cells are re-implanted in the periodontal defect (Taba et al. 2005; Neel et al. 2014). In spite of the great potential of these techniques, there are several concerns about the safety of the adenovirus vector that still currently limit the clinical applicability of the gene therapy approach (Nussenbaum and Krebsbach 2006; Einhorn 2003).

Nanobiomaterials and Functionally-Graded Implants: The Last Frontiers in Periodontal Tissue Engineering

This valuable strategy is in its beginning in the entire field of tissue engineering—not restricted to only dentistry—and shows great promise for the 21st-century regenerative medicine. The existing examples of multi-layer scaffolds for multi-tissue regeneration often involve the use of nanomaterials, which can more closely reproduce the fine characteristics of the structure to regenerate (Kargozar and Mozafari 2018).

Sowmya et al. (Sowmya et al. 2017) recently reported the development of a 3-layer scaffold for the simultaneous regeneration of cementum, alveolar bone and periodontal ligament. Specifically, this scaffold was structured with chitosan/poly(lactic-co-glycolic acid) (PLGA)/nano-sized bioactive glass layer loaded with cementum protein 1 (CEMP1) for the cementum regeneration, chitosan/ PLGA layer loaded with fibroblast growth factor 2 for periodontal ligament regeneration, and chitosan/PLGA /nano-sized bioactive glass layer loaded with Platelet-Rich Plasma (PRP) growth factors for bone regeneration. Histological and tomographic evaluations showed that the implantation of this scaffold in rabbits led to complete periodontal healing and new alveolar bone deposition after three months.

Incorporation of growth factors further stimulating tissue regeneration in nanomaterial-based constructs was also investigated. Zhang et al. (Zhang et al. 2015) fabricated MBG/silk fibroin scaffolds incorporating BMP-7 or PDGF-B adenovirus and implanted them in dogs. The scaffolds loaded with PDGF-B adenovirus were able to partially regenerate the periodontal ligament while those loaded with BMP-7 primarily improved new alveolar bone formation. The synergistic combination of these two growth factors promoted periodontal healing by allowing up to two times greater regeneration of the periodontal ligament, alveolar bone and cementum as compared to each adenovirus used alone.

Chen et al. (Chen et al. 2016) used a poly(caprolactone) (PCL)-based multiphasic scaffold, enriched with type I collagen, as a carrier to release CEMP1-loaded poly(ethylene glycol) (PEG)-stabilized amorphous calcium phosphate nanoparticles. After implantation in rats for 8 weeks, these scaffolds were shown to be able to stimulate cementum-like tissue formation but little new bone regeneration.

None of these scaffolds have been experimented in humans as yet, probably due to the lack of clear evidence about the effects of growth factors in the long term and the fate of nanomaterials used *in vivo*, as well as the high cost of the devices.

Scaffolds and their Key Characteristics

The fundamental concept underlying tissue engineering is to combine a scaffold with living cells and/or biologically active molecules to form a 'tissue engineering construct', which promotes the repair and/or regeneration of tissues (Bartold et al. 2000; Bartold and Raben 1996). The scaffold is expected to perform various functions, including the support of cell colonization, migration, growth and differentiation. The design of these scaffolds also needs to consider physico-chemical properties, morphology and degradation kinetics. The external size and shape of the construct are of importance, particularly if the construct is customized for an individual patient. Most importantly, clinically successful constructs should stimulate and support both the onset and the continuance of tissue in-growth as well as subsequent remodeling and maturation by providing optimal stiffness and external and internal geometrical orientation. Scaffolds must provide sufficient initial mechanical strength and stiffness to substitute for the loss of mechanical function of the diseased, damaged or missing tissue. Continuous cell and tissue remodeling is important for achieving stable biomechanical conditions and vascularization at the host site. In addition to these essentials of mechanics and geometry, a suitable construct will (a) possess a three-dimensional and highly porous interconnected pore network with surface properties that are optimized for the attachment, migration, proliferation and differentiation of cell types of interest and enable flow transport of nutrients and metabolic waste; and (b) be biocompatible and biodegradable with a controllable rate to complement cell/tissue growth and maturation (Hutmacher and Cool 2007). It is essential to understand and control the scaffold degradation process in order to achieve successful tissue formation, remodeling and maturation at the defect site. Initially, it was believed that scaffolds should be degraded as the tissue is growing. Yet, tissue in-growth and maturation differ temporally from tissue to tissue, and simply achieving tissue in-growth does not necessarily equate to tissue maturation and remodeling. Indeed, many scaffold-based strategies have failed in key scaffold characteristics. The fundamental concept underlying tissue engineering is to combine a scaffold with living cells and/or biologically active molecules to form a 'tissue engineering construct', which promotes the repair and/or regeneration of tissues (Bartold et al. 2000; Bartold and Raben 1996). The scaffold is expected to perform various functions, including the support of cell colonization, migration, growth and differentiation. The design of these scaffolds also needs to consider physico-chemical properties, morphology and degradation kinetics. The external size and shape of the construct are of importance, particularly if the construct is customized for an individual patient. Most importantly, clinically successful constructs should stimulate and support both the onset and the continuance of tissue in-growth as well as subsequent remodeling and maturation by providing optimal stiffness and external and internal geometrical orientation. Scaffolds must provide sufficient initial mechanical strength and stiffness to substitute for the loss of mechanical function of the diseased, damaged or missing tissue. Continuous cell and tissue remodeling is important for achieving stable biomechanical conditions and vascularization at the host site. In addition to these essentials of mechanics and geometry, a suitable construct will (a) possess a three-dimensional and highly porous interconnected pore network with surface properties that are optimized for the attachment, migration, proliferation and differentiation of cell types of interest and enable flow transport of nutrients and metabolic waste; and (b) be biocompatible and biodegradable with a controllable rate to complement cell/tissue growth and maturation (Hutmacher and Cool 2007). It is essential to understand and control the scaffold degradation process in order to achieve successful tissue formation, remodeling, and maturation at the defect site. Initially, it was believed that scaffolds should be degraded as the tissue is growing. Yet, tissue in growth and maturation differ temporally from tissue to tissue, and simply achieving tissue in growth does not necessarily equate to tissue maturation and remodeling. Indeed, many scaffold-based strategies have failed in the past as the scaffold degradation was more rapid than tissue remodeling and/or maturation. It is now recognized that the onset of degradation should only occur after the regenerated tissue within the scaffold has been remodeled at least once in the natural remodeling cycle. Thus, it is important that the scaffold remain intact as the tissue matures within the scaffold, with bulk degradation occurring later (Woodruff and Hutmacher 2010).

Scaffolds and Carriers for Periodontal Tissue Engineering

There are different scaffold materials that could be used as a carrier for cells in tissue engineering. Generally, they could be divided into natural (for example, collagen, gelatin and chitosan) and synthetic scaffolds (for example, β tricalcium phosphate, poly glycolic acid and polycarpolactone), and also could be classified into resorbable and nonresorbable scaffolds (Bartold and Raben 1996; Shue et al.

2012). Three dimensional electrospun nanofibrous scaffolds are receiving widespread interest in bone and cartilage regeneration, and have also been utilized for periodontal regeneration (Li et al. 2014). Studies have shown good attachment and proliferation of periodontal ligament cells on electrospun gelatin scaffolds (Zhang et al. 2009) and a variety of multilayered electrospun polymeric membranes and scaffolds (Inanc et al. 2009; Bottino et al. 2011; Vaquette et al. 2012).

Advanced Tissue-Engineered Construct Design With Multiphasic Scaffolds

A multiphasic scaffold can be defined by the variation within the architecture (porosity, pore organization, etc.), and/or the chemical composition of the resulting construct, which usually recapitulates to some extent the structural organization and/or the cellular and biochemical composition of the native tissue. Multiphasic scaffolds aimed at imparting biomimetic functionality to tissue-engineered bone and soft tissue grafts have been recognized for some time as having significant potential to enable clinical translation in the field of orthopaedic tissue engineering, and are more recently emerging in the field of periodontal tissue regeneration. Multiphasic scaffolds represent an attractive option for facilitating periodontal regeneration because of the requirement to temporally and spatially control the interaction between multiple soft and hard tissues. Recently, the use of multiphasic scaffolds for periodontal tissue engineering purposes has been used by several groups. Carlo Reis and colleagues developed a semirigid PLGA (polylactide co glycolide acid)/CaP (calcium phosphate) bilayered biomaterial construct with a continuous outer barrier membrane obtained by solvent casting and an inner topographically complex and porous component fabricated by solvent casting sugar leaching. The scaffold was tested in Class II furcation defects in dogs and was shown to promote cementum, bone and periodontal ligament fiber insertion. This bilayered construct approach represents a modification of the traditional 'guided tissue regeneration', whereby the construct acts as both a barrier and an enhanced space maintainer. Despite these promising results, the authors noted that the periodontium was not fully regenerated in the most coronal regions of the defect. They hypothesized that the space maintenance properties of the bilayered construct were decreased over time as the scaffold was gradually degraded. Indeed, at 120 days postimplantation, no traces of the polymeric material were found. This study highlights the importance of appropriate material selection and demonstrates that a polymer undergoing a slow *in vivo* degradation might be more suited for periodontal regeneration. The approach proposed by Carlo-Reis and colleagues relies solely on the regenerative performance of the host progenitors residing in the vicinity of the damaged area, and it can be hypothesized that the combination with exogenous progenitor cells could enhance the regenerative process (Carlo-Reis et al. 2011). Another interesting approach involves the use of polycaprolactone (PCL)-polyglycolic acid constructs for controlling fiber orientation and facilitating morphogenesis of the periodontal tissue complexes (Park et al. 2010; Park et al. 2012). This approach utilized multicompartmental scaffold architecture using computational scaffold design and manufacturing by 3D printing. When combined with BMP-7 transfected gingival cells, newly formed tissues demonstrated the interfacial generation of parallel and obliquely oriented fibers that formed human tooth dentin ligament bone complexes in an *in vivo* ectopic mouse periodontal regeneration model (Park et al. 2010). Subsequently, biomimetic fiber-guiding scaffolds using similar 3D wax/solvent casting methods combined with BMP-7 transduced PDL cells were tested in an athymic rat periodontal defect model, and resulted in perpendicularly oriented micro-channels that guided the periodontal fiber orientation at the root ligament interface (Park et al. 2012). The authors advocated the manufacture of individualized multiphasic scaffolds via computational design and 3D printing (Park et al. 2013). A biphasic tissue-engineered construct has also been utilized, which comprises an electrospun membrane for the delivery of a periodontal cell sheet attached to a three-dimensional porous scaffold for bone regeneration (Vaquette et al. 2012). The periodontal compartment was composed of a solution electrospun PCL membrane to facilitate the delivery of PDL cell sheets and improve the stability and the application of the cell sheets onto the dentine root surface. This study showed that the PCL membrane provided additional anchorage to the cell sheets, which resulted in enhanced adhesion and stability. *In vitro*, it was shown that the bone compartment supported cell growth and mineralization, and the periodontal component was suitable for supporting multiple PDL cell sheets. When applied onto a dentine block and implanted in a subcutaneous animal model, cementum deposition was seen on the surface of the dentine. This approach demonstrated that a biphasic scaffold combined with cell sheet technology could be beneficial for periodontal regeneration. The concept was further developed by enhancing the osteoconductive nature of the bone compartment by coating with a layer of calcium phosphate, as well as utilizing a periodontal compartment possessing a larger pore size that could enhance the integration of PDL tissue with the newly formed alveolar bone (Costa et al. 2014).

Another approach has been to utilize PCL-HA (90:10 wt%) scaffolds, which were fabricated using three dimensional printing in three phases: 100 μm microchannels in Phase A designed for cementum/dentine interface, 600 μm microchannels in Phase B designed for the PDL and 300 μm microchannels in Phase C designed for alveolar bone (Lee et al. 2014). Recombinant human amelogenin, connective tissue growth factor and bone morphogenetic protein-2 were delivered in Phases A, B, and C, respectively. On 4-weeks *in vitro* incubation with either dental pulp stem/progenitor cells, PDL stem/progenitor cells or alveolar bone stem/progenitor cells, distinctive tissue phenotypes were formed in each compartment. The strategy used for the regeneration of multiphase periodontal tissues in this study involved the spatiotemporal delivery of multiple proteins. Using this method, it was shown that a single stem/progenitor cell population appeared to differentiate into putative cementum, PDL and alveolar bone complex by using the scaffold's biophysical properties, combined with spatially released bioactive cues.

Tissue-engineered Decellularized Matrices and Periodontal Regeneration

The use of decellularized matrices as a biologic scaffold is gaining increasing attention in regenerative medicine. The rationale of using this approach is to produce three-dimensional scaffolds that mimic the natural tissue's composition, microstructure and biological and mechanical properties. The aim is to enhance the recruitment of host progenitor cells into these scaffolds and induce them to differentiate into the target tissue cell phenotype. Decellularized matrices can be obtained by decellularizing native tissues or organs, or by removing the cellular components of tissue-engineered constructs (Hoshiba et al. 2010). In a recent study, it was demonstrated that periodontal cell sheets placed on melt electrospun PCL membranes could be decellularized by bidirectional perfusion with NH_4OH/Triton X-100 and DNase solutions. The decellularized cell sheets demonstrated an intact extracellular matrix, retained growth factors and had the capacity to support the proliferation of allogenic PDL cells (Farag et al. 2014). Indeed, decellularized matrices have been obtained from various tissues and organs, such as heart valves, blood vessels, Small Intestinal Submucosa (SIS), lung, trachea, skin, nerves and the cornea (Hoshiba et al. 2010). It has also been demonstrated that the decellularization process not only results in preservation of the ECM microstructure but retention of biologically active components, such as growth factors, is also achievable (Badylak 2007). Importantly, tissue-engineered decellularized scaffolds did not elicit an immune response when implanted *in vivo* (Bloch et al. 2011).

Blood Supply: Vascularization and Endothelial Progenitors

Vascularization is an important part of any regeneration approach to avoid tissue necrosis, and the use of prevascularized tissue engineered scaffolds is receiving increasing attention in regenerative medicine (Baldwin et al. 2014). In the context of periodontal tissue engineering, Nagai and colleagues (2009) used a tissue engineering construct of human PDL fibroblasts (HPDLFs) co-cultured with or without Human Umbilical Vein Endothelial Cells (HUVECs). The HUVECs were found to form capillary-like structures when co-cultured with the HPDLFs. These cultures demonstrated longer survival, higher ALP activity and lower osteocalcin production than the HDPLF cultures alone. These findings suggest that the incorporation of endothelial progenitors into tissue-engineered constructs may be beneficial in maintaining adequate vascularization, which would, in turn, improve regenerative outcomes (Nagai et al. 2009).

Gene Therapy

One way to overcome the issue of the short half-life of growth factors and ensure a sustained local release is to deliver cells capable of producing the growth factor *in situ* within the periodontal defect. This can be achieved by gene therapy, which involves the genetic manipulation of cells to enhance their ability to produce a given protein, in this case, a growth or differentiation factor. More specifically, this strategy utilizes vectors to insert genetic material into cells that are subsequently inserted into the periodontal defect, eliciting transcription of these genes and subsequent growth and differentiation of surrounding host cells, leading to new attachment formation. Gene delivery of PDGF has been accomplished by the successful transfer of the gene into various periodontal cell types (Park et al. 2010; Jin et al. 2004; Chang et al. 2009). Animal studies have demonstrated that gene delivery of PDGF stimulated more cementoblast activity and improved regeneration compared with a single application of recombinant PDGF (Chang et al. 2009). Although our understanding of the *in vivo* effect of sustained growth factor activity has improved with experimental gene therapy studies, significant safety concerns remain in relation to this technology.

Strategies of Overlaying to Mimic Multiple Periodontal Tissues

Given that the veritable regeneration of periodontal tissue involves re-establishment of the bone-PDL-cementum apparatus, the simple combination of *in vitro*-cultured stem cells and biomaterials cannot realistically be used to achieve this goal (Fournier and Doillon 1996; Koop et al. 2012). However, it is possible to regain the periodontal hybrid tissues by layering materials and cells to mimic the different tissue layers involved in the periodontium (Giannobile and Somerman 2003). The vertical stacking of three-layered PDLSC sheets, woven poly(glycolic acid) and porous β-TCP in an orderly manner based on this concept and their placement into the three wall periodontal defects of a canine resulted in newly formed bone and cementum interspersed with the aligned collagen fibers (Vuong and Otto 2002). Similarly, cell sheets comprising PDLSCs and/or jaw BMMSCs have been multilayered to regenerate a complex periodontium-like architecture (Pihlstrom et al. 2005). Recently, a 'sandwich' tissue engineering complex was constructed by adding a layer of mineralized membrane on each side of the collagen membrane. After seeding with gingival fibroblasts, this complex was implanted into periodontal defect areas in dogs, and simultaneous neogenesis of ligamentous and osseous structures was achieved (Zimmerli et al. 2004). Although the development of bilayered cell constructs serve as a promising strategy to simultaneously regenerate multiple periodontal tissues, the orchestrated use of multiple regenerated tissues to reconstruct the periodontal complex with micron-scaled tissue compartmentalization as well as osseous and ligamentous interfacial structures with systematic tooth-supporting functions remains a major clinical challenge.

Engineering Approaches to Reconstruct Periodontal Complex Interfaces and Architectures

Recent advances in biomaterials technology have enabled the engineering of periodontal scaffolds with triphasic tissue interfaces and structures (e.g., see (Hojo et al. 2009)). In particular, computer aided design and three-dimensional (3D) printing now allow the generation of compartmentalized hybrid scaffolds with biomimetic interfaces in 3D space (e.g., see (Fürst et al. 2007)). These scaffolds with spatial constructs can specifically regulate cell behaviors and influence the orientations of multiple tissues and thus provide an essential foundation for the formation of oriented ligamentous tissues and their successful incorporation into the new bone and cementum (reviewed in (Lyngstadaas et al. 2009)). In this context, biomimetic hybrid scaffolds containing bone-specific and PDL-specific polymer compartments are used to engineer human tooth-ligament interfaces (Hojo et al. 2009), and the engineering of bone-ligament complexes has successfully been achieved with fiber-guiding scaffolds (Hojo et al. 2009). Biphasic scaffolds that mimic the bone compartment by using a fused deposition modeling scaffold and mimic the PDL compartment by using an electrospun membrane have been reported, and the combination of these scaffolds and multiple PDLSC sheets allows the *in vivo* regeneration of complex periodontal structures (Quirynen et al. 2002). Furthermore, the coating of the bone compartment with a calcium phosphate layer increased the osteoconductivity of seeded osteoblasts (Price et al. 1996).

Using 3D printing and directional freeze-casting techniques, a fiber-guiding hybrid scaffold has been designed to spatiotemporally control the morphogenesis, integration and functionalization of various tissues (Adeli and Parvizi 2012). Animal experiments have shown that this customized fiber-guiding scaffold can accurately adapt to defect sites and thus successfully guide cell/tissue directionality during regeneration and facilitate the formation of a more stable ligament-ligand complex with a rapidly maturing matrix (Shahi and Parvizi 2015). More recently, a similar '3D-patterned, periodontal mimic multiphasic architecture' approach was reported, taking this technology one step closer to clinical translation (Turgut et al. 2005). The aforementioned fiber-guiding scaffold has been approved by the U.S. Food and Drug Administration for clinical use (Adeli and Parvizi 2012). Collectively, 3D-patterned multiphasic complexes have enabled the reconstruction of periodontal complex architectures for periodontal tissue engineering strategies, but when translated to clinical use, the complexity of tissue-engineered constructs must be kept to a minimum to ensure cost-effectiveness and ease of production.

Crucial Barriers to Progress

In vitro stem cell-material designs can mimic the anatomy of the periodontium and different biomaterials, such as the bone graft materials and cell sheets, particularly barrier membranes could be applied for this strategy. The goal of periodontal tissue engineering is to reinstate the normal function of the diseased periodontium to support the teeth, whereas the triad would require stem cells, biomaterials and infection control. However, the efficacy of cell-based interventions in humans remains to be confirmed (Melcher et al. 1987). Although experimental data have allowed the initiation of clinical trials in periodontal cell therapy, proper consideration of the cell source, material type and regulatory concerns is crucial to facilitate clinical translation (Ketonis et al. 2010). For periodontal tissue engineering, regeneration in response to overengineered constructs should be investigated in larger animals, including rodents, because the anatomy of the dentoalveolar architecture in larger animals more closely resembles that of humans (Zhao et al. 2009). In addition, the experimentally generated defects need to accurately mimic the pathophysiology of periodontitis in humans. In many cases, the periodontal defects generated in the currently used animal studies do not sufficiently reflect the inflammatory status of human periodontitis.

Clinical Application and Closing Remarks

Regenerative therapies for periodontal disease that use patients' cells to repair the periodontal defect have been proposed in a number of preclinical and clinical studies. Periodontal tissue-derived stem cells, such as PDLSCs, are committed toward all the periodontal developmental lineages that contribute to cell turnover in the steady-state and would thus be useful cell sources for treating periodontally destructive diseases, such as periodontitis.

Treatments that partially regenerate damaged periodontal tissue through the localized administration of GFs have now been established. Although the clinical practice is not very successful, such regenerative therapies have provided very useful and feasible clinical study models for the future design of tissue engineering and stem cell therapies. Using currently available clinical strategies, partial regeneration of the periodontal tissue is becoming possible; however, methods to achieve the functional regeneration of large defects caused by severe periodontal diseases are still lacking. To address this, it is essential to better understand the cellular and molecular mechanisms underlying periodontal development and, thereby, identify the appropriate functional molecules that induce the differentiation of stem cells into periodontal lineage cells for the successful reconstruction of periodontal tissue. The field of periodontal bioengineering has entered an exciting new developmental phase that will make increasingly important contributions to the patient. Particularly, a number of biological technologies are being aggressively explored for clinical translation, signifying a veritable 'coming of age' of the field. However, such issues as appropriate delivery devices, immunogenicity, autologous cells vs. allogenic cells, identifying tissues that provide the most appropriate donor source, control of the whole process and cost-effectiveness are all important considerations that should not be overlooked. The future of periodontal bioengineering is undoubtedly driven by technology. New applications and improvement on current designs will largely depend on innovations in biomaterials engineering. Progress in stem cell biology will be imperative in dictating advances in stem cell-based regeneration. A better understanding of the molecular mechanisms by which substrate interactions impact stem cell self-renewal and differentiation is of paramount importance for targeted design of biomaterials. Discoveries in the fields of developmental biology and functional genomics should also be exploited for broadening the repertoire of biological molecules that can be incorporated into biomaterials for fine-tuning stem cell activities. With the merger between the two powerful disciplines biomaterials engineering and stem cell biology a new drawing board now lies before us to develop therapies that promise to revolutionize periodontal tissue engineering.

Future Challenges (Yanagawa 2016)

The goal of 3D tissue engineering is not only the fabrication of whole-organ structures but also the generation of functional engineered organs and tissues, to restore the sites of injury (Bhatia and Ingber 2014; Mao and Mooney 2015). Although the hydrogels are useful for fabricating and maintaining 3D structure, the engineered tissue for transplantation therapy should synchronize with the tissues of the recipient following the transplantation. Therefore the ideal transplantation scaffolds should be hydrolytically or enzymatically degradable. A number of studies have focused on the development of biodegradable hydrogels (Lee and Shin 2007; Nicodemus and Bryant 2008), designed to degrade by hydrolysis (Peister et al. 2009) reduction (Chien et al. 2012), enzymatic reaction (Zhang et al. 2016; Daemi et al. 2016) and oracombination. Since the hydrogel degradation can, in practice, be tuned by chemical moieties of the hydrogels, the degradation rate and profile are controllable as well.

The future challenge for 3D tissue engineering in the field of regenerative therapy is the development of 3D tissue constructs without hydrogels. One of the recently developed hydrogel-free approaches is the use of a decellularized Extracellular Matrix (dECM) as a native scaffold (Badylak et al. 2011). This approach was first reported by Cho et al. (Pati et al. 2014), who fabricated the complex channel structure with dECM, using 3D printing technology. Despite this success, numerous issues remain, the most significant being the potential removal of the various types of molecules in the ECM during the decellularization process. Therefore, the mechanical properties of fabricated tissues remain less strong as compared to those of native tissues. Another example of hydrogel-free approaches is the use of cellular aggregates, such as spheroids. Hydrogel-free tubular tissues have been created by the 3D printing of spheroids into a needle array (Itoh et al. 2015). These approaches may overcome the limitations of the hydrogels, but their potential is still being investigated. The understanding and methodology developed in the hydrogel microfabrication can contribute to the advancements in hydrogel free technologies.

The search for the ideal scaffold

Scaffolds provides a 3D framework for cells to attach to and on which to proliferate that can be implanted into a tissue-defected site. In general, two main types of scaffolds can be differentiated: synthetically derived polymer scaffolds and naturally occurring scaffolds.

Synthetically derived polymer scaffolds should be biocompatible and biodegradable and should ensure an optimal interaction with endothelial cells to promote angiogenesis. To develop scaffolds that fulfill these properties, it is of great importance to investigate how different biomaterials modulate endothelial cell function. For this purpose, a variety of endothelial cell culture systems have been established during recent years. They allow for the evaluation of new biomaterials in terms of endothelial cell attachment, cytotoxicity, growth, angiogenesis and gene regulation.

Additionally, co-culture systems can be used to study blood vessel development in tissue constructs on a higher level of cellular organization, considering the interaction between endothelial cells and other cell types.

In addition to the cellular interaction with biomaterials, the architecture of the scaffold seems to play an important role in adequate vascularization. Pinney et al. reported that the 3D structure of a scaffold in itself can change the angiogenic activity of incorporated cells.

When culturing fibroblasts on a lactate-glycolate copolymer scaffold to form a dermal-equivalent tissue, they observed that the cellular content of Vascular Endothelial Growth Factor (VEGF) messenger ribonucleic acid in these 3D cultures was 22 times greater than that in the same fibroblasts grown as monolayers. In addition, the pore size of the scaffolds has been shown to be a critical determinant of blood vessel in growth, which is significantly faster in pores with a size greater than 250 mm than in those less than 250 mm. Finally, molecular deteriorations that are induced during the incorporation process of the implant might influence the in growth of blood vessels into scaffolds. This may explain why currently used biomaterials partly fail to vascularize, independent of their material properties. Fibronectin, for example, is the only mammalian adhesion protein that binds and activates a 5b1-integrins, which are known to exert proangiogenic actions, although Vogel and Baneyx, who investigated the role of this adhesion protein in scaffold vascularization, could not demonstrate an acceleration or improvement of implant vascularization but documented an inhibitory action on the process of new vessel formation.

This inhibition was interpreted as the result of excessive tension generated by cells in contact with implanted biomaterials, which may have changed the molecular structure of fibronectin fibrils, deteriorating their capability to bind and activate a 5b1-integrins. In addition to synthetically derived polymer scaffolds, naturally occurring scaffolds composed of extracellular matrix proteins offer promising alternatives for tissue repair and regeneration. Important examples are small intestinal submucosa, acellular dermis, cadaveric fascia, the bladder acellular matrix graft and the amniotic membrane.

These types of scaffolds have been shown to promote rapid interaction with the surrounding host tissue, to induce the deposition of cells and additional extra cellular matrix, and to accelerate the process of angiogenesis. These naturally occurring scaffolds can be processed so that they retain growth factors, glycosaminoglycans and structural elements such as fibronectin, elastin and collagen, which are important regulatory factors of angiogenesis.

Conclusion

With recent advances in 3D tissue engineering, hydrogel microfabrication technologies, such as micromolding, 3D bioprinting, photolithography and stereolithography, are providing a more realistic approach to drug discovery and the development of alternative methods of organ transplantation. The production of well-defined architectures that mimic natural tissues and organs, without causing significant cell damage, is one of the key challenges during the fabrication process. Additionally, combining the hydrogel microfabrication technology with cell culture platform, such as microfluidics devices, to provide nutrients and oxygen to the cells within the hydrogels, has not been investigated thoroughly to date. The progress in the fabrication of hydrogels and the development of methodology for cell cultivation will offer long term improvement of the biological functions in 3D tissue constructs. However, many questions remain to be answered until naturally occurring scaffolds can be used for clinically relevant tissue engineering, including the immunologic response of the host to such implants and the methods to modify their mechanical and physical properties. Thus, the ideal scaffold, which promotes angiogenesis of engineered tissue sufficiently, has not yet been determined.

Acknowledgements

We thank all the researchers and authors for their contribution to this topic and ongoing work, without which this chapter would not have been possible.

References

1) Gottlow, J., Nyman, S., Lindhe, J., Karring, T. and Wennström, J. 1986 July. New attachment formation in the human periodontium by guided tissue regeneration. Case Reports. J. Clin. Periodontol. 13(6): 604–16.

Adeli, B. and Parvizi, J. 2012. Strategies for the prevention of periprosthetic joint infection. J. Bone Jt. Surg. Br. Vol. 94: 42–46.

Albrektsson, T., Brånemark, P.-I., Hansson, H.-A. and Lindström, J. 1981. Osseointegrated titanium implants: Requirements for ensuring a long-lasting, direct bone-to-implant anchorage in man. Acta Orthop. Scand. 52: 155–170.

Antoniades, H.N. and Owen, A.J. 1982. Growth factors and regulation of cell growth. Annu. Rev. Med. 33: 445–463.

Antoniades, H.N., Galanopoulos, T., Neville-Golden, J., Kiritsy, C.P., Lynch, S.E. 1991. Injury induces *in vivo* expression of platelet-derived growth factor (PDGF) and PDGF receptor mRNAs in skin epithelial cells and PDGF mRNA in connective tissue fibroblasts. Proc. Natl. Acad. Sci. USA 88(2): 565–569.

Anusaksathien, O., Webb, S.A., Jin, Q.-M. and Giannobile, W.V. 2003. Platelet-derived growth factor gene delivery stimulates *ex vivo* gingival repair. Tissue Eng. 9: 745–756.

Anusaksathien, O., Jin, Q., Zhao, M., Somerman, M.J. and Giannobile, W.V. 2004. Effect of sustained gene delivery of platelet-derived growth factor or its antagonist (PDGF-1308) on tissue-engineered cementum. J. Periodontol. 75: 429–440.

Aurer, A. and Jorgić-Srdjak, K. 2005. Membranes for periodontal regeneration. Acta Stomatol. Croat. 39: 107–112.

Badylak, S.F. 2007. The extracellular matrix as a biologic scaffold material. Biomaterials 28: 3587–93.

Badylak, S.F., Taylor, D. and Uygun, K. 2011. Whole-organ tissue engineering: Decellularization and recellularization of three-dimensional matrix scaffolds. Annu. Rev. Biomed. Eng. 13: 27e53.

Baldwin, J., Antille, M., Bonda, U., De-Juan-Pardo, E.M., Khosrotehrani, K., Ivanovski, S. et al. 2014. *In vitro* pre-vascularisation of tissue-engineered constructs: A co-culture perspective. Vasc. Cell 6: 13.

Barron, V. and Pandit, A. 2003. Combinatorial approaches in tissue engineering: progenitor cells, scaffolds, and growth factors. pp. 1–21. *In*: Ashammakhi, N. and Ferretti, P. (eds.). Topics in Tissue Engineering; University of Oulu: Oulu, Finland.

Bartold, P.M. and Raben, A. 1996. Growth factor modulation of fibroblasts in simulated wound healing. J. Periodontal. Res. 31: 205–16.

Bartold, P.M., Mcculloch, C.A., Narayanan, A.S. and Pitaru, S. 2000. Tissue engineering: a new paradigm for periodontal regeneration based on molecular and cell biology. Periodontol. 24: 253–69.

Bei, M. 2009. Molecular genetics of tooth development. Curr. Opin. Genet. Dev. 19: 504–510.

Bhatia, S.N. and Ingber, D.E. 2014. Microfluidic organs-on-chips. Nat. Biotechnol. 32: 760e72.

Bloch, O., Golde, P., Dohmen, P.M., Posner, S., Konertz, W. and Erdbrügger, W. 2011. Immune response in patients receiving a bioprosthetic heart valve: lack of response with decellularized valves. Tissue Eng. Part A 17: 399–405.

Bottino, M.C., Thomas, V. and Janowski, G.M. 2011. A novel spatially designed and functionally graded electrospun membrane for periodontal regeneration. Acta Biomater. 7: 216–24.

Burchardt, H. 1983. The biology of bone graft repair. Clin. Orthop. Relat. Res. 174: 28–42.

Carlo-Reis, E.C., Borges, A.P.B., Araújo, M.V.F., Mendes, V.C., Guan, L. and Davies, J.E. 2011. Periodontal regeneration using a bilayered PLGA/calcium phosphate construct. Biomaterials 32: 9244–53.

Carranza. 2014. Carranza's Clinical Periodontology. Michael Newman Henry Takei Perry Klokkevold Fermin Carran 12th Edition. Elsevier.

Caton, J., DeFuria, E., Polson, A. and Nyman, S. 1987. Periodontal regeneration via selective cell repopulation. J. Periodontol. 58: 546–552.

Chang, P.C., Cirelli, J.A., Jin, Q., Seol, Y.J., Sugai, J.V., D'silva, N.J. et al. 2009. Adenovirus encoding human platelet-derived growth factor-B delivered to alveolar bone defects exhibits safety and biodistribution profiles favorable for clinical use. Hum. Gene Ther. 20: 486–96.

Chen, X., Liu, Y., Miao, L., Wang, Y., Ren, S., Yang, X. et al. 2016. Controlled release of recombinant human cementum protein 1 from electrospun multiphasic scaffold for cementum regeneration. Int. J. Nanomed. 11: 3145.

Chien, H.W., Tsai, W.B. and Jiang, S.Y. 2012. Direct cell encapsulation in biodegradable and functionalizable carboxybetaine hydrogels. Biomaterials 33: 5706e12.

Costa, P.F., Vaquette, C., Zhang, Q., Reis, R.L., Ivanovski, S. and Hutmacher, D.W. 2014. Advanced tissue engineering scaffold design for regeneration of the complex hierarchical periodontal structure. J. Clin. Periodontol. 41: 283–94.

Dabra, S., Chhina, K., Soni, N. and Bhatnagar, R. 2012. Tissue engineering in periodontal regeneration: A brief review. Dent. Res. J. 9: 671–680.

Daemi, H., Rajabi-Zeleti, S., Sardon, H., Barikani, M., Khademhosseini, A. and Baharvand H. 2016. A robust super-tough biodegradable elastomer engineered by supramolecular ionic interactions. Biomaterials 84: 54e63.

Dennison, D.K., Vallone, D.R., Pinero, G.J., Rittman, B. and Caffesse, R.G. 1994. Differential effect of TGF-β1 and PDGF on proliferation of periodontal ligament cells and gingival fibroblasts. J. Periodontol. 65: 641–648.

Einhorn, T.A. 2003. Clinical applications of recombinant human BMPs: Early experience and future development. JBJS 85: 82–88.

Engler, W., Ramfjord, S. and Hiniker, J. 1966. Healing following simple gingivectomy. A tritiated thymidine radioautographic study. I. Epithelialization. J. Periodontol. 37: 298–308.

Esposito, M., Grusovin, M.G., Papanikolaou, N., Coulthard, P. and Worthington, H.V. 2005. Enamel matrix derivative (Emdogain) for periodontal tissue regeneration in intrabony defects. Cochrane Database Syst. Rev. CD003875.

Fang, J., Zhu, Y.-Y., Smiley, E., Bonadio, J., Rouleau, J.P., Goldstein, S.A. et al. 1996. Stimulation of new bone formation by direct transfer of osteogenic plasmid genes. Proc. Natl. Acad. Sci. USA 93: 5753–5758.

Farag, A., Vaquette, C., Theodoropoulos, C., Hamlet, S.M., Hutmacher, D.W. and Ivanovski, S. 2014. Decellularized periodontal ligament cell sheets with recellularization potential. J. Dent. Res. 93: 1313–9.

Fournier, N. and Doillon, C.J. 1996. Biological molecule-impregnated polyester: An *in vivo* angiogenesis study. Biomaterials 17: 1659–1665.

Frost, H. 1989a. The biology of fracture healing. An overview for clinicians. Part II. Clin. Orthop. Relat. Res. 248: 294–309.

Frost, H.M. 1989b. The biology of fracture healing. An overview for clinicians. Part I. Clin. Orthop. Relat. Res. 248: 283–293.

Fürst, M.M., Salvi, G.E., Lang, N.P. and Persson, G.R. 2007. Bacterial colonization immediately after installation on oral titanium implants. Clin. Oral. Implants Res. 18: 501–508.

Gansbacher, B. 2003. Report of a second serious adverse event in a clinical trial of gene therapy for X-linked severe combined immune deficiency (X-SCID) Position of the European Society of Gene Therapy (ESGT). J. Gene Med. 5: 261–262.

Giannobile, W. 1996. Periodontal tissue engineering by growth factors. Bone 19: S23–S37.

Giannobile, W., Whitson, S. and Lynch, S. 1997. Non-coordinate control of bone formation displayed by growth factor combinations with IGF-I. J. Dent. Res. 76: 1569–1578.

Giannobile, W.V., Hernandez, R.A., Finkelman, R.D., Ryarr, S., Kiritsy, C.P., D'Andrea, M. et al. 1996. Comparative effects of platelet derived growth factor-BB and insulin-like growth factor-I, individually and in combination, on periodontal regeneration in Macacafascicularis. J. Periodontal. Res. 31: 301–312.

Giannobile, W.V. and Somerman, M.J. 2003. Growth and amelogenin-like factors in periodontal wound healing. A systematic review. Ann. Periodontol. 8: 193–204.

Giorgio, I., Saeid, K. and Francesco, B. 2019. Biomaterials, current strategies, and novel nano-technological approaches for periodontal regeneration. J. Funct. Biomater. 10(1): 3.

Gristina, A.G. 1987. Biomaterial-centered infection: Microbial adhesion versus tissue integration. Science 237: 1588–1595.

Hammarström, L.J 1997. Enamel matrix, cementum development and regeneration. Clin. Periodontol. Sep. 24(9 Pt 2): 658–68.

Hauschka, S.D. 1988. Cell surface fibroblast growth factor and epidermal growth factor receptors are permanently lost during skeletal muscle terminal differentiation in culture. J. Cell Biol. Aug; 107(2): 761–9.

Hojo, K., Nagaoka, S., Ohshima, T. and Maeda, N. 2009. Bacterial interactions in dental biofilm development. J. Dent. Res. 88: 982–990.

Hoshiba, T., Lu, H., Kawazoe, N. and Chen, G. 2010. Decellularized matrices for tissue engineering. Expert. Opin. Biol. Ther. 10: 1717–28.

Hutmacher, D.W. and Cool, S. 2007. Concepts of scaffold-based tissue engineering: the rationale to use solid free-form fabrication techniques. J. Cell. Mol. Med. 1: 654–69.

Inanc, B., Arslan, Y., Seker, S., Elcin, A. and Elcin, Y. 2009. Periodontal ligament cellular structures engineered with electrospun poly(DL-lactide-coglycolide) nanofibrous membrane scaffolds. J. Biomed. Mater Res. A 90: 186–95.

Itoh, M., Nakayama, K., Noguchi, R., Kamohara, K., Furukawa, K., Uchihashi, K. et al. 2015. Scaffold-free tubular tissues created by a Bio-3D printer undergo remodeling and endothelialization when implanted in rat aortae. PLoS One 10: c0136681.

Iviglia, G., Kargozar, S. and Baino, F. 2019. Biomaterials, current strategies, and novel nano-technological approaches for periodontal regeneration. J. Funct. Biomater. Jan. 2: 10(1).

Izumi, Y., Aoki, A., Yamada, Y., Kobayashi, H., Iwata, T., Akizuki, T. et al. 2011. Current and future periodontal tissue engineering. Periodontol. 56: 166–187.

Jepsen, S., Eberhard, J., Herrera, D. and Needleman, I. 2002. A systematic review of guided tissue regeneration for periodontal furcationdefects. What is the effect of guided tissue regeneration compared with surgical debridement in the treatment of furcation defects? J. Clin. Periodontol. 29(Suppl. 3): 103–16; discussion 160–2.

Jin, Q., Anusaksathien, O., Webb, S.A., Printz, M.A. and Giannobile, W.V. 2004. Engineering of tooth-supporting structures by delivery of PDGF gene therapy vectors. Mol. Ther. 9: 519–526.

Jin, Q.M., Anusaksathien, O., Webb, S., Rutherford, R. and Giannobile, W. 2003. Gene therapy of bone morphogenetic protein for periodontal tissue engineering. J. Periodontol. 74: 202–213.

Jose, B., Antoci, V., Jr. Zeiger, A.R., Wickstrom, E. and Hickok, N.J. 2005. Vancomycin covalently bonded to titanium beads kills Staphylococcus aureus. Chem. Biol. 12: 1041–1048.

Kargozar, S. and Mozafari, M. 2018. Nanotechnology and nanomedicine: Start small, think big. Mater. Today Proc. 5: 15492–15500.

Ketonis, C., Adams, C.S., Barr, S., Aiyer, A., Shapiro, I.M., Parvizi, J. et al. 2010. Antibiotic modification of native grafts: Improving upon nature's scaffolds. Tissue Eng. Part A 16: 2041–2049.

Koop, R., Merheb, J. and Quirynen, M. 2012. Periodontal regeneration with enamel matrix derivative in reconstructive periodontal therapy: A systematic review. J. Periodontol. 83: 707–720.

Lee, S.H. and Shin, H. 2007. Matrices and scaffolds for delivery of bioactive molecules in bone and cartilage tissue engineering. Adv. Drug Deliv. Rev. 59: 339e59.

Lee, C.H., Hajibandeh, J., Suzuki, T., Fan, A., Shang, P. and Mao, J.J. 2014. Three-dimensional printed multiphase scaffolds for regeneration of periodontium complex. Tissue Eng. Part A 20: 1342–51.

Li, G., Zhang, T., Li, M., Fu, N., Fu, Y., Ba, K. et al. 2014. Electrospun fibers for dental and craniofacial applications. Curr. Stem Cell Res. Ther. 9: 187–95.

Linde, A., Alberius, P., Dahlin, C., Bjurstam, K. and Sundin, Y. 1993. Osteopromotion: A soft-tissue exclusion principle using a membrane for bone healing and bone neogenesis. J. Periodontol. 64: 1116–1128.

Lindhe, J., Westfelt, E., Nyman, S., Socransky, S. and Haffajee, A. 1984. Long-term effect of surgical/non-surgical treatment of periodontal disease. J. Clin. Periodontol. 11: 448–458.

Lynch, S.E., Buser, D., Hernandez, R.A., Weber, H., Stich, H., Fox, C.H. et al. 1991. Effects of the platelet-derived growth factor/insulin-like growth factor-I combination on bone regeneration around titanium dental implants. Results of a pilot study in beagle dogs. J. Periodontol. 62: 710–716.

Lyngstadaas, S., Wohlfahrt, J., Brookes, S., Paine, M., Snead, M. and Reseland, J. 2009. Enamel matrix proteins; old molecules for new applications. Orthodont. Craniofac. Res. 12: 243–253.

Mao, A.S. and Mooney, D.J. 2015. Regenerative medicine: Current therapies and future directions. Proc. Natl. Acad. Sci. USA 112: 14452e9.

Melcher, A. 1976. On the repair potential of periodontal tissues. J. Periodontol. 47: 256–260.

Melcher, A., McCulloch, C., Cheong, T., Nemeth, E. and Shiga, A. 1987. Cells from bone synthesize cementum-like and bone-like tissue *in vitro* and may migrate into periodontal ligament *in vivo*. J. Periodontal. Res. 22: 246–247.

Misch, C.E. and Dietsh, F. 1993. Bone-grafting materials in implant dentistry. Implant Dent. 2: 158–167.

Nagai, N., Hirakawa, A., Otani, N. and Munekata, M. 2009. Development of tissue-engineered human periodontal ligament constructs with intrinsic angiogenic potential. Cells Tissues Organs 190: 303–12.

Nanci, A. and Bosshardt, D.D. 2006. Structure of periodontal tissues in health and disease. Periodontol. 40: 11–28.

Needleman, I.G., Worthington, H.V., Giedrys-Leeper, E. and Tucker, R.J. 2006. Guided tissue regeneration for periodontal infra-bony defects. Cochrane Database Syst. Rev. Apr. 19(2): CD001724.

Neel, E.A.A., Chrzanowski, W., Salih, V.M., Kim, H.W. and Knowles, J.C. 2014. Tissue engineering in dentistry. J. Dent. 42: 915–928.

Nevins, M., Camelo, M., Nevins, M.L., Schenk, R.K. and Lynch, S.E. 2003. Periodontal regeneration in humans using recombinant human platelet-derived growth factor-BB (rhPDGF-BB) and allogenic bone. J. Periodontol. 74: 1282–1292.

Nicodemus, G.D. and Bryant, S.J. 2008. Cell encapsulation in biodegradable hydrogels for tissue engineering applications. Tissue Eng. Part B Rev. 14: 149e65.

Niklaus P. Lang and Jan Linde. 2015. Clinical Periodontology and Implant Dentistry. 2 Volume Set, 6th Edition. Willey Blackwell.

Nussenbaum, B. and Krebsbach, P.H. 2006. The role of gene therapy for craniofacial and dental tissue engineering. Adv. Drug Deliv. Rev. 58: 577–591.

Nyman, S., Gottlow, J., Lindhe, J., Karring, T. and Wennstrom, J. 1987. New attachment formation by guided tissue regeneration. J. Periodontal. Res. 22: 252–254.

Oates, T.W., Rouse, C.A. and Cochran, D.L. 1993. Mitogenic effects of growth factors on human periodontal ligament cells *in vitro*. J. Periodontol. 64: 142–148.

Palumbo, A. 2011. The anatomy and physiology of the healthy periodontium. In Gingival Diseases-their Aetiology, Prevention and Treatment; Intech.

Park, C.H., Rios, H.F., Jin, Q., Bland, M.E., Flanagan, C.L., Hollister, S.J. et al. 2010. Biomimetic hybrid scaffolds for engineering human tooth-ligament interfaces. Biomaterials 31: 5945–52.

Park, C.H., Rios, H.F., Jin, Q., Sugai, J.V., Padialmolina, M., Taut, A.D. et al. 2012. Tissue engineering bone-ligament complexes using fiber-guiding scaffolds. Biomaterials 33: 137–45.

Park, C.H., Rios, H.F., Taut, A.D., Padial-Molina, M., Flanagan, C.L., Pilipchuk, S.P. et al. 2013. Imagebased, fiber guiding scaffolds: a platform for regenerating tissue interfaces. Tissue Eng. Part C Methods 20: 533–42.

Pati, F., Jang, J., Ha, D.H., Kim, S.W., Rhie, J.W., Shim, J.H. et al. 2014. Printing three dimensional tissue analogues with decellularized extracellular matrix bioink. Nat. Commun. 5: 3935.

Peister, A., Deutsch, E.R., Kolambkar, Y., Hutmacher, D.W. and Guldberg, R.E. 2009. Amniotic fluid stem cells produce robust mineral deposits on biodegradable scaffolds. Tissue Eng. Part A 15: 3129e38.

Pihlstrom, B.L., Michalowicz, B.S. and Johnson, N.W. 2005. Periodontal diseases. Lancet 366: 1809–1820.

Pledger, W.J., Stiles, C.D. and Antonaides, H.N. 1977. Induction of DNA synthesis in BALB/C3T3 by serum components. re-evaluation of the commitment process. Pro. Natl. Acad. Sci. USA 74: 4481–5.

Price, J., Tencer, A., Arm, D. and Bohach, G. 1996. Controlled release of antibiotics from coated orthopedic implants. J. Biomed. Mater. Res. 30: 281–286.

Prichard, J. 1957. The infrabony technique as a predictable procedure. J. Clin. Periodontol. 28: 202–216.

Quirynen, M., De Soete, M. and Van Steenberghe, D. 2002. Infectious risks for oral implants: A review of the literature. Clin. Oral Implants Res. Rev. Artic. 13: 1–19.

Ross, R., Raines, E.W. and Bowen, P. 1986. The biology of platelet derived growth factor. Cell 46: 155–69.

Schek, R.M., Hollister, S.J. and Krebsbach, P.H. 2004. Delivery and protection of adenoviruses using biocompatible hydrogels for localized gene therapy. Mol. Ther. 9: 130–138.

Shahi, A. and Parvizi, J. 2015. Prevention of periprosthetic joint infection. Arch. BoneJt. Surg. 3: 72–81.

Sheikh, Z., Sima, C. and Glogauer, M. 2015. Bone replacement materials and techniques used for achieving vertical alveolar bone augmentation. Materials 8: 2953–2993.

Shimauchi, H., Nemoto, E., Ishihata, H. and Shimomura, M. 2013. Possible functional scaffolds for periodontal regeneration. Jpn. Dent. Sci. Rev. 49: 118–130.

Shue, L., Yufeng, Z. and Mony, U. 2012. Biomaterials for periodontal regeneration: a review of ceramics and polymers. Biomatter 2: 271–7.

Sowmya, S., Mony, U., Jayachandran, P., Reshma, S., Kumar, R.A., Arzate, H. et al. 2017. Tri-layered nanocomposite hydrogel scaffold for the concurrent regeneration of cementum, periodontal ligament, and alveolar bone. Adv. Healthc. Mater. 6: 1601251.

Springer, B.D., Lee, G.-C., Osmon, D., Haidukewych, G.J., Hanssen, A.D. and Jacofsky, D.J. 2004. Systemic safety of high-dose antibiotic-loaded cement spacers after resection of an infected total knee arthroplasty. Clin. Orthop. Relat. Res. 427: 47–51.

Stigter, M., Bezemer, J., De Groot, K. and Layrolle, P. 2004. Incorporation of different antibiotics into carbonated hydroxyapatite coatings on titanium implants, release and antibiotic efficacy. J. Controll. Release. 99: 127–137.

Taba, M., Jr. Jin, Q., Sugai, J. and Giannobile, W. 2005. Current concepts in periodontal bioengineering. Orthodont. Craniofac. Res. 8: 292–302.

Tabata, Y. 2003. Tissue regeneration based on growth factor release. Tissue Eng. 9: 5–15. [CrossRef] [PubMed]

Tencate. 2017. Development, Structure, and Function. Ten Cate's Oral Histology. Antonio Nanci. 9th Edition. Elsevier.

Trombelli, L. and Farina, R. 2008. Clinical outcomes with bioactive agents alone or in combination with grafting or guided tissue regeneration. J. Clin Periodontol. Sep. 35(8 Suppl.): 117–35.

Turgut, H., Sacar, S., Kaleli, I., Sacar, M., Goksin, I., Toprak, S. et al. 2005. Systemic and local antibiotic prophylaxis in the prevention of *Staphylococcus epidermidis* graft infection. BMC Infect. Dis. 5: 91.

Vaquette, C., Fan, W., Xiao, Y., Hamlet, S., Hutmacher, D.W. and Ivanovski, S. 2012. A biphasic scaffold design combined with cell sheet technology for simultaneous regeneration of alveolar bone/periodontal ligament complex. Biomaterials 33: 5560–73.

Vuong, C. and Otto, M. 2002. *Staphylococcus epidermidis* infections. Microbes Infect. 4: 481–489.

Whang, K., Tsai, D., Nam, E., Aitken, M., Sprague, S., Patel, P. et al. 1989. Ectopicbone formation via rhBMP-2 delivery from porous bioabsorbable polymer scaffolds. J. Biomed. Mater. Res. 42: 491–499.

Wikesjö, U.M., Nilvéus, R.E. and Selvig, K.A. 1992. Significance of early healing events on period ontal repair: A review. J. Periodontol. 63: 158–165.

Woo, B.H., Fink, B.F., Page, R., Schrier, J.A., Jo, Y.W., Jiang, G. et al. 2001. Enhancement of bone growth by sustained delivery of recombinant human bone morphogenetic protein-2 in a polymeric matrix. Pharm. Res. 18: 1747–1753.

Woodruff, M.A. and Hutmacher, D.W. 2010. The return of a forgotten polymer: polycaprolactone in the 21st century. Prog. Pol. Sci. 35: 1217–56.

Yamamoto, M., Takahashi, Y. and Tabata, Y. 2003. Controlled release by biodegradable hydrogels enhances the ectopic bone formation of bone morphogenetic protein. Biomaterials 24: 4375–4383.

Zhang, S., Huang, Y., Yang, X., Mei, F., Ma, Q. and Chen, G. 2009. Gelatin nanofibrous membrane fabricated by electrospinning of aqueous gelatine solution for guided tissue regeneration. J. Biomed. Mater. Res. A 90: 671–9.

Zhang, Y., Miron, R.J., Li, S., Shi, B., Sculean, A. and Cheng, X. 2015. Novel meso porous bioglass/silk scaffold containing ad PDGF-B and ad BMP 7 for the repair of periodontal defects in beagle dogs. J. Clin. Periodontol. 42: 262–271.

Zhang, Y., Rossi, F., Papa, S., Violatto, M.B., Bigini, P., Sorbona, M. et al. 2016. Non-invasive *in vitro* and *in vivo* monitoring of degradation of fluorescently labeled hyaluronan hydrogels for tissue engineering applications. Acta Biomater. 30: 188e98.

Zhao, L., Chu, P.K., Zhang, Y. and Wu, Z. 2009. Antibacterial coatings on titanium implants. J. Biomed. Mater. Res. Part B Appl. Biomater. 91: 470–480.

Zilberman, M. and Elsner, J.J. 2008. Antibiotic-eluting medical devices for various applications. J. Controll. Release. 130: 202–215.

Zimmerli, W., Trampuz, A. and Ochsner, P.E. 2004. Prosthetic-joint infections. N. Engl. J. Med. 351: 1645–1654.

11

Dentin-Pulp Complex Regeneration

Amaury Pozos-Guillén and *Héctor Flores**

Introduction

In this chapter we will focus on presenting new strategies for local regeneration therapy of the dentin-pulp complex; such a process is mediated by odontoblast and relates to preservation of pulp vitality, and clinical management of deep caries, necrotic pulp with apical periodontitis necrotic and immature permanent teeth; also, to understand factors involved in repairing of the damaged pulp and to review the current knowledge of the potential beneficial effects derived from the interaction of dental materials with the dentin-pulp complex as well as potential future developments.

Recent studies in the area of prevention, diagnosis and treatment of pulpal and periradicular disease has led to an increasing interest into the role of the dentin-pulp complex and its ability to repair itself and regenerate tissues. Regenerative endodontics should be considered as two concerns: local dentin-pulp complex regeneration and regenerative endodontics; the first, also called dentin/odontoblast complex regeneration, relates to preservation of pulp vitality and pulp capping; the latter relates to regeneration of vital tissue within an empty root canal space.

Historically, pulp capping and dentin bridge formation induction were reported since the 30s, with the first experimental studies performed by Zander in 1939 (Zander 1939).

Attempts to regenerate pulp tissue were carried out in the 60s and 70s without success. Studies focusing on the formation of fibrous connective tissue inside the root canal space have been reported by Ostby and Nygaard-Ostby and Hjortdal. They determined that filling the root canal space with a blood clot could lead to regeneration of pulp tissue. Generation of a disorganized soft connective tissue was observed. Histological examination of extracted teeth revealed that fibrous connective tissue and cellular cementum were formed in the apical portion of the root canal space when the teeth previously contained vital pulp tissue (Ostby 1961; Nygaard-Ostby and Hjortdal 1971).

Publications related to regenerative endodontics have increased significantly in the last decade. In an electronic search in PubMed, with appropriate MeSh terms including 'regenerative endodontics', 259 studies of potential relevance were identified (18 April 2019). The first case reports of 'revascularization' were reported in 2001 and 2004. Successful clinical outcomes in teeth with

Basic Sciences Laboratory, Faculty of Dentistry, Autonomous University of San Luis Potosí. 2 Manuel Nava, Zona Universitaria, 78290; San Luis Potosí, S.L.P., México.
  Email: apozos@uaslp.mx
* Corresponding author: heflores@uaslp.mx

pulp necrosis were reported without the conventional obturation of the root canal with gutta-percha or bioceramic materials. These studies defined the direction of the investigation in this topic. From these statistics, it becomes immediately clear that these two conditions remain a significant public health problem and require better strategies for disease prevention and clinical management (Iwaya et al. 2001; Banchs and Trope 2004).

Clinical Setting

Clinically, there are two scenarios that clinicians must handle correctly and that are the most common diseases of the pulp. First, when the dental pulp is still vital and potentially inflamed; in these cases, the main objective is to maintain pulp vitality. The treatment strategy will be focused on locally regenerate dentin and promote reorganization of the underlying connective tissue. In the second clinical scenario, there is a complete loss of pulp tissue, due to cell and tissue death in response to infection and subsequent bacterial invasion and uncontrolled inflammation. In this condition, the strategy aims to generate new vital connective tissue, imitating the original dental pulp.

Caries is the most common disease worldwide. The Global Burden of Disease Study 2016 estimated that oral diseases affected at least 3.58 billion people worldwide, and 2.4 billion people suffer from caries of permanent teeth and 486 million children suffer from caries of primary teeth. Data in USA estimate that 92% of adults between 20 to 64 years have had dental caries in their permanent teeth and 26% have untreated decay with an average of 3.28% decayed or missing permanent teeth (National Institute of Dental and Craniofacial Research, 18 April 2019). Epidemiological studies estimate the annual incidence of dental trauma at about 4.5%. Approximately one-fifth of adolescents and adults (permanent teeth) sustained a traumatic dental injury (commonly involved the maxillary central incisors) (Lam 2016). The American Association of Endodontists has estimated that 22.3 million endodontic procedures were performed annually by different causes (American Association of Endodontists, 18 April 2019).

This condition has a clinical impact since an early loss of a permanent tooth in young patients has consequences ranging from aesthetic problems, alterations of the function and bone development of the jaws, problems with phonetics, respiration and mastication, to severe effects of the psychosocial development of the patients. Significant advances have been made in the field of caries management, leading to a better understanding of the mineralization process of the teeth and the biological behavior of the dentin-pulp complex. It is evident that the dentin-pulp complex is able to adapt to a variety of stimuli that generate defense responses to maintain its vitality, and the main role of the dentin-pulp complex is to form defense dentin. It is a new paradigm advocating the complete replacement of compromised tissue, based on tissue engineering rather than traditional restoration.

Basic Concepts

Dentin and pulp tissues are specialized connective tissues derived from ectomesenchymal cells, formed from the dental papilla of the tooth bud.

Dental Pulp

Dental pulp is a highly innervated and vascularized connective tissue of mesenchymal origin, confined within dentin and enamel; it is located in the center of a tooth and is mainly made of living pulp cells, odontoblasts, immune system cells, neurons, endothelial cells and extracellular matrix. Dental pulp has a blood vessel system to deliver nutrients and clear waste products, and a special neural system to offer protection against harmful stimuli. The immune system made by dendritic cells, macrophages and T-lymphocytes are responsible to protect teeth from microorganisms and other foreign antigens (Pashley 2002).

Dentin

Dentin is a mineralized tissue that forms the bulk of the crown and root of the tooth, giving the root its form; it surrounds coronal and radicular pulp, forming the walls of the pulp chamber and root canals; its composition is approximately 67% inorganic, 20% organic and 13% water (Glossary of Endodontic Terms 2016).

Dentin can be classified as primary, secondary or tertiary, depending on when it was formed. Primary dentin is the regular tubular dentin formed before tooth eruption, including mantle dentin. Secondary dentin is the regular circumferential dentin formed after tooth eruption, whose tubules remain continuous with that of the primary dentin. Tertiary dentin is the irregular dentin that is formed in response to abnormal stimuli, such as excess tooth wear, cavity preparation, restorative materials and caries (Cox et al. 2002).

Dentin-pulp Complex

The main internal part of the tooth under the enamel layer in the crown is the dentin-pulp complex, involving the whole tooth root covered by a thin layer of cement. The pulp is a unique tissue, which is a soft tissue of mesenchymal origin with specialized cells, the odontoblasts, arranged peripherally in direct contact with dentin matrix. The close relationship between odontoblasts and dentin, is referred to as the dentin-pulp complex. The structural integrity and isolative characteristics of the tooth are kept by highly mineralized dentin that encloses the pulp chamber and root canals; thus dentin and pulp should be considered as a functional entity made up of histologically distinct constituents (Goldberg and Lasfargues 1995).

Different stimuli and aggressions to dental tissues represent a special situation, since the pulp tissue is surrounded by non-flexible, mineralized walls. Also, the tubular structure of dentine allows permeability, which grants diffusion of bacterial metabolites, and the degradation products of the matrix produced by carious processes, which causes a response of the pulp cells, originating molecular events in response to this damage.

Dentin-pulp Complex Damage Response

When the dentin-pulp complex is injured by caries, trauma, chemical stimuli or other aggressors, the dentin-pulp complex responds dynamically as a functional unit to protect the pulp tissue, either through the formation of sclerotic dentin, calcification of the dentin tubules and/or promoting the formation of reparative dentine by the pulpal odontoblasts, which rapidly deposit dentin (Pashley 2002).

Dentin has a strong relationship with pulp tissue through the odontoblastic process. Dental caries and trauma generate cellular and molecular responses in the pulp that can cause inflammatory and/or regenerative events at tissue and cellular levels (Goldberg et al. 2011).

The biological response of the dentin-pulp complex to different harmful stimuli is a complex interrelation between the aggression, defense mechanisms and the regeneration process. These factors are addressed independently, and the relation between factors and their relative balance is essential to determine tooth vitality. The balance of these events could directly impact the nature and capacity of any regeneration process mediated by the dentin-pulp complex (Smith 2003).

Pulp cells and odontoblasts play dynamic roles in the regeneration of damaged dentin, as a protective physical defense in the removal of exogenous stimuli by depositing tertiary dentin on the pulp chamber surface. When dentine is invaded by pathogens and their products, the first pulp cells to act are odontoblasts. These cells, located at the dentin-pulp interface with their long cellular process embedded in dentin tubules, represent the first line of defense (Durand et al. 2006).

Odontoblasts may be involved in combating bacterial invasion and activating innate and adaptive aspects of dental pulp immunity. This recognition happens through the detection of molecular structures shared by pathogens and are essential for microorganism survival (Veerayutthwilai et al. 2007).

Dentin-pulp Complex Response to Dental Caries

Dental caries is a chronic infectious disease mediated by a complex and dynamic bacterial biomass that affects the mineralized tissues of the tooth. Bacterial invasion of dentinal tubules has been described as the main cause of the inflammatory response of dental pulp (Cooper et al. 2010). The proliferation and metabolic activity of these microorganisms lead to the release of bacterial components into dentinal tubules and their diffusion towards the peripheral pulp (Cooper et al. 2011). Recognition of bacterial components by host cells at the dentin-pulp interface generates host defensive events including antibacterial, immune and inflammatory reactions and may eliminate early stages bacterial infection and block the route of its progression when accompanied by dentin formation at the dentin-pulp interface.

Dental caries may end up in pulpal necrosis and potential tooth loss if not treated. Three basic reactions protect the pulp against caries: (i) a decrease in dentin permeability, (ii) tertiary dentin formation, and (iii) inflammatory and immune reactions. These responses occur concomitantly, and their robustness is highly dependent on the aggressive nature of the advancing lesion (Smith 2002). There are different factors about the dental pulp response that distinguish it from other tissues in the body. Clinically and histologically, the pulpal response to aggressions has been well studied.

Different factors limit the possibility of pulp tissue regeneration. The dental pulp has the least collateral blood supply because of the anatomical features of the pulp chamber, and this leads to a disruption in the function of the immune system for infection control. Odontoblasts, also as post mitotic cells, have a restricted ability to proliferate. Stimulating odontoblast cells in promoting their secretory activity causes losses due to superficial caries, and this leads to dental restorative competence. Only pulp tissue can lead to the regeneration of dentin; however, the regeneration of pulp tissue is difficult, because the tissue is recapped in dentin with no collateral blood supply except from the root apical foramen. Overall, the dentin-pulp complex is responsible for dental health (Farges et al. 2015).

Deep Caries Management

Minimally invasive caries excavation techniques are not new. Different excavation methods to avoid pulp exposure have been previously proposed (Marending et al. 2016). The traditional 'invasive' approach is to fully excavate caries. When the pulp is not exposed during complete caries excavation, there is a high probability of success, and for the pulp to remain vital (Fitzgerald and Heys 1991).

Different studies describe minimally invasive techniques. The main clinical objectives are stopping caries progression and maintaining pulp vitality. Three different options of treatments have been recognized:

1) Caries-sealing method: caries is only removed from the enamel, leaving caries in the dentin.
2) Partial caries removal: a portion of caries close to the pulp is left, where two different techniques are described:
   - stepwise caries excavation, where the remaining caries is chemically treated and after a period of temporization (few weeks), it is excavated and completely removed;
   - indirect capping with an immediate definitive restoration, where the cavity is filled with a permanent restorative material.
3) Complete caries removal: the softened dentin is completely removed. In case of pulp exposure, there are three methods to maintain full or partial pulp vitality: direct capping; partial pulpotomy and complete pulpotomy. For all clinical procedures, the cavity is treated with a permanent restoration, which ensures peripheral sealing.

There are no objective clinical parameters to determine how much carious dentin should be removed; the question arises as to whether to cap the exposed pulp or perform a root canal treatment directly. A systematic review reported success in different longitudinal studies when a complete root canal treatment is performed (Ng 2007). Tooth with pulp exposure subsequent to caries excavation, the cost-benefit relation between a capping procedure, and root canal treatment could still be balanced or even favor pulpectomy (Schwendicke and Stolpe 2014). Completed root formation is a prerequisite for pulpectomy after pulp exposure. As an alternative to pulpectomy, pulpotomy offered a viable alternative to root canal treatment for teeth with vital pulps in short terms (Simon et al. 2013).

Some authors report that the maintenance of pulp vitality and the promotion of biologically based management strategies are the core of deep caries management. Pulp exposure can be avoided in radiographically deep caries and asymptomatic or mildly symptomatic teeth by selective removal of caries and restoration in one or two visits (Bjørndal et al. 2019; Duncan et al. 2019).

For all cases, disinfection of the dental tissue is mandatory for the health of the tooth, the subsequent interaction between dental tissue defense and repair is complex and the fine-tuning of the regulation of these processes is important for ensuring which response predominates when vital pulp tissue can be clinically retained or regenerated.

Local Response

After primary dentinogenesis and tooth formation, the odontoblasts are responsible for regeneration or healing of injuries in the form of tertiary dentinogenesis, which may continue after injuries and, provides the basis for the development of dentine bridges in pulp exposure sites. The formation of the dentinal bridge occurs after the death of the odontoblast near the site of the exposure, and the following differentiation of cells called odontoblast-like cells derived from pulpal stem cells or progenitor cells. This event requires a sequence of cellular actions, which include the recruitment of stem cells, cytodifferentiation and the activation and up-regulation of secretory activity of cells (Choung et al. 2016).

Different factors are involved in the initiation of tertiary dentinogenesis and could be related to harmful agents such as acids and bacterial metabolic products, or by leakage from the restorative material used to fill a cavity. Tertiary dentinogenesis has been described in relation to the nature of the injury. This has led to adoption of terms like 'reactionary' and 'reparative' to subdivide tertiary dentinogenesis into the responses seen after survival and death of the primary odontoblast population, respectively. The reactionary dentinogenesis represents the focal up-regulation of a group of primary odontoblasts surviving injury to the tooth, while reparative dentinogenesis represents the response of tertiary dentin secretion by a new generation of odontoblast-like cells after death of the primary odontoblast cells (Smith et al. 2001).

Local regeneration of the dentin-pulp complex from residual dental pulp has been developed by researchers who are involved in clinical practice. Induction of appropriate pulp wound healing and formation of new dentin in tooth defects are mandatory for local repair of the dentin-pulp complex and to maintain vital pulp.

Tissue Engineering in Endodontics

A challenging problem for endodontists and pediatric dentists is the clinical management of immature permanent teeth with necrotic pulp resulting from infection or trauma (Albuquerque et al. 2014).

Here we will review the current knowledge of regenerative dentistry as an emerging concept that challenges modern dentistry to step up basic dental research and translate scientific knowledge into the future for clinical scenarios. This methodology is based on the knowledge of the essential mechanisms of tooth development and the biological processes of healing, repair or regenerating (engineering) the damaged tissue or organ (Angelova Volponi et al. 2018).

Since the beginning of the 20s, Davis was the first to recognize the importance of the integrity of the apical/periapical tissues in endodontic therapy (Gutmann 2016). As above mentioned (Ostby

1961; Nygaard-Ostby and Hjortdal 1971) produced the experimental evidence for Davis' clinically based observation and called attention to the need for biocompatible endodontic materials.

The traditional endodontic treatment for necrotic immature permanent teeth is calcium hydroxide apexification, which has antibacterial properties and can stimulate enzyme pyrophosphatase, facilitating repair mechanisms. However, this therapy requires multiple visits over an extended period; which results in a delay of root canal obturation and placement of permanent restoration and may predispose the tooth to increased susceptibility of root fracture (Andreasen et al. 2002; Giuliani et al. 2002).

Another type of apexification named 'apical MTA plug' was described using Mineral Trioxide Aggregate (MTA). MTA is a repair material made of fine hydrophilic particles of tri/dicalcium silicate, tricalcium aluminate, tricalcium oxide and silicate oxide (Parirokh and Torabinejad 2010). MTA is placed into the root canal space and acts as a mechanical barrier to prevent coronal leakage and penetration of microorganisms. Some disadvantages of this material are difficulty to manipulate, the possibility of tooth discoloration and difficulty to remove from the root canal. However, neither calcium hydroxide nor MTA barrier technique allow further root growth in length, maturation of the apex or root wall thickening. New calcium silicate-based materials have recently been developed with the purpose of improving clinical use and overcoming MTA limits. Biodentine™ is a bioceramic made of tricalcium silicate, dicalcium silicate, zirconium oxide, calcium carbonate, calcium oxide and iron oxide. It is mixed with a hydrosoluble polymer and calcium chloride to decrease the setting time (Rajasekharan et al. 2014). This biomaterial has shown reduced setting time with interesting physical and biological properties as a dentine restorative material (Koubi et al. 2013; Topçuoğlu and Topçuoğlu 2016).

It has described a procedure that allows complete root development of immature permanent teeth with necrotic pulp/apical periodontitis. This procedure suggests the use of a combination of an antimicrobial paste and irrigants, no canal walls instrumentation, induced apical bleeding to form a blood clot, and a tight seal into the root canal to promote healing, offering results in terms of penetration through dentine and antibacterial efficacy of the drug combination when the drugs are placed in root canals (Takushige et al. 2004; Hwang et al. 2018; Arruda et al. 2018; Zancan et al. 2019; Zargar et al. 2019; Fundaoğlu Küçükekenci et al. 2019; McIntyre et al. 2019; Sadek et al. 2019). The application of antibacterial drugs may represent a way to eradicate bacteria during root canal treatment; however, this local application has no effect on tissue regeneration by itself.

Regenerative Endodontics

The main goal of regenerative endodontics is the use of biologic-based procedures to arrest the disease process, preventing its recurrence while favoring the repair or replacement of damaged structures of the dentin-pulp complex (Diogenes and Ruparel 2017). The term regenerative endodontics was introduced in clinical endodontics, which also included revascularization and revitalization to describe the treatment of immature permanent teeth with necrotic pulp.

There is considerable discussion on the use of the term 'regeneration' because there is convincing evidence from histologic studies that the newly formed tissue following current forms of regenerative endodontic procedures does not resemble the lost dentin-pulp complex. Repaired tissue that promotes resolution of the disease and re-establishment of some or all the original tissue functions should be an acceptable goal (Simon et al. 2014; Diogenes and Ruparel 2017; Song et al. 2017).

Regenerative endodontics comprises both vital and non-vital pulp treatments. Specifically, non-vital treatments include procedures to promote new vital tissue formation after necrosis following infection. The regenerative endodontic treatment is an alternative to a conventional endodontic treatment (Murray et al. 2007).

Some terms have been used to identify clinical procedures in this field:

Regenerative endodontics. Biologically-based procedures designed to physiologically replace damaged tooth structures, including dentin and root structures, as well as cells of the dentin-pulp complex.

Revascularization. The restoration of blood supply (Glossary of Endodontic Terms 2016).

Revitalization. An ingrowth of tissue that may not be the same as the original lost tissue (Wang et al. 2010).

The clinical procedures and results of regenerative endodontics are very different from conventional endodontic therapy; that has created an interest in the field of endodontics in recent years. Immature permanent teeth with necrotic pulp/apical periodontitis are traditionally treated with apexification treatment using calcium hydroxide or apical MTA plugs to induce formation of an apical hard tissue barrier before root canal filling. The calcium hydroxide apexification procedure usually takes multiple visits over an extended period, which could increase the risk of root fracture. However, an apexification procedure has no potential to restore the vitality of damaged tissue in the canal space and promote root maturation of immature permanent teeth with necrotic pulp. In the year 2001, Iwaya et al. reported a clinical case in a necrotic immature mandibular second premolar with periapical involvement in a 13-year-old patient; as an alternative to conventional root canal treatment protocol and apexification, antimicrobial agents were used in the root canal. Radiographic examination showed the start of apical closure 5 months after treatment and a thickening of the canal wall and complete apical closure was confirmed 30 months after the procedure, suggesting a possible revascularization potential into the root canal space (Iwaya et al. 2001). Other reports on this new technique showed induced root maturation in infected immature teeth and described the use of a blood clot into the root canal as 'revascularization' (Branchs and Trope 2004). On the other hand, the term 'revitalization' has been suggested as it describes non-specific vital tissue rather than just blood vessels as implied by the term 'revascularization' (Lenzi and Trope 2012).

Regenerative Endodontic Procedures (REPs)

Regeneration is defined as reconstitution of damaged tissues by a tissue similar to the original tissue and restoration of biological functions. Repair is the replacement of the damaged tissue by tissue different from the original tissue and consequently, the loss of the original biological functions. The dental pulp has a limited potential of regeneration (Shi et al. 2005).

REPs are defined as "biologically based procedures designed to replace damaged structures, including dentin and root structures, as well as cells of the dentin-pulp complex" (Murray et al. 2007). This term has been widely accepted and refers to all procedures that aim to achieve organized repair of the dental pulp and include future therapies yet to evolve in the field of regenerative endodontics (Diogenes et al. 2017).

REPs are bioengineering treatments that aim to restore the normal physiological functions of the dental pulp, including innate pulp immunity, pulp repair through mineralization (tertiary dentin) and pulp sensibility (sensation of occlusal pressure and pain). These therapeutic techniques include a triad of elements that is integrated by stem cells, growth factors and biomaterials or scaffolds. The successful endodontic regeneration requires synchronized effects of infection control, biomaterials and stem cells (Cao et al. 2015). In regenerative endodontic procedures, growth factors embedded in the dentin matrix are released into the canal space after a cleaning protocol. However, the mesenchymal stem cells introduced into the canal space during REPs do not appear to be able to differentiate into odontoblast-like cells and produce the dentin-pulp complex in animal and human studies (Lovelace et al. 2011). Regenerative endodontics implies that further root maturation results in reestablishment of the dentine-pulp complex. REPs, suggesting repair rather than regeneration (Wang et al. 2010; Becerra et al. 2014).

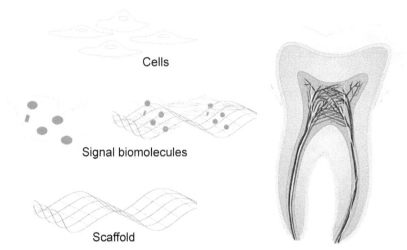

Cells

Signal biomolecules

Scaffold

Figure 11.1. Local regeneration of the dentin-pulp complex by tissue engineering strategies: Stem cells, signal biomolecules and scaffolds.

In recent years, regenerative therapies in endodontics have gained attention as a treatment for teeth without a complete root development affected by caries or trauma. The success of these clinical procedures has offered new treatment alternatives to traditional endodontic therapy. With advances in tissue engineering, the use of stem cells, and the progress made in biomaterials, it is possible to consider including regeneration procedures in daily endodontic clinical practice (Kim et al. 2016) (Fig. 11.1).

Tissue Engineering Strategies

In general, tissue engineering strategies include the evaluation of an appropriate scaffold for the regulation of cell differentiation, selection of growth factors that can promote stem cell differentiation, and an appropriate source of stem cell/progenitor cells (Hargreaves et al. 2013; Albuquerque et al. 2014) (Table 11.1). The success of tissue engineering in combination with tissue regeneration depends on the behavior and cellular activity in the biological processes developed within a structure that functions as a support, better known as scaffolds or directly at the site of the injury. The cell-cell and cell-biomaterial interaction are key factors for the induction of a specific cell behavior, together with bioactive factors that allow the formation of the desired tissue (Ortiz et al. 2019) (Fig. 11.2).

Stem Cell/Progenitor Cells

Stem cells are defined as clonogenic cells capable of both self-renewal and multilineage differentiation since they are thought to be undifferentiated cells with varying degrees of potency and plasticity (Gronthos et al. 2002).

All tissues originate from stem cells, which play an indispensable role in embryonic development and tissue regeneration. These cells are capable of self-renewal, proliferation, and differentiation into multiple mature cell types. Stem cell potency describes the potential of the cell to divide and express different cell phenotypes. Totipotent stem cells are able to divide and produce all the cells in an individual. Pluripotent stem cells have not completely divided and can become many cells, but not all lineages. They are able to differentiate into any of the three germ layers: endoderm, mesoderm or ectoderm, where the progeny has multiple distinct phenotypes, whilst multipotent stem cells can differentiate into cells from multiple, but a limited number of lineages (Robey 2000).

Table 11.1. Current techniques in regenerative endodontic.

Technique	Objective
Root canal revascularization	Open up the tooth apex to allow bleeding into root canals
Stem cells therapy	Stem cells are delivered to teeth via injectable matrix
Injectable scaffold	Pulp cells are seeded into a 3D scaffold made of polymers and injectable implanted
3D cell printing scaffold	Ink-jet like device dispenses layers of cells in a hydrogel which is surgically implanted
Gene therapy	Mineralizing genes are transfected into the vital pulp cells of necrotic and symptomatic teeth

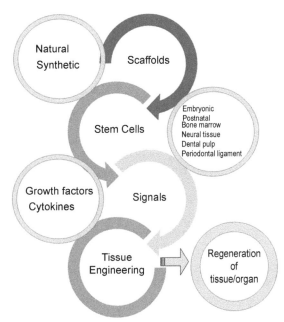

Figure 11.2. Key components in tissue engineering: Scaffold, stem cells, and signal biomolecules.

There are two types of stem cells: embryonic and postnatal. Embryonic stem cells are pluripotent cells capable of differentiating into any cell type as well as maintaining an undifferentiated state. These cells are plastic and have the capacity to develop into various specialized cell types with an enormous potential for tissue regeneration. Postnatal stem cells have been isolated from various tissues including bone marrow, neural tissue, dental pulp and periodontal ligament. These are multipotent stem cells capable of differentiating into more than one cell type, but not all cell types (Antoniou 2001).

Studies of stem cells of dental pulp has led to a significant change in our understanding of the mechanisms involved in the preservation of dental pulp homeostasis in health and in the pulp response to damage. These cells are related to the physiology of the dental pulp tissue. Also, it has been suggested that stem cells are involved in the regulation of pulp angiogenesis in response to caries. Recently, the potential use of stem cells in dentin-pulp complex engineering has increased the interest in regenerative endodontics.

The first cellular lines established of Mesenchymal Stem Cells (MSC) obtained from different structures of teeth were Dental Pulp Stem Cells (DPSC) and Stem cells from Human Exfoliated Deciduous teeth (SHED) reported by Gronthos et al., at the National Institutes of Dental and Craniofacial Research, both with high potential for differentiation into other cell lineages (Gronthos et al. 2000; Gronthos et al. 2002; Miura et al. 2003). To date, four types of human dental stem cells

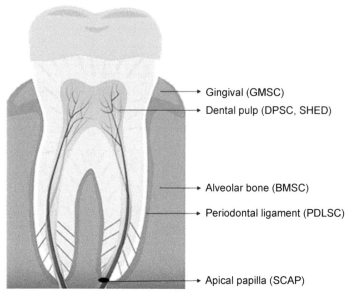

Figure 11.3. Types of human dental stem cells and sites of isolated.

have been isolated; DPSC, SHED, Stem Cells from Apical Papillae (SCAP), and Periodontal Ligament Stem Cells (PDLSC). DPSC, SHED, SCAP are derived from neural crest mesenchyme (Fig. 11.3).

DPSC show a higher proliferation capacity compared to osteogenic cells and have the ability to differentiate into odontoblast-like cells which express the early odontoblast cell marker, dentine sialophosphoprotein and can form a dentine–pulp complex when transplanted *in vivo*. DPSC are capable of generating new stem cells or multilineage differentiation into odontoblasts, adipocytes and neural-like cells, suggesting a hierarchy of progenitors within the pulp, including a small population of stem cells amongst a larger population of more committed cells (Gronthos et al. 2002).

The fraction of DPSC in the dental pulp is small, approximately 1% of the total cells (Smith et al. 2005), and the effect of aging reduces the cell pool available to participate in regeneration which reflects the better healing outcomes seen in younger patients (Huang et al. 2008).

SHED, mesenchymal stem cells isolated from the dental pulp of exfoliated deciduous teeth, were recognized to have a high proliferative rate and capability of differentiating into various cell types, including neural cells, adipocytes and odontoblasts. These cells are distinct from DPSC regarding their higher proliferation rate, increased cell population doublings, viability, osteoinductive capacities, failure to reconstitute a dentin-pulp complex and markers (Miura et al. 2003; Koyama et al. 2009; Nakamura et al. 2009) (Table 11.2).

The transplantation of MSC obtained from dental tissues for therapeutic purposes in the dental area is a method that has been performed experimentally in animal models, demonstrating high predictability of success; however, there is still little evidence to bring these experimental protocols to controlled clinical trials in humans. Although results show that dental pulp cells have the capacity to repair damaged dental tissue, when it comes to *in vivo* tooth repair, the dental reparative capacity is limited (Gronthos et al. 2000). It has been shown that dental pulp stem cells can regenerate a pulp-like tissue in root canals *in vivo*. The tissue formed had functional odontoblast-like cells able to deposit a mineralized matrix on the root canal walls (Huang et al. 2010). The advances on dental pulp tissue engineering are geared towards the generation of a viable and healthy pulp throughout the entire root canal length. SHED cells are able to attach to the dentin walls and proliferate inside root canals *in vitro* (Gotlieb et al. 2008).

The use of stem cells for the regeneration of pulpal tissues is promising, their use in a clinical situation to induce apexification is incipient and unpredictable; however, advances in the study

Table 11.2. Postnatal dental stem cells studied.

Type	Isolation	Marker	Possible applications
Dental pulp stem cells (DPSC)	Dental pulp	STRO-1 CD44 CD46	Osteogenesis and dentinogenesis inductors; without ability to produce dentin-pulp complex regeneration
Stem cells from human exfoliated deciduous teeth (SHED)	Exfoliated deciduous teeth and coronal pulp	CD146 STRO-1 CD44	Osteoinductive capacities; but failure to reconstitute a dentin-pulp complex regeneration
Periodontal ligament stem cells (PDLSC)	Root of extracted teeth	CD90 CD73 CD105 CD29	Capacity for tissue regeneration and periodontal repair
Stem cells from the apical papilla (SCAP)	Impacted third molars	CD146 STRO-1 CD29	Higher proliferative rate and effective for tooth formation
Stem cells from the dental folicle (DFSC)	Follicles of impacted third molars	CD13 CD29 CD59 CD90 CD105	Capacity to differentiate into osteoblasts and cementoblasts, adipocytes and neurons

and understanding of pulp regeneration processes could make this clinical procedure a viable and predictable treatment in a few years.

Growth Factors

Growth factors are proteins that regulate many aspects of cellular function, including survival, morphogenesis, proliferation, migration, apoptosis, differentiation and secretory processes of cells. Growth factors and cytokines are polypeptides or proteins that bind to specific receptors on the surface of target cells; they may act as signaling molecules that modulate cell behavior by mediating intracellular communication. It has been reported that growth factors, as well as bone morphogenic proteins, are essential for tissue engineering in endodontics. Growth factors may be released from the dentin matrix as a result of both injury events to the tissues and clinical restorative procedures. Also, growth factors may be molecules in the signaling of reactionary and reparative dentinogenesis processes (Lind 1996; Smith 2003).

In vivo and *in vitro* studies report that placement of exogenous growth factors, particularly TGF-β and Bone Morphogenetic Proteins (BMPs), on exposed pulps, have demonstrated the potential of these molecules to signal reparative dentinogenic events. In application of growth factors to exposed pulps in capping conditions, growth factors stimulated reparative responses, but the reparative dentin matrix secreted showed variable structure ranging from a tubular matrix like physiological dentin, to atubular osteodentin-like matrices (Rutherford et al. 1994; Nakashima 1994a; Hu et al. 1998). The TGF-β family of growth factors as well as different components of the matrix have been demonstrated as chemotactocos for mesenchymal cells and the migration of these cells to the sites of damage. TGF-β1 regulates a wide range of cellular activities, such as cell migration, cell proliferation, cell differentiation and extracellular matrix (ECM) synthesis. TGF-β1 has been shown to increase cell proliferation and production of the ECM in dental pulp tissue culture, and promotes odontoblastic differentiation of dental pulp cells (Massagué et al. 2000; Verrecchia and Mauviel 2002; Lambert et al. 2011). BMPs include a subgroup of the TGF-β superfamily and are involved in biological activities such as cell proliferation, differentiation and apoptosis. BMPs have osteoinductive and chondrogenic effects. More than 20 BMPs have been identified and characterized; its activity is regulated by the antagonists of BMPs such as noggin and chordin; this modulation has a critical role in tooth development (Chen et al. 2004). BMP2, BMP4, BMP7 and BMP11 are of clinical significance due to their role in inducing mineralization. Bovine dental pulp cells treated with BMP2 and BMP4 differentiate into pre-odontoblasts (Nakashima et al. 1994b). Dentin sialophosphoprotein expression and odontoblastic differentiation are regulated via BMP2. It also stimulates the differentiation of dental pulp stem/progenitor cells into odontoblasts *in vivo* and *in vitro* (Iohara et al. 2004; Chen et al. 2008). BMP7 or osteogenic protein-1, promotes dentin formation when placed over amputated dental pulp in animal models (Rutherford et al. 1994; Rutherford and Gu 2000; Six et al. 2002). Dental pulp cells transfected with BMP11 or GDF11, promotes mineralization. Dentin matrix protein 1, ALP, DSPP, enamelysin and phosphate-regulating gene are expressed in BMP11-transfected cells. Transplantation of BMP11-transfected cell pellets induces formation of dentin-like tissue on amputated dental pulp in an animal model (Nakashima et al. 2004).

The spatial proximity between odontoblasts and blood vessels suggests the existence of an active exchange of signaling molecules during dentinogenesis. In this sense, Vascular Endothelial Growth Factor (VEGF) is a heparin-binding protein with specific affinity to endothelial cells and plays an essential role in angiogenesis. The functions of VEGF include the proliferation of endothelial cells and their enhanced survival, stimulating neovascularization in the area of injury (Nör et al. 1999). Also, VEGF control the vascular permeability during physiological and pathological events and is expressed in dental pulp tissues of teeth undergoing caries-induced pulpitis (Ferrara et al. 2003; Guven et al. 2007). In addition, VEGF increases the proliferation and osteogenic differentiation of dental pulp cells under osteogenic conditions (D' Alimonte et al. 2011).

VEGF participates in the dentinogenesis by inducing the vascularization required to sustain the high metabolic demands of odontoblastic cells in active processes of dentin matrix secretion. Further studies about the effect of VEGF in the dental pulp tissue are necessary in order to understand the impact of this growth factor to tissue damage and tissue regeneration.

Growth factors regulate either transplanted cells or endogenously cells in dentin-pulp complex regeneration. Three therapeutic approaches have been proposed for the regeneration of dental tissue where these growth factors are used, the complete regeneration of a tooth, the local regeneration of the dentin-pulp complex in a dentin defect and the regeneration of the dental pulp by the apical pulp or periapical tissues, including the periodontal ligament and bone (Kitamura et al. 2012).

Scaffolds

According to the American Association of Endodontists, scaffold is defined as "a lattice that provides a framework for stem cells to grow for pulpal regeneration" (Glossary of Endodontic Terms 2016). The American Society for Testing Materials defines a scaffold as "the support, delivery vehicle, or matrix for facilitating the migration, binding, or transport of cells or bioactive molecules used to replace, repair, or regenerate tissues" (ASTM Standard F2312-11).

To avoid immune responses, scaffolds should be synthesized from biocompatible and biodegradable materials. The adequate design of scaffolds is necessary in order to ensure mechanical integrity and functionality and, the surface needs to have the precise morphology for promoting cell adhesion and differentiation. Conditions for an ideal scaffold include chemical stability and physical properties, matching the surrounding tissues for cell compatibility, adhesion function, cell proliferation, controlled degradation and mechanical strength. A critical factor for an appropriate scaffold is the selection of biomaterials that fully or partially mimic the ECM of the tissue to be replaced (Zhang et al. 2013).

Different biomaterial scaffolds have been developed as ECM analogs with the ability to keep cell attachment, and producing engineered tissue or organs. Scaffold materials in endodontic regeneration include natural polymers such as collagen, protein, chitosan, alginate, hyaluronic acid and their derivatives. These scaffolds have been applied for restoring and regenerating dental tissue due to their properties, such as biocompatibility, bioactivity and tissue construct for cell growth and differentiation. Furthermore, there are some synthetic polymers among scaffold materials, such as polyglycolic acid (PGA), polylactic acid (PLA), polyglycolic acid-poly-l-lactic acid (PGA-PLLA) and poly-lactic polyglycolic acid (PLGA) (Hashemi-Beni et al. 2017).

Biological (natural) Polymer Scaffolds

Collagen is a major protein that can be found in sinew, cartilage, bone and skin. One of the most important advantages of collagen is that it can be processed into a variety of set-ups, as porous sponges, gels and sheets, and can also be crosslinked with chemicals to make it stronger or to alter its degradation rate (Nasir et al. 2006). Collagen sponges have demonstrated to be similar to the structure of an extracellular matrix, they have low immunogenicity and cytotoxicity, and also, the ability of forming several shapes and stimulating the differentiation of osteoblasts (Silver and Pins 1991; Chevallay et al. 2000). Collagen sponge scaffolds and gels have been reported for tooth regeneration; results suggest that not only does collagen retain and support cell proliferation and differentiation, but also help in the production of calcified tissues (Sumita et al. 2006). Seeding Dental Pulp Stem Cells (DPSC) on a collagen scaffold, the collagen scaffold could stimulate a systematized comparable matrix formation similar to that of pulp tissue (Prescott et al. 2008). Collagen scaffold allows easy placement of cells and growth factors. It also allows for substitution by natural tissues after suffering degradation (Sumita et al. 2006; Yamauchi et al. 2011). However, the results are not always consistent; consequently, the characteristics of collagen scaffolds and gels require further investigation before being applied in human trials.

Chitosan is a natural polymer derived from chitin. It is a biodegradable natural carbohydrate biopolymer that has shown to improve wound healing and bone formation. It is non-toxic and non-immunogenic. Depending on the type of polymer used, once the percentage of deacetylation of chitin gets to approximately 50%, chitin transforms to chitosan, which is soluble in aqueous acidic media. Chitosan is biocompatible and biodegradable and is currently used with other polymers in a variety of tissue engineering applications (Madihally and Matthew 1999; Suh and Matthew 2000; Seol et al. 2004). The uniqueness of this scaffold is its layered macroscale bio-mimetic structure with tunable mechanical characteristics that supports movement of the two cell types in all directions (Ravindran et al. 2010). Viability by using BMP-7 gene-activated chitosan/collagen scaffolds for human dental pulp stem cells has been previously demonstrated. One of the potential clinical applications of this chitosan scaffold may be the regeneration of the dentin–pulp complex. The elasticity of chitosan/collagen scaffold supports the application of DPSC in pulp cavities to allow tissue generation. However, to progress to potential therapeutic application toward dentin regeneration, improved scaffold design in composition, as well as morphology, are essential.

Alginate is a natural polysaccharide obtained from brown seaweed and has several attractive physical properties such as biocompatibility, mildness of gelation conditions and low immunogenicity, making it a desirable material for many uses in biomedical engineering. Purified alginate has been extensively used in the food and pharmaceutical industries, as well as various biomedical, biomaterial and therapeutic applications. For example, alginate hydrogels have been considered in wound healing, drug delivery and tissue engineering applications to date. They are used to maintain structural similarity to extracellular matrices in tissues and other important applications with manipulation (Lee and Mooney 2012). The combination of alginate microspheres of calcium with soluble alginate solutions led to the construction of an injectable self-gelling alginate gel with macropores (pores in micrometer range) to use *in vivo* immunotherapy, which encouraged cellular penetration and provided ready access to microspheres, extending therapeutic factors implanted in the matrix (Hori et al. 2009). It also demonstrated to induce pulp cells differentiation into odontoblast-like cells and secreted tubular dentin matrix (Dobie et al. 2002). A simple technique to produce a collagen/alginate composite scaffold with geometry similar to a guttapercha point was developed to be used for endodontic regeneration following root canal treatment. The scaffold was seeded with stem cells from the apical papilla (SCAPs). This construct was characterized in terms of elastic modulus, shape preservation over time, cell viability and biocompatibility. This composite scaffold can be populated by MSC derived from the apical papilla. These cells are able to secrete a calcified ECM under osteogenic stimulation (Devillard et al. 2017).

Hyaluronic Acid (HA) is a naturally occurring polysaccharide that belongs to the glycosaminoglycan family and contains a basic unit of two sugars, D-glucuronic acid and N-acetylglucosamine. HA is synthesized in the inner cell membrane by hyaluronan synthases, a class of transmembrane proteins. After synthesis, it is removed through the membrane ECM, where it is degraded by the hyaluronidase enzymes family, after 3 to 5 days. This polysaccharide allows transportation of key metabolites and conserves tissue structure by binding to water; it also causes metalloproteinase inhibitors to activate, suppressing tissue decomposition, like inflammatory cytokines (Schwartz et al. 2007). HA is found in the ECM of all living tissues, with different concentrations and molecular weights, being most abundantly present in tissues subjected to mechanical loads, such as cartilages, dermis and vocal folds (Walimbe et al. 2017). HA and its derivatives can obviously help preserve tissue because of chemical and structural modifications under different circumstances. Dental pulp regeneration with sound dentin can be developed by mixing growth factors and a hyaluronic acid sponge as an implant for dental pulp regeneration because of its appropriate physical structure, biocompatibility and biodegradation. However, HA has some disadvantages, which include poor mechanical properties and rapid degradation *in vivo*. HA and HA-based materials are extensively used for the preparation of scaffolds for tissue engineering. These scaffolds offer the advantages of increased biocompatibility, controllable degradation rate by using a crosslinker, suitable porosity for cell encapsulation, differentiation and proliferation (Jafarkhani et al. 2016; de Sant'Anna et al. 2017; Şapte et al. 2017; Faisal and Kumar 2017).

Synthetic Polymer Scaffolds

A number of synthetic polymers such as polylactic acid (PLA), poly-l-lactic acid (PLLA), polyglycolic acid (PGA), PLGA and polyepsiloncaprolactone (PCL) have been used as scaffolds for pulp regeneration. The FDA accepts all materials as the biocompatible polymeric materials for drug delivery systems (Jain 2000; Mauth et al. 2007). Vacanti et al. reported for the first time the use of these polymers as matrices for cell transplantation (Vacanti et al. 1998). The synthetic polymers are nontoxic, biodegradable, and allow precise manipulation of the physicochemical properties such as mechanical stiffness, degradation rate, porosity and microstructure (Mao et al. 2012). Synthetic polymers are generally degraded by simple hydrolysis, when natural polymers are mainly degraded enzymatically (Gunatillake and Adhikari 2003). Due to their biocompatibility and wide range of reproducibility, they are candidates for tissue engineering protocols (Bohl et al. 1998; Nakashima 2005; Rezwan et al. 2006). The biocompatibility and biodegradability of these polymers have been shown by many in *in vitro* and *in vivo* studies. It can be said that the material degradation level leads to moderate inflammatory responses *in vivo* after a large amount of acidic degradation (Hutmacher 2000; Grayson et al. 2004; Tomson et al. 2007). PLLA is a very strong polymer and has found many applications where structural strength is important. Experiments were carried out by Cordeiro et al. and Sakai et al. showing PLLA scaffolds promoted dental pulp cell differentiation into endothelial cells and odontoblasts (Cordeiro et al. 2008; Sakai 2010). PGA has been used as an artificial scaffold for cell transplantation, and degrades as the cells excrete ECM. PLGA was used as a scaffold to demonstrate that dentin-like tissue was formed, and pulp-like tissue could be regenerated after 3–4 months (Huang et al. 2010). There are many advantages in using reproducible and biocompatible synthetic materials to control the mechanical and chemical characteristics, such as the structure, size, viscosity, porosity and degradation rate. The controlled release of the biodegradable PLA, PGA or PLGA system could locally affect the use of incorporated bioactive molecules and the expression of cell phenotype. The main disadvantage is that the synthetic polymers can cause a chronic or acute inflammatory host response, and localized pH decrease due to relative acidity of hydrolytically degraded byproducts (Chan and Mooney 2008).

Ceramic Scaffolds

This group of scaffolds refers to calcium/phosphate materials, bioactive glasses and glass ceramics (Sharma et al. 2014). Most common biomaterials in use are calcium phosphate-based (Ca/P) bioceramics. Many scaffold systems have been proposed for hard tissue engineering such as bone substitutes and in dental tissue repair as pulp-capping agents. Ceramics such as calcium phosphates (Ca/P) and bioactive glasses or glass ceramics are natural choices. Hydroxyapatite was one of the first biomaterial to be used as a scaffold. It may be derived from bovine bone or coralline or made out of a pure synthetic material. TCP is a naturally occurring material made of calcium and phosphorous and is used as a ceramic bone substitute. The Ca/P scaffolds contain b-tricalcium phosphate (b-TCP) or hydroxyapatite (Foroughi et al. 2012; Foroughi et al. 2013). The advantages of TCP and/or HA scaffold usage are related to produce mineral matrix of the bone and tooth; moreover, they can be manufactured synthetically and are used in medical applications; for example, in the therapeutic healing processes of bone defects in dental regeneration and maxillofacial surgery. In addition, since these materials are similar to naturally mineralized tissue, they are suitable as biocompatible and osteoinductive materials (Mauth et al. 2007). Ca/P scaffolds include β-TCP or HA and have been widely tested for bone regeneration due to their properties of resorption, biocompatibility, low immunogenicity, osteoconductivity, bone bonding and similarity to mineralized tissues. 3D Ca/P porous granules have proved to be useful in dental tissue engineering by providing favorable 3D substrate conditions for human Dental Pulp Stem Cell (hDPSC) growth and odontogenic differentiation. Addition of SiO_2 and ZnO dopants to pure TCP scaffolds increases its mechanical strength as well as cellular proliferation properties. In tissue engineering, the utilization of biocompatible and biodegradable synthetic scaffold reduced the negative side effect of foreign materials where the materials are easy to handle, and no immunogenic reaction has been reported so far as the materials are degradable.

Various techniques have been used to manufacture two- and three-dimensional scaffolds. The main technique to fabricate two-dimensional scaffolds is electrospinning; whereas the main techniques to fabricate three-dimensional scaffolds include solvent casting, freeze drying, particle/salt leaching, chemical/gas foaming, thermally induced phase separation and the foam-gel technique (Loh and Choong 2013; Lu et al. 2013; Park et al. 2015; Gómez-Lizárraga et al. 2017; Ortiz et al. 2017; Del Bakhshayesh et al. 2018; Granados-Hernández et al. 2018; Vázquez-Vázquez et al. 2018; Xu et al. 2018). These techniques have some limitations to yield scaffolds with specific micro-architectures in terms of porosity, pore size, pore geometry and interconnectivity. They are still being used because of their low cost and minimal equipment complexity. Besides, due to their manufacturing conditions, these techniques do not allow including living cells or soluble factors within the process. Additive manufacturing techniques have arisen as a solution to these disadvantages. The most accepted of these techniques include: stereolithography, selective laser sintering, fused deposition modeling and three-dimensional printing (Moreno Madrid et al. 2019).

The appropriate combination of the growth factors, the delivery system of the growth factor and well-designed scaffolds inducing stem cells and blood vessels are necessary to establish local regeneration of the dentin-pulp complex. In addition, the facility to introduce the cells into the root canal using a biocompatible and biodegradable scaffold with appropriated strategies will be critical for the future use of tissue engineering therapies in clinical regenerative endodontics.

Perspectives and Clinical Implications

The objective of regenerative endodontics is to re-establish normal pulp function in necrotic and infected teeth that would result in re-instatement of protective functions. This is achieved by the clinical application of tissue engineering principles with the goal of achieving maximum disinfection while creating the most conducive environment for stem cells to direct the repair and regeneration of the target tissue (Cao et al. 2015).

One crucial step of clinical evidence is to critically consider and use the primary articles about therapy, i.e., randomized clinical trials, the study design that best addresses the questions related to this clinical field; a process that involves assessing the reliability of results, risk of bias (internal validity) and applicability of reported clinical findings (external validity). So, controlled clinical trials would provide the best evidence to try and answer the question; however to date, there are no well-designed randomized clinical trials with sufficient follow-up, with an appropriate sample size that demonstrates the total success of the so-called regenerative endodontic procedures in necrotic and immature permanent teeth. There are a few randomized clinical trials with short periods of follow-up. Most studies are case reports and case series; however, these designs represent the lowest level of evidence.

There are different factors that may hinder the design and implementation of a controlled clinical trial, and that should be considered:

- Adequate diagnosis with well-defined selection criteria. Regenerative endodontic procedures are recommended for teeth with a necrotic pulp and an immature apex. However, some immature permanent teeth with necrotic pulps may be suitable for regenerative endodontic procedures, whilst others suitable for apical MTA sealing and root canal filling.
- Adequate and standardized treatment that can be applied in the same way in all patients included with appropriate controls; stem cell with/without growth/differentiation factors, etc.
- Adequate and standardized response variable that can be measured in the same way in all the patients included. Studies need to quantitatively report the defined outcomes (pulpal regeneration; dentinal regeneration; vascular regeneration; neuronal regeneration). The American Association of Endodontists (2016) clinical considerations for regenerative endodontic procedures define success by three measures: (i) Primary goal: The elimination of symptoms and the evidence of bone healing; (ii) Secondary goal: Increased root wall thickness and/or increased root length (desirable); (iii) Tertiary goal: positive response to vitality testing (which if achieved, could indicate a more organized vital pulp tissue).
- Appropriately manage the basic design elements such as the initial selection of patients, randomization, follow-up and blinding, so that the results can be attributed to the treatment and not susceptible to bias. Ideally split-mouth designs to reduce inter-individual variability.
- Compliance with international bioethical guidelines.
- In general, fully comply with the guidelines suggested for this type of design and that are included in the Consolidated Standards of Reporting Trials (CONSORT). This statement comprises a checklist of essential items in order to enhance the quality of the reporting of randomized clinical trials (Moher et al. 2010).

According to published articles, many unique clinical cases with a short follow-up time, are very varied in the technique, the use of biomaterials, follow-up time, among others, and there is no consensus for the clinical management of the damaged dentin-pulp complex; so it can be said that, there is currently no gold standard treatment and randomized controlled clinical trials are required to evaluate and develop predictable clinical regenerative procedures.

Conclusions

In the last two decades, advances in biomaterials and clinical research have made different modalities of treatment to treat the dentin-pulp complex possible, allowing manipulation of reactionary and reparative dentinogenesis. The resolution of infection and the disease process remains the primary goal of any endodontic therapy. A repaired tissue that promotes the resolution of disease and re-establishment of some or all of the original tissue functions should be a desirable goal. The goal of regenerative endodontics is the use of biologic-based procedures to arrest the disease process, preventing its recurrence while favoring the repair or replacement of damaged structures of the dentin-pulp complex.

The European Society of Endodontology and the American Association for Endodontists have released position statements and clinical considerations for regenerative endodontics. In general, they consider that the degree of success of regenerative endodontics depends on three factors: (i) resolution of clinical signs and symptoms and bone healing; (ii) further root maturation; and (iii) return of neurogenesis, positive response to vitality testing. The results are variable for these objectives, and a true regeneration of the pulp/dentine complex is not reached (Feigin and Shope 2017; Kim et al. 2018).

The American Association of Endodontists has suggested the "AAE Clinical Considerations for a Regenerative Procedure. Revised 6-8-16" to help clinicians manage immature permanent teeth with necrotic pulp/apical periodontitis. These considerations should be seen as one possible source of information and, given the rapid evolving nature of this field, clinicians should also actively review new findings elsewhere as they become available.

Significant advances in regenerative endodontics are permitting a better understanding of factors that mediated regeneration and repair of the damaged dentin-pulp complex. However, the evidence from diverse histologic and clinical results suggests that the radiographic findings derivate from successful cases of regenerative endodontic only imitate ectopic tissue development in the root canal space. Clinically, positive responses of vascularized teeth to pulp testing do not indicate that a more organized vital pulp tissue is formed and do not imply regeneration of the pulp-like tissue in the root canal space (Lin et al. 2014). No studies have shown that normal regeneration of human tissue or organ is possible if it is totally damaged; thus, tissue or organ transplantation is necessary. Based on the published evidence, true regeneration of the pulp and dentin-pulp complex has not been supported consistently.

Tissue engineering strategies for dental dentin-pulp complex regeneration in preclinical studies include transplantation of stem/progenitor cells or use of biological molecules (Kim 2017). Regenerative endodontic clinical management uses the endogenous stem/progenitor cells from periapical tissues and biological molecules released from dentin or evoked bleeding (Smith et al. 2016). On the other hand, numerous case reports, and series of regenerative endodontic cases have been published (Kim et al. 2013). Considering global evidence, although the pre-clinical animal models have provided significant scientific bases of the possible benefits associated with the use of techniques based on tissue engineering, these days, important challenges remain for these methodologies and results to be extrapolated to humans.

Continuous laboratory and clinical studies are necessary in order to elucidate the potential benefits of clinical application of tissue engineering principles for repair/regeneration of the damage dentin-pulp complex. Complete regeneration of pulp–dentin complex in empty root canal space with all laboratory or clinical approaches has not been achieved as yet. All current approaches of pulp regeneration are still in the development phase. Significant concerns should be considered, such as treatments, the education of clinicians and students, facilities for obtaining dental stem cells, production of scaffolds, laboratory and clinical studies (Hashemi-Beni et al. 2017). Long-term, well-conducted and controlled randomized clinical trials with adequate sample size are necessary to achieve a high level of evidence to demonstrate the success rate to regenerate the dentine-pulp complex in the root canal space.

References

Albuquerque, M.T., Valera, M.C., Nakashima, M., Nör, J.E. and Bottino, M.C. 2014. Tissue-engineering-based strategies for regenerative endodontics. J. Dent. Res. 93: 1222–1231.

American Association of Endodontists (AAE). 2016. Clinical Considerations for a Regenerative Procedure. Revised 6-8-16, https://www.aae.org/specialty/wp-content/uploads/sites/2/2017/06/currentregenerativeendodonticconsiderations.pdf.

American Association of Endodontists (AAE). https://www.aae.org/specialty/about-aae/news-room/endodontic-treatment-statistics/, Accessed on 18 April 2019.

Andreasen, J.O., Farik, B. and Munksgaard, E.C. 2002. Long-term calcium hydroxide as a root canal dressing may increase risk of root fracture. Dent. Traumatol. 18: 134–137.

Angelova Volponi, A., Zaugg, L.K., Neves, V., Liu, Y. and Sharpe, P.T. 2018. Tooth repair and regeneration. Curr. Oral Health Rep. 5: 295–303.

Antoniou, M. 2001. Embryonic stem cell research: The case against. Nat. Med. 7: 397–399.

Arruda, M.E.F., Neves, M.A.S., Diogenes, A., Mdala, I., Guilherme, B.P.S., Siqueira, J.F. Jr et al. 2018. Infection control in teeth with apical periodontitis using a triple antibiotic solution or calcium hydroxide with chlorhexidine: A randomized clinical trial. J. Endod. 44: 1474–1479.

ASTM Standard F2312-11. Standard Terminology Relating to Tissue Engineered Medical Products. West Conshocken, PA: ASTM International, 2011.

Banchs, F. and Trope, M. 2004. Revascularization of immature permanent teeth with apical periodontitis: new treatment protocol? J. Endod. 30: 196–200.

Becerra, P., Ricucci, D., Loghin, S., Gibbs, J.L. and Lin, L.M. 2014. Histologic study of a human immature permanent premolar with chronic apical abscess after revascularization/revitalization. J. Endod. 40: 133–139.

Bjørndal, L., Simon, S., Tomson, P.L. and Duncan, H.F. 2019. Management of deep caries and the exposed pulp. Int. Endod. J. 52: 949–973.

Bohl, K.S., Shon, J., Rutherford, B. and Mooney, D.J. 1998. Role of synthetic extracellular matrix in development of engineered dental pulp. J. Biomater. Sci. Polym. Ed. 9: 749–764.

Cao, Y., Song, M., Kim, E., Shon, W., Chugal, N., Bogen, G. et al. 2015. Pulp-dentin regeneration: Current state and future prospects. J. Dent. Res. 94: 1544–1551.

Chan, G. and Mooney, D.J. 2008. New materials for tissue engineering: towards greater control over the biological response. Trends Biotechnol. 26: 382–392.

Chen, D., Zhao, M. and Mundy, G.R. 2004. Bone morphogenetic proteins. Growth Factors 22: 233–241.

Chen, S., Gluhak-Heinrich, J., Martinez, M., Li, T., Wu, Y. and Chuang, H.H. et al. 2008. Bone morphogenetic protein 2 mediates dentin sialophosphoprotein expression and odontoblast differentiation via NF-Y signaling. J. Biol. Chem. 283: 19359–19370.

Chevallay, B., Abdul-Malak, N. and Herbage, D. 2000. Mouse fibroblasts in long-term culture within collagen three-dimensional scaffolds: influence of cross linking with diphenylphosphorylazide on matrix reorganization, growth, and biosynthetic and proteolytic activities. J. Biomed. Mater. Res. A 49: 448–459.

Choung, H.W., Lee, D.S., Lee, J.H., Shon, W.J., Lee, J.H., Ku, Y. et al. 2016. Tertiary dentin formation after indirect pulp capping using protein CPNE7. J. Dent. Res. 95: 906–912.

Cooper, P.R., Takahashi, Y., Graham, L.W., Simon, S., Imazato, S. and Smith, A.J. 2010. Inflammation-regeneration interplay in the dentine-pulp complex. J. Dent. 38: 687–697.

Cooper, P.R., McLachlan, J.L., Simon, S., Graham, L.W. and Smith, A.J. 2011. Mediators of inflammation and regeneration. Adv. Dent. Res. 23: 290–295.

Cordeiro, M.M., Dong, Z., Kaneko, T., Zhang, Z., Miyazawa, M., Shi, S. et al. 2008. Dental pulp tissue engineering with stem cells from exfoliated deciduous teeth. J. Endod. 34: 962–969.

Cox, C.F., Bogen, G., Kopel, H.M. and Ruby, J.D. 2002. Repair of pulpal injury by dental materials. pp. 325–344. *In*: Hargreaves, K.M., Goodis, A.E. and Seltzer, S. (eds.). Seltzer and Bender's Dental Pulp. Quintessence Pub. Co., Chicago, USA.

D' Alimonte, I., Nargi, E., Mastrangelo, F., Falco, G., Lanuti, P., Marchisio, M. et al. 2011. Vascular endothelial growth factor enhances *in vitro* proliferation and osteogenic differentiation of human dental pulp stem cells. J. Biol. Regul. Homeost Agents 25: 57–69.

de Sant'Anna, M.M.S., Batista, L.A., da Silva, T.C.M., Oliveira, L.P. and de Carvalho, J.J. 2017. Spatiotemporal expression of extracellular matrix components during the chondrogenic and osteogenic phases of bone healing. Rom. J. Morphol. Embryol. 58: 1201–1216.

Del Bakhshayesh, A.R., Mostafavi, E., Alizadeh, E., Asadi, N., Akbarzadeh, A. and Davaran, S. 2018. Fabrication of three-dimensional scaffolds based on nano-biomimetic collagen hybrid constructs for skin tissue engineering. ACS Omega. 3: 8605–8611.

Devillard, R., Rémy, M., Kalisky, J., Bourget, J.M., Kérourédan, O., Siadous, R. et al. 2017. *In vitro* assessment of a collagen/alginate composite scaffold for regenerative endodontics. Int. Endod. J. 50: 48–57.

Diogenes, A. and Ruparel, N.B. 2017. Regenerative endodontic procedures: Clinical outcomes. Dent. Clin. North Am. 61: 111–125.

Dobie, K., Smith, G., Sloan, A. and Smith, A. 2002. Effects of alginate hydrogels and TGF-1 on human dental pulp repair *in vitro*. Connect Tissue Res. 43: 387–390.

Duncan, H.F., Galler, K.M., Tomson, P.L., Simon, S., El-Karim, I., Kundzina, R. et al. 2019. European society of endodontology position statement: Management of deep caries and the exposed pulp. Int. Endod. J. 52: 923–934.

Durand, S.H., Flacher, V., Roméas, A., Carrouel, F., Colomb, E., Vincent, C. et al. 2006. Lipoteichoic acid increases TLR and functional chemokine expression while reducing dentin formation in *in vitro* differentiated human odontoblasts. J. Immunol. 176: 2880–2887.

Faisal, N. and Kumar, K. 2017. Polymer and metal nanocomposites in biomedical applications. Biointerface Res. Appl. Chem. 7: 2286–2294.

Farges, J.C., Alliot-Licht, B., Renard, E., Ducret, M., Gaudin, A., Smith, A.J. et al. 2015. Dental pulp defence and repair mechanisms in dental caries. Mediators Inflamm. 2015: 230251.

Feigin, K. and Shope, B. 2017. Regenerative endodontics. J. Vet. Dent. 34: 161–178.

Ferrara, N., Gerber, H.P. and LeCouter, J. 2003. The biology of VEGF and its receptors. Nat. Med. 9: 669–676.

Fitzgerald, M. and Heys, R.J. 1991. A clinical and histological evaluation of conservative pulpal therapy in human teeth. Oper Dent. 16: 101–112.

Foroughi, M.R., Karbasi, S. and Ebrahimi-Kahrizsangi, R. 2012. Physical and mechanical properties of a poly-3-hydroxybutyrate-coated nanocrystalline hydroxyapatite scaffold for bone tissue engineering. J. Porous. Mater. 19: 667–675.

Foroughi, M.R., Karbasi, S. and Ebrahimi-Kahrizsangi, R. 2013. Mechanical evaluation of nHAp scaffold coated with Poly-3-Hydroxybutyrate. J. Nanosci. Nanotechnol. 13: 1555–1562.

Fundaoğlu Küçükekenci, F., Küçükekenci, A.S. and Çakici, F. 2019. Evaluation of the preventive efficacy of three dentin tubule occlusion methods against discoloration caused by triple-antibiotic paste. Odontology 107: 186–189.

GBD 2016 Disease and Injury Incidence and Prevalence Collaborators. 2017. Global, regional, and national incidence, prevalence, and years lived with disability for 328 diseases and injuries for 195 countries, 1990–2016: A systematic analysis for the Global Burden of Disease Study 2016. Lancet 390: 1211–1259.

Giuliani, V., Baccetti, T., Pace, R. and Pagavino, G. 2002. The use of MTA in teeth with necrotic pulps and open apices. Dent. Traumatol. 18: 217–221.

Glossary of Endodontic Terms. 2016. American Association of Endodontists, Ninth Edition. https://www.aae.org/specialty/clinical-resources/glossary-endodontic-terms/.

Goldberg, M. and Lasfargues, J.J. 1995. Pulpo-dentinal complex revisited. J. Dent. 23: 15–20.

Goldberg, M., Kulkarni, A.B., Young, M. and Boskey, A. 2011. Dentin: structure, composition and mineralization. Front Biosci. (Elite Ed.) 3: 711–735.

Gómez-Lizárraga, K.K., Flores-Morales, C., Del Prado-Audelo, M.L., Álvarez-Pérez, M.A., Piña-Barba, M.C. and Escobedo, C. 2017. Polycaprolactone- and polycaprolactone/ceramic-based 3D-bioplotted porous scaffolds for bone regeneration: A comparative study. Mater. Sci. Eng. C Mater. Biol. Appl. 79: 326–335.

Gotlieb, E.L., Murray, P.E., Namerow, K.N., Kuttler, S. and Garcia-Godoy, F. 2008. An ultrastructural investigation of tissue-engineered pulp constructs implanted within endodontically treated teeth. J. Am. Dent. Assoc. 139: 457–465.

Granados-Hernández, M.V., Serrano-Bello, J., Montesinos, J.J., Alvarez-Gayosso, C., Medina-Velázquez, L.A., Álvarez-Fregoso, O. et al. 2018. *In vitro* and *in vivo* biological characterization of poly(lactic acid) fiber scaffolds synthesized by air jet spinning. J. Biomed. Mater. Res. B Appl. Biomater. 106: 2435–2446.

Grayson, A.C., Voskerician, G., Lynn, A., Anderson, J.M., Cima, M.J. and Langer, R. 2004. Differential degradation rates *in vivo* and *in vitro* of biocompatible poly (lacticacid) and poly (glycolic acid) homo- and co-polymers for a polymeric drug-delivery microchip. J. Biomater. Sci. Polym. Ed. 15: 1281–1304.

Gronthos, S., Mankani, M., Brahim, J., Robey, P.G. and Shi, S. 2000. Postnatal human dental pulp stem cells (DPSCs) *in vitro* and *in vivo*. Proc. Natl. Acad. Sci. 97: 13625–13630.

Gronthos, S., Brahim, J., Li, W., Fisher, L.W., Cherman, N., Boyde, A. et al. 2002. Stem cell properties of human dental pulp stem cells. J. Dent. Res. 81: 531–535.

Gunatillake, P.A. and Adhikari, R. 2003. Biodegradable synthetic polymers for tissue engineering. Eur. Cell Mater. 5: 1–16.

Gutmann, J.L. 2016. Apical termination of root canal procedures—ambiguity or disambiguation? Evidence-Based Endodontics 1: 4.

Guven, G., Altun, C. and Günhan, O. 2007. Co-expression of cyclooxygenase-2 and vascular endothelial growth factor in inflamed human pulp: an immune histochemical study. J. Endod. 33: 18–20.

Hargreaves, K.M., Diogenes, A. and Teixeira, F.B. 2013. Treatment options: biological basis of regenerative endodontic procedures. J. Endod. 39(suppl. 3): S30–S43.

Hashemi-Beni, B., Khoroushi, M., Foroughi, M.R., Karbasi, S. and Khademi, A.A. 2017. Tissue engineering: Dentin-pulp complex regeneration approaches (A review). Tissue Cell 49: 552–564.

Hori, Y., Winans, A.M. and Irvine, D.J. 2009. Modular injectable matrices based on alginate solution/microsphere mixtures that gel *in situ* and co-deliver immunomodulatory factors. Acta Biomater. 5: 969–982.

Hu, C.C., Zhang, C., Qian, Q. and Tatum, N.B. 1998. Reparative dentin formation in rat molars after direct pulp capping with growth factors. J. Endod. 24: 744–751.

Huang, A.H., Chen, Y., Lin, L., Shieh, T. and Chan, A.W. 2008. Isolation and characterization of dental pulp stem cells from a supernumery tooth. J. Oral. Pathol. Med. 37: 571–574.

Huang, G.T., Yamaza, T., Shea, L.D., Djouad, F., Kuhn, N.Z., Tuan, R.S. et al. 2010. Stem/progenitor cell-mediated *de novo* regeneration of dental pulp with newly deposited continuous layer of dentin in an *in vivo* model. Tissue Eng. Part A 16: 605–615.

Hutmacher, D. 2000. Scaffolds in tissue engineering bone and cartilage. Biomaterials 21: 2529–2543.

Hwang, D., Fong, H., Johnson, J.D. and Paranjpe, A. 2018. Efficacy of different carriers for the triple antibiotic powder during regenerative endodontic procedures. Aust. Endod. J. 44: 208–214.

Iohara, K., Nakashima, M., Ito, M., Ishikawa, M., Nakasima, A. and Akamine, A. 2004. Dentin regeneration by dental pulp stem cell therapy with recombinant human bone morphogenetic protein 2. J. Dent. Res. 83: 590–595.

Iwaya, S.I., Ikawa, M. and Kubota, M. 2001. Revascularization of an immature permanent tooth with apical periodontitis and sinus tract. Dent. Traumatol. 17: 185–187.

Jafarkhani, M., Salehi, Z. and Ghelich, P. 2016. An overview on the experimental and mathematical modelings of angiogenesis and vasculogenesis. Biointerface Res. Appl. Chem. 6: 1190–1199.

Jain, R.A. 2000. The manufacturing techniques of various drug loaded biodegradable poly (lactide-co-glycolide) (PLGA) devices. Biomaterials 21: 2475–2490.

Kim, S.G., Zheng, Y., Zhou, J., Chen, M., Embree, M.C., Song, K. et al. 2013. Dentin and dental pulp regeneration by the patient's endogenous cells. Endod. Topics 28: 106–117.

Kim, S.G., Kahler, B. and Lin, L.M. 2016. Current developments in regenerative endodontics. Curr. Oral Health Rep. 3: 293–301.

Kim, S.G. 2017. Biological molecules for the regeneration of the pulp-dentin complex. Dent. Clin. North Am. 61: 127–141.

Kim, S.G., Malek, M., Sigurdsson, A., Lin, L.M. and Kahler, B. 2018. Regenerative endodontics: a comprehensive review. Int. Endod. J. 51: 1367–1388.

Kitamura, C., Nishihara, T., Terashita, M., Tabata, Y. and Washio, A. 2012. Local regeneration of dentin-pulp complex using controlled release of fgf-2 and naturally derived sponge-like scaffolds. Int. J. Dent. 2012: 190561.

Koubi, G., Colon, P., Franquin, J.C., Hartmann, A., Richard, G., Faure, M.O. et al. 2013. Clinical evaluation of the performance and safety of a new dentine substitute, Biodentine, in the restoration of posterior teeth—a prospective study. Clin. Oral. Investig. 17: 243–249.

Koyama, N., Okubo, Y., Nakao, K. and Bessho, K. 2009. Evaluation of pluripotency in human dental pulp cells. J. Oral. Maxillofac. Surg. 67: 501–506.

Lam, R. 2016. Epidemiology and outcomes of traumatic dental injuries: a review of the literature. Aust. Dent. J. 61(Suppl. 1): 4–20.

Lambert, K.E., Huang, H., Mythreye, K. and Blobe, G.C. 2011. The type III transforming growth factor-β receptor inhibits proliferation, migration, and adhesion in human myeloma cells. Mol. Biol. Cell 22: 1463–1472.

Lee, K.Y. and Mooney, D.J. 2012. Alginate: properties and biomedical applications. Prog. Polym. Sci. 37: 106–126.

Lenzi, R. and Trope, M. 2012. Revitalization procedures in two traumatized incisors with different biological outcomes. J. Endod. 38: 411–414.

Lin, L.M., Ricucci, D. and Huang, G.T. 2014. Regeneration of the dentine-pulp complex with revitalization/revascularization therapy: challenges and hopes. Int. Endod. J. 47: 713–724.

Lind, M. 1996. Growth factors: possible new clinical tools. A review. Acta Orthop. Scand. 67: 407–417.

Loh, Q.L. and Choong, C. 2013. Three-dimensional scaffolds for tissue engineering applications: role of porosity and pore size. Tissue Eng. Part B Rev. 19: 485–502.

Lovelace, T.W., Henry, M.A., Hargreaves, K.M. and Diogenes, A.J. 2011. Evaluation of the delivery of mesenchymal stem cells into the root canal space of necrotic immature teeth after clinical regenerative endodontic procedure. J. Endod. 37: 133–138.

Lu, T., Li, Y. and Chen, T. 2013. Techniques for fabrication and construction of three-dimensional scaffolds for tissue engineering. Int. J. Nanomedicine 8: 337–350.

Madihally, S. and Matthew, H. 1999. Porous chitosan scaffolds for tissue engineering. Biomaterials 20: 1133–1142.

Mao, J.J., Kim, S.G., Zhou, J., Ye, L., Cho, S., Suzuki, T. et al. 2012. Regenerative endodontics: Barriers and strategies for clinical translation. Dent. Clin. North Am. 56: 639–49.

Marending, M., Attin, T. and Zehnder, M. 2016. Treatment options for permanent teeth with deep caries. Swiss. Dent. J. 126: 1007–1027.

Massagué, J., Blain, S.W. and Lo, R.S. 2000. TGFbeta signaling in growth control, cancer, and heritable disorders. Cell 103: 295–309.

Mauth, C., Huwig, A., Graf-Hausner, U. and Roulet, J. 2007. Restorative applications for dental pulp therapy. Topics Tissue Eng. 3: 1–32.

McIntyre, P.W., Wu, J.L., Kolte, R., Zhang, R., Gregory, R.L., Bruzzaniti, A. et al. 2019. The antimicrobial properties, cytotoxicity, and differentiation potential of double antibiotic intracanal medicaments loaded into hydrogel system. Clin. Oral. Investig. 23: 1051–1059.

Miura, M., Gronthos, S., Zhao, M., Lu, B., Fisher, L.W., Robey, P.G. et al. 2003. SHED: stem cells from human exfoliated deciduous teeth. Proc. Natl. Acad. Sci. USA 100: 5807–5812.

Moher, D., Hopewell, S., Schulz, K.F., Montori, V., Gøtzsche, P.C., Devereaux, P.J. et al. 2010. CONSORT 2010 explanation and elaboration: Updated guidelines for reporting parallel group randomised trials. J. Clin. Epidemiol. 63: e1–e37.

Moreno Madrid, A.P., Vrech, S.M., Sanchez, M.A. and Rodriguez, A.P. 2019. Advances in additive manufacturing for bone tissue engineering scaffolds. Mater. Sci. Eng. C Mater. Biol. Appl. 100: 631–644.

Murray, P.E., Garcia-Godoy, F. and Hargreaves, K.M. 2007. Regenerative endodontics: a review of current status and a call for action. J. Endod. 33: 377–390.

Nakamura, S., Yamada, Y. and Katagiri, W. 2009. Stem cell proliferation pathways comparison between human exfoliated deciduous teeth and dental pulp stem cells by gene expression profile from promising dental pulp. J. Endod. 35: 1536–1542.

Nakashima, M. 1994a. Induction of dentin formation on canine amputated pulp by recombinant human bone morphogenetic proteins (BMP)-2 and -4. J. Dent. Res. 73: 1515–1522.

Nakashima, M., Nagasawa, H., Yamada, Y. and Reddi, A.H. 1994b. Regulatory role of transforming growth factor-beta, bone morphogenetic protein-2, and protein-4 on gene expression of extracellular matrix proteins and differentiation of dental pulp cells. Dev. Biol. 162: 18–28.

Nakashima, M., Iohara, K., Ishikawa, M., Ito M., Tomokiyo, A., Tanaka, T. et al. 2004. Stimulation of reparative dentin formation by *ex vivo* gene therapy using dental pulp stem cells electrotransfected with growth/differentiation factor 11 (Gdf11). Hum. Gene Ther. 15: 1045–1053.

Nakashima, M. 2005. Bone morphogenetic proteins in dentin regeneration for potential use in endodontic therapy. Cytokine Growth Factor Rev. 16: 369–376.

Nasir, N.M., Raha, M.G., Kadri, K.N., Rampado, M. and Azlan, C.A. 2006. The study of morphological structure, phase structure and molecular structure of collagen-PEO 600K blends for tissue engineering application. Am. J. Biochem. Biotechnol. 2: 175–179.

National Institute of Dental and Craniofacial Research, https://www.nidcr.nih.gov/research/data-statistics/dental-caries/adults, Accessed on 18 April 2019.

Ng, Y.L., Mann, V., Rahbaran, S., Lewsey, J. and Gulabivala, K. 2007. Outcome of primary root canal treatment: systematic review of the literature—Part 1. Effects of study characteristics on probability of success. Int. Endod. J. 40: 921–939.

Nör, J.E., Christensen, J., Mooney, D.J. and Polverini, P.J. 1999. Vascular endothelial growth factor (VEGF)-mediated angiogenesis is associated with enhanced endothelial cell survival and induction of Bcl-2 expression. Am. J. Pathol. 154: 375–384.

Nygaard-Ostby, B. and Hjortdal, O. 1971. Tissue formation in the root canal following pulp removal. Scand. J. Dent. Res. 79: 333–349.

Ortiz, M., Rosales-Ibáñez, R., Pozos-Guillén, A., De Bien, C., Toye, D., Flores, H. et al. 2017. DPSC colonization of functionalized 3D textiles. J. Biomed. Mater. Res. B Appl. Biomater. 105: 785–794.

Ortiz, M., Romo, A., Romo, F., Escobar, D., Flores, H. and Pozos, A. 2019. Behavior of mesenchymal stem cells obtained from dental tissues: A review of the literature. ODOVTOS-Int. J. Dent. Sc. 21: 31–40.

Ostby, B.N. 1961. The role of the blood clot in endodontic therapy. An experimental histologic study. Acta Odontol. Scand. 19: 324–353.

Parirokh, M. and Torabinejad, M. 2010. Mineral trioxide aggregate: a comprehensive literature review—part I: chemical, physical, and antibacterial properties. J. Endod. 36: 16–27.

Park, H.J., Lee, O.J., Lee, M.C., Moon, B.M., Ju, H.W., Lee, J.M. et al. 2015. Fabrication of 3D porous silk scaffolds by particulate (salt/sucrose) leaching for bone tissue reconstruction. Int. J. Biol. Macromol. 78: 215–223.

Pashley, D. 2002. Pulpodentin complex. pp. 63–93. *In*: Hargreaves, K.M., Goodis, A.E. and Seltzer, S. (eds.). Seltzer and Bender's Dental Pulp. Quintessence Pub. Co., Chicago, USA.

Prescott, R., Alsanea, R., Fayad, M., Johnson, B., Wenckus, C., Hao, J. et al. 2008. *In vivo* generation of dental pulp-like tissue by using dental pulp stem cells, a collagen scaffold, and dentin matrix protein I after subcutaneous transplantation in mice. J. Endod. 34: 421–426.

Rajasekharan, S., Martens, L.C., Cauwels, R.G. and Verbeeck, R.M. 2014. Biodentine™ material characteristics and clinical applications: a review of the literature. Eur. Arch. Paediatr. Dent. 15: 147–158.

Ravindran, S., Song, Y.Q. and George, A. 2010. Development of three-dimensional biomimetic scaffold to study epithelial–mesenchymal interactions. Tissue Eng. Part A 16: 327–342.

Rezwan, K., Chen, Q., Blaker, J. and Boccaccini, A.R. 2006. Biodegradable and bioactive porous polymer/inorganic composite scaffolds for bone tissue engineering. Biomaterials 27: 3413–3431.

Robey, P.G. 2000. Stem cells near the century mark. J. Clin. Invest. 105: 1489–1491.

Rutherford, R.B., Spångberg, L., Tucker, M., Rueger, D. and Charette, M. 1994. The time-course of the induction of reparative dentine formation in monkeys by recombinant human osteogenic protein-1. Arch. Oral. Biol. 39: 833–838.

Rutherford, R.B. and Gu, K. 2000. Treatment of inflamed ferret dental pulps with recombinant bone morphogenetic protein-7. Eur. J. Oral. Sci. 108: 202–206.

Sadek, R.W., Moussa, S.M., El Backly, R.M. and Hammouda, A.F. 2019. Evaluation of the efficacy of three antimicrobial agents used for regenerative endodontics: An *in vitro* study. Microb. Drug. Resist. 2: 761–771.

Sakai, V.T., Zhang, Z., Dong, Z., Neiva, K.G., Machado, M.A., Shi, S. et al. 2010. SHED differentiate into functional odontoblasts and endothelium. J. Dent. Res. 89: 791–796.

Şapte, E., Costea, C.F., Cărăuleanu, A., Dancă, C., Dumitrescu, G.F., Dimitriu, G. et al. 2017. Histological, immunohistochemical and clinical considerations on amniotic membrane transplant for ocular surface reconstruction. Rom. J. Morphol. Embryol. 58: 363–369.

Schwartz, Z., Goldstein, M., Raviv, E., Hirsch, A., Ranly, D.M. and Boyan, B.D. 2007. Clinical evaluation of demineralized bone allograft in a hyaluronic acid carrierfor sinus lift augmentation in humans: a computed tomography and histomorphometric study. Clin. Oral. Implants Res. 18: 204–211.

Schwendicke, F. and Stolpe, M. 2014. Direct pulp capping after a carious exposure versus root canal treatment: a cost-effectiveness analysis. J. Endod. 40: 1764–1770.

Seol, Y., Lee, J., Park, Y., Lee, Y., Ku, Y., Rhyu, I. et al. 2004. Chitosan sponges as tissue engineering scaffolds for bone formation. Biotechnol. Lett. 26: 1037–1041.

Sharma, S., Srivastava, D., Grover, S. and Sharma, V. 2014. Biomaterials in tooth tissue engineering: a review. J. Clin. Diagn. Res. 8: 309–115.

Shi, S., Bartold, P.M., Miura, M., Seo, B.M., Robey, P.G. and Gronthos, S. 2005. The efficacy of mesenchymal stem cells to regenerate and repair dental structures. Orthod. Craniofac. Res. 8: 191–199.

Silver, F.H. and Pins, G. 1991. Cell growth on collagen: a review of tissue engineering using scaffolds containing extracellular matrix. J. Long Term Eff. Med. Implants 2: 67–80.

Simon, S., Perard, M., Zanini, M., Smith, A.J., Charpentier, E., Djole, S.X. et al. 2013. Should pulp chamber pulpotomy be seen as a permanent treatment? Some preliminary thoughts. Int. Endod. J. 46: 79–87.

Simon, S.R., Tomson, P.L. and Berdal, A. 2014. Regenerative endodontics: regeneration or repair? J. Endod. 40(4 Suppl.): S70–S75.

Six, N., Lasfargues, J.J. and Goldberg, M. 2002. Differential repair responses in the coronal and radicular areas of the exposed rat molar pulp induced by recombinant human bone morphogenetic protein 7 (osteogenic protein 1). Arch. Oral. Biol. 47: 177–187.

Smith, A.J., Murray, P.E., Sloan, A.J., Matthews, J.B. and Zhao, S. 2001. Trans-dentinal stimulation of tertiary dentinogenesis. Adv. Dent. Res. 15: 51–54.

Smith, A.J. 2002. Pulpal responses to caries and dental repair. Caries Res. 36: 223–232.

Smith, A.J. 2003. Vitality of the dentin-pulp complex in health and disease: Growth factors as key mediators. J. Dent. Educ. 67: 678–689.

Smith, A.J., Patel, M., Graham, L., Sloan, A.J. and Cooper, P.R. 2005. Dentine regeneration: The role of stem cells and molecular signaling. Oral. Biosc. Med. 2: 127–132.

Smith, A.J., Duncan, H.F., Diogenes, A., Simon, S. and Cooper, P.R. 2016. Exploiting the bioactive properties of the dentin-pulp complex in regenerative endodontics. J. Endod. 42: 47–56.

Song, J.S., Takimoto, K., Jeon, M., Vadakekalam, J., Ruparel, N.B. and Diogenes, A. 2017. Decellularized human dental pulp as a scaffold for regenerative endodontics. J. Dent. Res. 96: 640–646.

Suh, J. and Matthew, H. 2000. Application of chitosan-based polysaccharide biomaterials in cartilage tissue engineering: a review. Biomaterials 21: 2589–2598.

Sumita, Y., Honda, M., Ohara, T., Tsuchiya, S., Sagara, H., Kagami, H. et al. 2006. Performance of collagen sponge as a 3-D scaffold for tooth-tissue engineering. Biomaterials 27: 3238–3248.

Takushige, T., Cruz, E.V., Asgor Moral, A. and Hoshino, E. 2004. Endodontic treatment of primary teeth using a combination of antibacterial drugs. Int. Endod. J. 37: 132–138.

Tomson, P.L., Grover, L.M., Lumley, P.J., Sloan, A.J., Smith, A.J. and Cooper, P.R. 2007. Dissolution of bio-active dentine matrix components by mineral trioxide aggregate. J. Dent. 35: 636–642.

Topçuoğlu, G. and Topçuoğlu, H.S. 2016. Regenerative endodontic therapy in a single visit using platelet-rich plasma and biodentine in necrotic and asymptomatic immature molar teeth: A report of 3 cases. J. Endod. 42: 1344–1346.

Vacanti, J.P., Morse, M.A., Saltzman, W.M., Domb, A.J., Perez-Atayde, A. and Langer, R. 1988. Selective cell transplantation using bioabsorbable artificial polymers as matrices. J. Pediatr. Surg. 23: 3–9.

Vázquez-Vázquez, F.C., Hernández-Tapia, L.G., Chanes-Cuevas, O.A. Álvarez-Pérez M.A. and Pozos-Guillén, A.J. 2018. Airflow electrofluidodynamics. pp. 123–138. *In*: Guarino, V. and Ambrosio, L. (eds.). Electrofluidodynamic Technologies (EFDTs) for Biomaterials and Medical Devices. Principles and Advances. Woodhead Publishing Elsevier, Duxford, UK.

Veerayutthwilai, O., Byers, M.R., Pham, T.T., Darveau, R.P. and Dale, B.A. 2007. Differential regulation of immune responses by odontoblasts. Oral. Microbiol. Immunol. 22: 5–13.

Verrecchia, F. and Mauviel, A. 2002. Transforming growth factor-beta signaling through the Smad pathway: Role in extracellular matrix gene expression and regulation. J. Invest. Dermatol. 118: 211– 5.

Walimbe, T., Panitch, A. and Sivasankar, P.M. 2017. A review of hyaluronic acid and hyaluronic acid-based hydrogels for vocal fold tissue engineering. J. Voice 31: 416–423.

Wang, X., Thibodeau, B., Trope, M., Lin, L.M. and Huang, G.T. 2010. Histologic characterization of regenerated tissues in canal space after the revitalization/revascularization procedure of immature dog teeth with apical periodontitis. J. Endod. 36: 56–63.

Xu, C., Bai, Y., Yang, S., Yang, H., Stout, D.A., Tran, P.A. et al. 2018. A versatile three-dimensional foam fabrication strategy for soft and hard tissue engineering. Biomed. Mater. 13: 025018.

Yamauchi, N., Yamauchi, S., Nagaoka, H., Duggan, D., Zhong, S., Lee, S.M. et al. 2011. Tissue engineering strategies for immature teeth with apical periodontitis. J. Endod. 37: 390–397.

Zancan, R.F., Calefi, P.H.S., Borges, M.M.B., Lopes, M.R.M., de Andrade, F.B., Vivan, R.R. et al. 2019. Antimicrobial activity of intracanal medications against both *Enterococcus faecalis* and *Candida albicans* biofilm. Microsc. Res. Tech. 82: 494–500.

Zander, H. 1939. Reaction of the pulp to calcium hydroxide. J. Dent. Res. 18: 373–379.

Zargar, N., Rayat Hosein Abadi, M., Sabeti, M., Yadegari, Z., Akbarzadeh Baghban, A. and Dianat, O. 2019. Antimicrobial efficacy of clindamycin and triple antibiotic paste as root canal medicaments on tubular infection: An *in vitro* study. Aust. Endod. J. 45: 86–91.

Zhang, L., Morsi, Y., Wang, Y., Li, Y. and Ramakrishna, S. 2013. Review scaffold design and stem cells for tooth regeneration. Jap. Dent. Sci. Rev. 49: 14–26.

12

Gene Therapy in Oral Tissue Regeneration

Fernando Suaste Olmos,[1,]* *Patricia González-Alva,*[2]
Alejandro Luis Vega-Jiménez[2] and *Osmar Alejandro Chanes-Cuevas*[2]

Introduction to Gene Therapy

Definition

Gene therapy is an experimental strategy initially developed for the treatment of a variety of genetic disorders. This approach pursues the re-establishment of a defective gene copy (which could be affected by single nucleotide substitutions, insertions or deletions) and its products by the introduction of a functional copy *in trans* (Siddique et al. 2016).

Tissue Regeneration (TR) is defined as a process of reproduction and reconstitution of the architecture of tissue loss, which requires three main elements: scaffolds, cell sources and tissue-inducing factors. Stem cells are the main element in tissue regeneration because its self-renewable and totipotential capacity to differentiated into multiple cell lineages. Additionally, growth factors are molecular signals that control the fate of these mesenchymal stem cells under differentiation. In tissue engineering, gene therapy has emerged as an alternative, supporting both cell therapy and therapy that relies on the use of signaling factors.

History

In tissue engineering, different approaches have been conducted to carry out tissue reconstruction and repair. These treatments aim to guide and induce the regeneration of damaged tissue, using specialized

---
[1] Physiology Research Institute, National Autonomous University of Mexico (IFC-UNAM), Circuito Exterior s/n. Col. Copilco el Alto, Alcaldía de Coyoacán, C.P. 04510, CDMX, México.
[2] Tissue Bioengineering Laboratory, Division of Graduate Studies and Research, Faculty of Dentistry, The National Autonomous University of Mexico (UNAM), Circuito Exterior s/n, Col. Copilco el Alto, Alcaldía de Coyoacán, C.P. 04510, CDMX, México.
Emails: pgonzalezalva@comunidad.unam.mx; dr.vegalex@gmail.com; osale_89@hotmail.com
* Corresponding author: fersuaste15@gmail.com

scaffolds that allow the rearrangement and proliferation of the cells that reshape the affected region. Likewise, scaffolds can function as triggered vectors that could delivery progenitor cells, growth factors, genes, or more recently, RNA molecules (Intini 2010).

Cell therapy has been constituted as a base strategy for the reconstruction of the tissues of the oral cavity, due to the cellular plasticity that progenitor cells possess to differentiate to multiple cellular lineages. Within the oral cavity, different types of mesenchymal cells have been isolated: Dental Pulp stem Cells (DPSCs), Stem cells from Human Exfoliated Deciduous teeth (SHED), Stem Cells from Apical Papilla (SCAP), Periodontal Ligament Stem Cells (PDLSCs), Dental Follicle Progenitor Cells (DFPCs) and Gingiva-derived Mesenchymal Stem/stromal Cells (GMSCs) (Botelho et al. 2017).

All these cell types are able to self-renew, as well as to differentiate into osteoblasts, adipocytes, odontoblasts and neural cells. While this methodology proposes restoring the mesenchymal cells derived from the bone marrow, oral cavity or other tissue in the wound area could trigger the repair of damaged tissues. The process of differentiation into a particular lineage is highly complex since it involves the space-time expression of growth and transcriptional factors that act in a hierarchical manner triggering signaling pathways, which in turn activates a pattern of gene expression that results in determining cell identity fate.

In addition to this process, the deposition of specific components of the extracellular matrix also contributes synergistically to the establishment and differentiation of each of these cell types. All these elements generate a limiting factor in cell therapy because not all the factors necessary for the differentiation of mesenchymal cells are produced simultaneously at the site of the injury.

The growth factors are a group of small polypeptides involved in the stimulation of different cellular signaling pathways through its association with specific membrane receptors and promoting its phosphorylation in tyrosine, threonine or serine amino acid residues, which in turn activates a complex system of transcriptional regulation inside of the cell.

Skeletal cells produce a variety of biological factors associated with bone remodeling, i.e., PDGF, BMPs, VEGF, FGF, TGF-beta, IGF, which are not exclusive from skeletal cells. In oral cavity regeneration, different signaling factors have been tested to improve the clinical outcomes: Platelet-Derived Growth Factor (PDGF), Bone Morphogenetic Proteins (BMPs), Fibroblast Growth Factor (FGF), Transforming Growth Factor-beta (TGF-beta) (Ohba et al. 2012).

Gene therapy has generated an alternative parallel to cell therapy, since its main component is the delivery of genes, which encode mainly for growth and transcription factors, allowing the activation of the differentiation process of undifferentiated mesenchymal cells present at the site of injury to the corresponding cell lineage. The primary limitations that this methodology faces are associated with the type of vector used for the release of the gene or genes of interest.

The genes used in gene therapy are released through the use of viral vectors such as adenovirus, adeno-associated viruses, retroviruses and lentiviruses. These vectors can integrate stably and in a random way to the host genome (lentivirus and retrovirus), generating mutations that can derive in severe genetic disorders. Alternatively, in the adeno-associated viruses and adenoviruses that have a very low integration efficiency to the host genome, they generate a robust immune response when producing viral proteins (Jooss and Chirnule 2003).

Currently, methodologies such as PCR (Polymerase Chain Reaction) and genetic cloning have made it possible to manipulate DNA sequences *in vitro* both coding regions (genes) and regulating elements of gene expression (promoters, enhancers, insulators, etc.). Through this method it has been possible to determine the role of specific genes on the cell homeostasis. Genetic manipulation is also associated with the reprogramming the functioning of a cell.

In prokaryotes, this approach has made it possible to generate organisms that contribute to the synthesis of biomolecules for therapeutic purposes, or to modify organisms that carry out bioremediation processes (Davies 2019).

In bacteria, the delivery of genetic material has been done through the use of plasmids, which are circular DNA molecules that replicate and segregate independently on the host bacterial chromosome (episomes); this allows that the sequences cloned in these vectors are expressed stably in the bacterial cytosol. Bacterial transformation emerged as one of the first approaches to introduce genetic material into a cell. In this strategy the permeability of the cell membrane is disturbed either by thermal shock or by the use of an electric field (electroporation). Additionally, cellular mechanisms such as conjugation and phage-mediated transduction were used to transfer genetic elements from one cell to another.

Vectors

In mammalian cells, unlike prokaryotes, there are no naturally occurring episomal vectors. The strategy to express target DNA sequences has been based on the development of viral vectors. The delivery of genetic material by this process is known as 'Transfection', from which two types are derived: stable transfection and transient transfection (Kim and Eberwine 2010).

In stable transfection the viral vector along with the target DNA sequence is integrated into the genome, and its expression is sustained for an extended period of time; whereas in transient transfection there is no integration into the host cell genome and its expression of the target DNA sequence occurs for a limited period of time (Kim and Eberwine 2010).

Several limitations have been generated from both types of transfection. For example, in stable transfection, the integration of the viral vector is random, which can create insertions in coding DNA sequences (ORFs) resulting in the silencing of a gene or genes and altering cell viability. Otherwise, the gene integrated into the genome is subject to transcriptional and epigenetic regulation controls, which can alter the chromatin state of surrounding genes, causing defects in its expression (Anguela and High 2019).

Transfection is, therefore, a method designed for the introduction of genetic elements (coding and non-coding DNA sequences) that control the expression of a gene, either by increasing its expression or suppressing it. In the former, transfection can be used in gene therapy for the treatment of diseases or induce changes in cell reprogramming by introducing transcription factors into mesenchymal cells (cell renewal). Moreover, transfection can be used to reduce the expression of a gene, through RNAi, or by performing genetic editing to correct errors in the sequence or eliminating them through the CRISPR/Cas system (Glorioso and Lemoine 2017; Anguela and High 2018).

Different types of viral vectors have been investigated for gene therapy, and several characteristics have to be considered for their use: the type of genome (RNA or DNA), its ability to infect dividing cells, the amount of genetic material that can be packaged in its capsid, activation of the immune response (toxicity), its ability to integrate into the genome and the expression of transgene in a long or short term (Anguela and High 2018).

Transfection Methods

In mammal cells, three types of strategies have been developed for the introduction of genetic material into a cell (transfection methods): Biological, chemical and physical. In the first instance, viral vectors have been the primary strategy due to their ability to infect different cell types by passing through the cell membrane and reaching the core by the endocytic pathway, and then releasing their genetic material into the nucleus, where its expression is controlled by endogenous transcriptional machinery (Table 12.1).

Another characteristic that distinguishes these type of vectors is their ability to evade the immune response of the cell. Chemical transfection is based on the use of cationic polymers that aim to change the net loaded charge of DNA, this results in an interaction with the plasmatic membrane, and then its introduction has been suggested to mediate by cellular processes such as endocytosis/phagocytosis. Finally, physical methods involve the use of electric current pulse (electroporation), direct microinjection of the genetic material to the target cell or the use of nanoparticles introduced by bio-ballistics (Kim and Eberwine 2010).

Table 12.1. Viral Vectors for Gene Therapy.

Viral vector	Viral genome	Packing range	Target cell infection	Transfection
Adenovirus (Ads)	DNA	8 kb	Dividing and non-dividing cells	Transiently transfection
Adeno-associated Virus (AAV)	DNA	5 kb	Dividing and non-dividing cells	Transiently and stable transfection
Lentivirus	RNA	8 kb	Dividing and non-dividing cells	Stable transfection
Retrovirus	RNA	8 kb	Dividing cells	Stable transfection
Herpes simplevirus	DNA	30–40 Kb	Dividing cells	Transiently transfection

The features of suitable vectors for gene therapy, are associated with several, types of the treatment to which it is directed (genetic disorders—hereditary and acquiring—or tissue engineering); the transfection method (biological, chemical and physical), and the mechanism through which the vector is delivered to the target cell (*in vivo* and *ex vivo*). For *in vivo* gene transfer, the vector is introduced directly into the region of the affected tissue, while in the *ex vivo* mechanism, cells in cell culture are transfected with the vector of interest and then transplanted into the organism with the affected tissue (Siddique et al. 2016) Figure 12.1.

Viral vectors for genetic manipulation in human cells are designed with defects in their replication mechanisms, and this is done by separating their structural and functional components into different plasmids (Van Tendeloo et al. 2001). On the one hand, a plasmid has the structural components of the virus, but the packaging signal for the encapsulation of the viral RNA is excluded.

On the other hand, the gene of interest is found in another vector with the packaging signal and the LTR (Long Terminal Repeats) sequence (for lentivirus), necessary for the integration of the virus in the genome. In practice, this process has been made more efficient by establishing a protocol for the generation of viral particles. In this process, the plasmid that carries the gene of interest along with the viral packaging signals, as well as the LTR integration sequences, is introduced to a cell line that expresses the structural components of the viral vector (capsid envelope proteins, transcriptase, integrase). Obtaining these viral particles ensures that a single transduction event is generated (Seow and Wood 2009).

Regarding the mechanisms that could be used by viral vectors to cross through the cell membrane, three examples have been well documented with adenovirus, adeno-associated virus and lentivirus (Waehler et al. 2007; Warnock et al. 2011). Adenoviruses which are non-envelop viruses, their capsid is structured in the form of an icosahedron, formed by a set of proteins denoted as hexon and penton, which make up the sides of the icosahedron and its vortices respectively.

Besides, the pentons extend a structure in the form of fiber ending in a knob domain. The genome of the adenoviruses is divided by a set of genes of early-stage, and late-stage comprising 40 kb, which contain Inverted Terminal Repeats (ITR) at their terminal ends. The introduction of this type of viral vector is generated by the interaction of the knob domain with the CAR receptor, followed by the union of a second receptor, the integrin that interacts with the penton and then the integration of the AVs is directed by endocytosis (Waehler et al. 2007; Warnock et al. 2011).

The Adeno-Associated Viruses (AVV) are DNA viruses without envelopes, surrounded by a protein cover that forms an icosahedron; its genome is constituted by two open reading frames that encode four proteins involved in its replication denoted Rep and three structural proteins of the capsid (VP1-VP3). The VP3 capsid protein binds to the membrane receptor proteoglycan heparan-sulfate (HSPG) and then to a coreceptor such as integrin, and finally the virus is internalized via endocytosis.

Finally, lentiviruses, which are a subcategory of retroviruses, are enveloped viruses. They have an RNA genome, with three common genes: gag, pol, and env; and six additional genes that fulfill both regulatory and structural functions. The most common lentivirus is the human immunodeficiency virus type I (VIH-I). The introduction of this type of vector to the host cell is given by the fusion

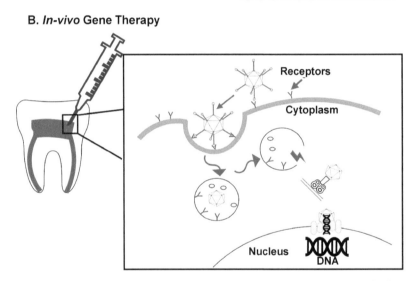

Figure 12.1. Strategies used for the introduction of viral vectors. (A) In *ex vivo* transfection, mesenchymal cells are removed from the tissue damage and then the culture in specialized scaffolds. The cells are then reprogrammed by introducing transcription factors employing viral vectors and subsequently transferred to the original niche; (B) *in vivo* transfection, the delivery vehicle is directly injected in the affected zone. In both strategies, the principal route to be internalized into the cell is the endocytic pathway.

of membranes, which is determined by the surface receptors of the virus envelope (glycoproteins) (Waehler et al. 2007; Warnock et al. 2011).

Gene therapy in the field of dental sciences, conducted its first studies on salivary gland cells, keratinocytes and cancer cells, from these early attempts, research in this field has been growing constantly (Baum et al. 2002; Siddique et al. 2016).

Target Genes, Growth Factors for Regeneration in the Oral Cavity

Gene therapy possesses numerous advantages over traditional treatments, such as greater sustainability than that of a single protein or compound application (Padial-Molina and Rios 2014).

In this regard, gene therapy is a two-step process; the first one involves a human genetic coding for the therapeutic protein; the coding must be cleaved and inserted into the genome of a carrier vector. The second part of the process includes the entry of the modified vector to the target human cells; then, the DNA sequence is released and becomes integrated within the chromosome. Ideally, the cells with the new genetic design begin forming the required therapeutic proteins (Siddique et al. 2016).

Coupled with tissue engineering strategies, gene therapy offers strong potential for three-dimensional tissue regeneration at the tooth-ligament-bone interface (Padial-Molina and Rios 2014).

An expanding area of clinical importance in dentistry which applies gene therapy is related to bone repair and regeneration, regeneration of tooth-supporting structures, implants, treatment of salivary gland disorder, among others. In this regard, gene therapy represents an ideal approach towards enhancing bone regeneration (Gupta et al. 2015). However, in the periodontium, regenerative treatment has been confronted with the morphological and functional specificities of each component of tooth-supporting tissues (Kaigler et al. 2006).

Notably, the healing of osseous tissues is a highly regulated process, in which growth factors and cytokines participate in a sequence of overlapping events similar to cutaneous wound repair (Kaigler et al. 2006). Systems for regulating the magnitude and temporal expression of the osteoblastic phenotype has been studied as a tool of genetic engineering in dentistry (Gupta et al. 2015).

The replacement of damaged bone with new bone mimics embryonic bone development. This process is driven by a cellular and molecular mechanism controlled by the Transforming Growth Factor-beta (TGF-β) superfamily of genes, also encoding a large number of extracellular signaling growth factors (Kaigler et al. 2006).

Next we provide a brief description of the most common study growth factor used in gene therapy (Table 12.2).

Bone Morphogenic Proteins

A unique family within the TGF-β superfamily are the Bone Morphogenic Proteins (BMPs). The BMPs play an essential role in the regulation of bone formation and repair; they can induce the formation of both bone and cartilage by stimulating the cellular events of mesenchymal progenitor cells (MSCs) (Kaigler et al. 2006; Rao et al. 2013).

Also, BMP-2, -4, -7 and -9 are the only growth factors that can individually induce *de novo* bone formation both *in vitro* and at heterotopic sites (Gupta et al. 2015). Moreover, BMPs -2, -4, -7 and -12 have been studied for periodontal and peri-implant bone regeneration. The most extensively research BMP for bone and periodontal regenerative treatment is BMP-2 (Kaigler et al. 2006).

In a direct gene therapy application, Chang et al. provided evidence that membranous bone repair with autologous tissue-engineered strategy could be achieved when using MSCs with adenovirus-BMP-2 gene transfer (Chang et al. 2003).

Moreover, Park et al. demonstrated that *ex vivo* BMP-2 gene delivery using periodontal ligament stem cell enhance new bone formation and re-osteointegration in peri-implantitis defect (Park et al. 2015).

Preclinical studies with BMP-2 have been conducted, and they demonstrated improvement of alveolar bone regeneration in different types of periodontal defects after treatment with BMP-2 via different vectors. However, ankylosis between tooth and alveolar bone is also occasionally reported, apparently associated with rapid osteogenesis (Kaigler et al. 2006).

In the case of BMP-7, direct *in vivo* gene delivery of adenovirus-BMP-7 transduced cells in a collagen gel carrier showed successful regeneration of alveolar bone defects around dental implants (Franceschi et al. 2000). However, preclinical and pilot human clinical studies made with rhBMP-7 and bone grafts for maxillary bone defects showed no significant differences between conventional treatments (Kaigler et al. 2006; Van den Bergh et al. 2000).

Table 12.2. Applications of growth factors and the vector use for their delivery.

Growth factors	Applications	Vector	Reference
BMP's	• Repair critical-sized mandibular defects. • Inducers of bone and bone marrow regeneration. • The osteoinductive potential makes them clinically valuable as alternatives to bone grafts. • Bone and cement regeneration.	• Adenoviral vector. • Biological matrices: demineralized bone matrix, collagen, fibrin. • Synthetic matrices: polylactic acid or polyglycolic acid. • Inorganic matrices: hydroxyapatite, tricalcium phosphate, other bioceramics. • Autologous bone marrow.	Akeel et al. 2000; Franceschi et al. 2000; Chang et al. 2003
PDGF	• Only PDGF has FDA-approval and commercialized as Regranex. • Enhance wound closure kinetics and re-epithelialization on full-thickness excisional skin wounds. • Lesions treated with Ad-PDGF-B tended to display increased evidence of new blood vessel formation, especially in the PDL region. • Ad-PDGF-B gene transfer not only enhanced the alveolar bone fill by two- to threefold but also increased cementum formation by higher than threefold, compared to controls.	• Adenoviral vector.	Anusaksathien et al. 2003; Cho et al. 1995; Jin et al. 2004
VEGF	• Plays an essential role in bone growth via the endochondral ossification pathway. • More bone was produced in calvaria defects that were treated with containing both VEGF and BMP genes than when either VEGF or BMP-transduced cells were implanted alone.	• Adenoviral vector VEGF + BMP.	Scheller et al. 2012; Passineau 2017
TGF-b	• Improvement wound closure kinetics, re-epithelialization and angiogenesis on full-thickness excisional skin wounds. • Treatment of critical size, supra-alveolar periodontal defects in beagle dogs may be of limited clinical benefit.	• Plasmid electroporation. • Hydroxy-ethyvisco-elastic gel with $CaCO_3$ particles.	Lee et al. 2012; Wikesjö et al. 2004; Nakashima et al. 2004
bFGF	• The increased periodontal ligament, bone formation on class II furcation on dog. • Dose-dependent bone and cement regeneration on class II furcation; nonhuman primate. • Enhanced alveolar bone height, periodontal ligament regeneration on 74 patients, Phase II randomized controlled trial, 2–3 wall intrabony defects; clinical.	• Topical application. • Gelatin hydrogel. • Hydroxypropyl cellulose hydrogel.	Kaigler et al. 2003; Shi et al. 2010
IGF	• No periodontal regeneration was shown in chronic periodontitis in a primate model. • PDGF-B/IGF showed increased bone fill on 38 patients, phase I/II randomized controlled trial, bilateral intrabony, and furcation defects; clinical.	• Methylcellulose gel. • Gelatin.	Giannobile et al. 2001; Shi et al. 2010

Platelet-Derived Growth Factor

The first growth factor to be evaluated in preclinical periodontal and peri-implant regenerative studies was Platelet-Derived Growth Factor (PDGF). The PDGF family is composed of four growth factors PDGF-A, -B, -C and D. Although, all PDGF factors participate in the wound-healing process, only the AA, BB and AB isoforms have been evaluated in periodontal therapy (Kaigler et al. 2006).

Periodontal ligament cells transfected with PDGF gene were able to stimulate gingival fibroblasts, periodontal ligament fibroblast, cementoblasts and osteoblast-like cells proliferation, migration and matrix synthesis, compared with PDGF continuous administration (Gupta et al. 2015; Kaigler et al. 2006).

PDGF has a significant impact on soft tissues and skeletal tissues. In this regard, FDA approved PDGF for the treatment of neurotrophic diabetic ulcers and for promoting bone repair of periodontal osseous defects (Gupta et al. 2015).

A wound repaid model using human gingival fibroblast showed that adenovirus-mediated PDGF-B gene transfer accelerates gingival soft tissue wound healing (Anusaksathien et al. 2003).

Transforming growth factor-β

The Transforming Growth Factor-b (TGF-β) is a multifunctional growth factor structurally related to BMPs; however, its function is quite different. The TGB- β1, the most abundant isoform of the TGF-β family, is a homodimeric protein synthesized and secreted by T CD4+ lymphocytes, macrophages. TGB-β1 is also found primarily in the platelets and osseous tissue (Li et al. 2014).

Previously reported data suggest that TGF-β1 the cellular proliferation secondary to topical administration is related to the differentiation of predetermined mesenchymal osteoprogenitor cells into osteoblast or chondroblasts. Also, in uncommitted precursor cells, TGF-β is not able to induce cells to become bone-forming cells; therefore, it is not considered to be osteoinductive (Clokie and Bell 2003).

Moreover, Li and Niyibizy (Li and Niyibizi 2012) used induced Pluripotent Stem Cells (iPSC) derived from mice; and exposed iPSC to TGB-β1 or TGF-β3. Their results demonstrated that TGB-β1 or TGF-β3 have the potential to induce iPSC to become cells that display osteoblast features *in vitro* and *in vivo*.

Vascular Endothelial Growth Factor

Bone formation requires nutritional supply from near blood vessels; therefore, angiogenic factors are involved in the deposit and maintenance of healthy bones (Gupta et al. 2015).

Vascular Endothelial Growth Factor (VEGF) has an angiogenic effect *in vitro* and *in vivo*, and has been studied for craniofacial regeneration. In particular, the proximity between osteoblast and endothelial cells during bone formation and repair suggest a cross-talk between these cells. Moreover, previously reported data suggest that VEGF and BMPS are involved in the cross-talk between bone cells and endothelial cells (Akeel et al. 2012).

Fibroblast Growth Factor 2

Belonging to the family of heparin-binding growth factors, that also have stimulatory properties is the basic Fibroblast Growth Factor 2 (FGF-2). The effects of FGF-2 include the induction of chemotaxis, angiogenesis and mitogenic activity; it is also involved in early differentiation and developmental processes (Lee et al. 2012).

The angiogenic stimulatory properties of FGF-2, particularly in fibroblast during wound healing, and its chemotactic and proliferative effect showed in periodontal ligament cells, suggested its potential use for periodontal regenerative therapeutic approaches (Kaigler et al. 2006).

However, the effect of FGF-2 in differentiation has shown contradictory results. For example, prolonged treatment with high concentrations of FGF-2 inhibits the differentiation of osteoblast by suppressing the synthesis of collagen type I and other osteogenic proteins (Lee et al. 2012).

In this respect, previous reports that evaluate more than one concentration of FGF-2 suggested that its effects are dose-dependent (Kaigler et al. 2006).

Insulin-like Growth Factor-I

Insulin-like Growth Factor-I (IGF-I) is a 7.6 kDa polypeptide capable of stimulating matrix synthesis and cell proliferation of chondrocytes. Similar to FGF-2, IGF-I is anabolic and mitogenic for articular chondrocytes. It also increases chondrocyte production of proteoglycan and collagen type II, the two main constitutes of cartilage (Madry et al. 2005).

In this respect, traumatic cartilage lesions have a limited capacity to heal. Therefore, the use of IGF-I as a therapeutic agent is often restrained by its short intra-articular residence time and the intrinsic scarcity of articular chondrocytes to serve as target cells.

Shi et al. (2010) demonstrated that gene transfer using a plasmid of an adeno-associated virus vector enables adult articular chondrocytes to produce appropriate amounts of IGF-I and FGF-2. Their results imply that genetically engineered chondrocytes, capable of secrete and supply IGF-I represent a potential approach for the application of gene therapy to articular cartilage repair (Shi et al. 2010).

Enamel Matrix Derivative

Enamel Matrix Derivative (EMD) is derived from embryonic enamel matrix. Moreover, amelogenis compose the significant fraction of the enamel matrix proteins, this protein family is hydrophobic proteins that account for more than 90% of the organic constituent of the enamel matrix (Venezia et al. 2004). Ameloblastin, MMP-20 and EMSP1 are also part of the enamel matrix proteins. The enamel matrix proteins have been purified and the cDNA cloned for developing porcine teeth and are known as EMD.

The commercial EMD matrix approved by the FDA (Emdogain®, Bira AB, Malmö, Sweden), a purified acidic extract of developing embryonal EMD from six month-piglets, is use for the treatment of periodontal defects. The proposed mechanisms that EMD aid in mimicking the specific events that occur in the development of periodontal tissues (Venezia et al. 2004; Kaigler et al. 2006).

Interestingly, EMD uses propylene glycol alginate as a vehicle in a viscous formulation, which is reduced under physiological conditions and facilitates EMD release and precipitation (Kaigler et al. 2006).

Gene Therapy and Oral Tissues

Salivary Glands

Salivary Glands (SGs) produce saliva, and several macromolecules involved in oral homeostasis. Moreover, saliva aids in lubrication, mastication, digestion of food and possesses antimicrobial properties that are a vital component of local immunity (Gupta et al. 2015; Nanci 2012).

Saliva is rich in antimicrobial peptides that are a vital component of local immunity (Khurshid et al. 2016).

Oral cancer treatment includes chemotherapy and radiation therapy, and both treatment modalities have a profound effect on salivary glands morphology and function.

For example, SGs are frequently used in the radiation field, and commonly affected by this treatment modality. Most complications of radiation therapy in head and neck cancer are related to SGs dysfunction.

The complications include acute toxicity, xerostomia (dry mouth), dysphagia (difficulty to swallow), erythema, transitory loss of taste, increased dental caries and other oral infections, local pain and discomfort. Consequently, a significant decline occurs in the quality of life of head and neck cancer patients (Huber et al. 2001; Khurshid et al. 2016).

Although salivary gland damage can be transitory in some patients, or responsive to pharmacological therapy; many patients experience a lifelong reduction in their ability to secrete saliva, and with that multiple oral morbidities (Khurshid et al. 2016).

In this respect, salivary glands were first targeted for intended clinical gene transfer applications in the early 1990s. However, gene-therapy for SGs could not be limited to their dysfunction. For example, the transfer of a new gene via retroductal cannulation of the main excretory ducts of a major SG could lead to the production of a cellular therapeutic protein or its secretion in saliva or the bloodstream.

Also, SGs can secrete proteins into the bloodstream making them potentially useful target sites for gene transfer in a minimally invasive manner with the help of intraductal cannulation (Passineau 2017).

Thus, SGs gene therapy represents an attractive approach for the treatment of major systemic pathologies such as diabetes and upper gastrointestinal tract diseases (Gupta et al. 2015; Yin et al. 2006).

Dentin

Current research on tissue engineering has reported promising outcomes for the regeneration of different oral and dental tissues. Moreover, in practice, scaffold-based approaches are generally used in combination with cells and growth factors (Boehler et al. 2011).

As previously described, the use of growth factors released at the site of bone defects or injury to activate endogenous stem cells constitutes a valid and efficient method for tissue regeneration. Although it is charged by high production costs and a short duration over time. Therefore, gene therapy is an alternative approach to overcome limitations of protein therapy (Mele et al. 2016).

Gene therapy for restoring dental tissues lost due to caries, periodontal disease and trauma is an attractive concept. Moreover, the dental pulp contains mesenchymal stem cells with the potential to differentiate into dentin-forming odontoblasts (Kichenbrand et al. 2019). Thus, stem cell-based genetically engineered cells could be cultured, modified or transfected, and then re-implanted into the host recipient (Siddique et al. 2016).

The healing potential of tissues, such as dentine pulp complex, is also enhanced by genes stimulating dentine formation. For example, Nakashima et al. (2004) applied *ex vivo* gene therapy; they used autogenous transplantation of Gdf11-transfected cells culture as pellet on amputated pulp in a canine model. They concluded that transfer of Gdf11 gene by electroporation to pulp cells *in vitro*, and their transplantation into amputated *in vivo* induced reparative dentin, particularly tubular dentin.

In the case of *in vivo* gene therapy, the healing potential of tissues such as dentine pulp complex could be enhanced by genes transfected into mesenchymal stem cells and stimulated dentine formation after being applied directly on the exposed dental pulp. Moreover, a tooth germ could be engineered *in vitro* or *ex vivo* with a culture of epithelial and mesenchymal stem cells, resulting into a non-metal tooth implant (Siddique et al. 2016; Bluteau et al. 2008).

Pulp

The dental pulp is a soft connective tissue located at the center cavity of the tooth; this is composed of nervous and vascular cells. The principal factor from tissue damage occurs as a result of caries. The primary treatment to protect the tissue from damage is the pulp capping procedure; through this treatment, the dental pulp is protected from necrosis through the deposition of calcium hydroxide. Other methods have employed the use of growth factors that have been installed directly at the site of the injury (OP-1, TGF-β, BMP-4, and BMP-2). Through the use of Gene Therapy Dental Pulp Stem Cells DSPCs transfected with Vascular Endothelial Growth Factor (VEGF) and Stromal cell-Derived Factor-1α (SDF-1α), have proven to be an effective treatment that allows an increase in pulp regeneration *in vivo* (Goldberg et al. 2001; Zhu et al. 2018).

Cementum

Cementum is a thin layer of mineralized tissue covering the tooth root surface, which provides a mineralized interface where the soft-tissue attachment has to be re-established. Also, the cementum matrix is a rich source of many growth factors that influence the activities of various periodontal cell types (Han et al. 2015; Grzesik and Narayanan 2002).

In humans, different types of cementum are classified according to the presence of cells and collagen fibers (Bosshardt and Selvig 1997).

Gene therapy has been another approach for the regeneration of dental support structures, including cementum. Various *in vitro* and *ex vitro* studies have suggested that PDGF (Platelet-Derived Growth Factor), FGF (Fibroblast Growth Factor), IGF (Insulin Growth Factor), BMP (Bone Morphogenic Protein) and other growth factors are able to strongly stimulate periodontal regeneration, including both bone and cementum (Sood et al. 2012).

For example, Giannobile et al. (2001) showed in their study that the administration of Ad/PGDF-A genes is equal to or greater than the continuous application of rhPDGF-AA.

Also, Okawa et al. (2017) reported significant alveolar bone formation and a moderate improvement in cement formation after gene therapy using marrow cells bone transduced with a lentiviral vector that expresses a form of TNSALP in a murine model.

Finally, Jin et al. (2003) reported that bone lesions treated by the supply of the Ad-BMP-7 gene showed a rapid chronogenesis, with subsequent osteogenesis, cementogenesis and predictable bridge of periodontal bone defects.

Periodontal Ligament

The periodontal ligament (PDL) supports and attaches the tooth to its alveolar socket made up of various collagen fibers around the tooth. These include gingival, trans-septal, alveolar crest, horizontal, oblique and apical fibers. It is essential to know that the periodontal ligament (PDL) is a very vascular soft connective tissue with the presence of blood vessels, that surrounds the root of the teeth, joining the root cement with the alveolar bone. Previously reported data showed the potential of gene therapy to introduce nucleotides into the cells to compensate for mutated genes or to restore the normal protein in specific tissues like bone, muscle, nerve, skin and mucosa (Gulabivala and Ng 2014).

Periodontitis is a periodontal disease that includes gingival inflammation and destruction of periodontal tissues including cementum, alveolar bone and PDL caused in most cases tooth loss. The therapy used is directed to stop the progression of the disease process, searching the regeneration of alveolar bone, periodontal ligament and the cementum that is lost. Finally to prevent the recurrence after treatment (Williams 1990).

Thus, this tissue, together with cementum and bone around the tooth root exists at an interface. So, the challenge of gene therapy is identified and deliver multiple signals in the correct order to achieve the regeneration or maintaining these kinds of complex tissues (Scheller et al. 2012).

Until now, the results have focused on finding the stimulation of fibroblasts, epithelial cells, osteoblasts, cementoblasts and mesenchymal stem cells included in PDL tissue widely accepted to have the potential to maintain or regenerate the periodontium (Maeda et al. 2013; Huri et al. 2015).

A part of these results is addressed to use the growth factors with gene therapy for periodontal regeneration that is defined as reproduction or reconstruction of a lost or injured part, so that form and function of lost structures are restored (Sood et al. 2012).

The regeneration of the periodontal tissues depends on appropriate signals, cells, blood supply and scaffold needed to target the fabric at the defect site (Taba et al. 2005). Thus, the objective of this therapy is to promote endogenous repair mechanisms and functional regeneration through the delivery of key Growth Factors (GFs) or cytokines that stimulate host cells to invade a tissue defect and direct robust extracellular matrix synthesis *in vivo* (Goker et al. 2019).

Platelet-Derived Growth Factor (PDGF) has demonstrated potent effects on the regeneration of tooth-supporting structures. Proliferation, migration and matrix synthesis has been observed in cultures of PDL (Sood et al. 2012). The three isoforms of PDGF, PDGF-AA, BB and AB have been evaluated in periodontal therapy. PDGF-BB is most effective on PDL cell mitogenesis and matrix biosynthesis (Jin et al. 2004).

Also, PDL cells transfected with PDGF gene have shown to stimulate gingival fibroblasts, periodontal ligament fibroblasts, cementoblasts and osteoblast-like cells proliferation, migration and matrix synthesis (Gupta et al. 2015; Kaigler et al. 2006).

Moreover, PDGF-BB demonstrated a complete regeneration in alveolar bone defects of critical size to Guided Tissue Regeneration (GTR) *in vivo* experiments. The results showed the formation of connective tissue in an early stage of repair, filling and stabilizing the wound. In a subsequent regenerative step, the connective tissue was substituted with new bone and PDL (Park et al. 1995; Sood et al. 2012; Cho et al. 1995).

On the other hand, the isoform 1 of the Transforming Growth Factor-β (TGF-β1) seems to play an essential role in inducing fibroblastic differentiation of PDL stem/progenitor cells and in maintaining the PDL apparatus under physiological conditions (Fujii et al. 2010; Kaigler et al. 2006).

BMPs have indicated significant alveolar bone regeneration in different types of periodontal defects after treatment with rhBMP-2 via various carriers. In the case of PDL the effect of BMP-12 to repair PDL tissue has been shown *in vitro* and *in vivo* studies. For example, Wikesjö et al. (2004) demonstrated that the BMP-12 treatment leads to less bone and more functionally oriented PDL between the new bone and new cementum by *in vivo* experiments. In another study BMP-7 gene transfers not only enhanced alveolar bone repair but also stimulated cementogenesis and PDL fiber formation (Sood et al. 2012).

Moreover, the use of biomaterials to achieve periodontal regeneration such as polymers serve as the delivery system for growth factors and DNA molecules, indicated by the retard apical migration of epithelial cells from the PDL region and alveolar bone. Diverse studies showed that bioresorbable recombinant human Growth/Differentiation Factor-5 (rhGDF-5) with poly(lactic-co-glycolic acid) (PLA) was applied in an *in vivo* experiment; bone formation was observed after 6 weeks of treatment in sites receiving rhGDF-5/PLGA exhibiting a significant increase in PDL, cementum and bone regeneration (Kwon et al. 2010).

Conclusion and Perspectives

Gene therapy has become a powerful tool in the treatment of clinical diseases of the oral cavity. The potential to manipulate the different cell types of the complex craniofacial system, by stimulating them with a combination of growth factors and later their reintegration into the damaged tissue, is a promising strategy to avoid surgical procedures. However, despite the advantages of this genetic strategy, severe limitations still prevent its use as a clinical therapy in humans. For example, to have sustained expression of a specific factor and promote specific cell reprogramming, the integration of transgene into the genome of the host cell is essentially required. Moreover, leading to devastating consequences for cell fate that can lead to the development of cancer. Future guidelines in gene therapy to avoid such reactions in the cell should aim the development of vectors whose transfection mechanism, whether transient or stable, is controlled, as well as improve specificity to infect a particular cell type.

References

Akeel, S., El-Awady, A., Hussein, K., El-Refaey, M., Elsalanty, M., Sharawy, M. et al. 2012. Recombinant bone morphogenetic protein-2 induces up-regulation of vascular endothelial growth factor and interleukin 6 in human pre-osteoblasts: Role of reactive oxygen species. Arch. Oral Biol. 57(5): 445–452.

Anguela, X.M. and High, K.A. 2019. Entering the modern era of gene therapy. Ann. Rev. Med. 70: 273–288.

Anusaksathien, O., Webb, S.A., Jin, Q.M. and Giannobile, W.V. 2003. Platelet-derived growth factor gene delivery stimulates *ex vivo* gingival repair. Tissue Eng. 9(4): 745–756.

Baum, B.J., Kok, M., Tran, S.D. and Yamano, S. 2002. The impact of gene therapy on dentistry: a revisiting after six years. J. Am. Dent. Assoc. 133(1): 35–44.

Bluteau, G., Luder, H.U., De Bari, C. and Mitsiadis, T.A. 2008. Stem cells for tooth engineering. Eur. Cell Mater. 16(1): 9.

Boehler, R.M., Graham, J.G. and Shea, L.D. 2011. Tissue engineering tools for modulation of the immune response. Biotechniques 51(4): 239–254.

Bosshardt, Dieter D. and Knut A. Selvig. 1997. Dental cementum: The dynamic tissue covering of the root. Periodontol. 2000 13(1): 41–75.

Botelho, J., Cavacas, M.A., Machado, V. and Mendes, J.J. 2017. Dental stem cells: recent progresses in tissue engineering and regenerative medicine. Ann. Med. 49: 644–651.

Chang, S.C., Chuang, H.L., Chen, Y.R., Chen, J.K., Chung, H.Y., Lu, L. et al. 2003. *Ex vivo* gene therapy in autologous bone marrow stromal stem cells for tissue-engineered maxillofacial bone regeneration. Gene Ther. 10(24): 2013–2019.

Cho, M.I., Lin, W.L. and Genco, R.J. 1995. Platelet-derived growth factor-modulated guided tissue regenerative therapy. J. Periodontol. 66(6): 522–530.

Clokie, C.M.L. and Bell, R.C. 2003. Recombinant human transforming growth factor β-1 and its effects on osseointegration. J. Craniofac. Surg. 14 (3): 268–277.

Davies, J.A. 2019. Real-world synthetic biology: Is it founded on an engineering approach, and should it be? Life (Basel) 9(1): 6.

Franceschi, R.T., Wang, D., Krebsbach, P.H. and Rutherford, R.B. 2000. Gene therapy for bone formation: *in vitro* and *in vivo* osteogenic activity of an adenovirus expressing BMP7. J. Cell Biochem. 78(3): 476–86.

Fujii, S., Hidefumi, M., Atsushi, T., Satoshi, M., Kiyomi, H., Naohisa, W. et al. 2010. Effects of TGF-β1 on the proliferation and differentiation of human periodontal ligament cells and a human periodontal ligament stem/progenitor cell line. Cell Tissue Res. 342(2): 233–242.

Giannobile, W.V., Lee, C.S., Tomala, M.P., Tejeda, K.M. and Zhu, Z. 2001. Platelet-derived growth factor (PDGF) gene delivery for application in periodontal tissue engineering. J. Periodontol. 72(6): 815–823.

Glorioso, J.C. and Lemoine, N. 2017. Gene therapy-from small beginnings to where we are now. Gene Ther. 24(9): 495–496.

Goker, F., Larsson, L., Del Fabbro, M. and Asa'ad, F. 2019. Gene delivery therapeutics in the treatment of periodontitis and peri-implantitis: A state of the art review. Inter. J. Mol. Sci. 20(14): 3551.

Goldberg, M., Six, N., Decup, F., Buch, D., Soheili Majd, E., Lasfargues, J.J. et al. 2001. Application of bioactive molecules in pulp-capping situations. Adv. Dent. Res. Aug; 15: 91–5. PubMed PMID: 12640750.

Grzesik, W.J. and Narayanan, A.S. 2002. Cementum and periodontal wound healing and regeneration. Crit. Rev. Oral Biol. Med. 13(6): 474–484.

Gulabivala, K. and Ng, Y.L. 2014. Endodontics E-Book: Elsevier Health Sciences.

Gupta, K., Singh, S. and Garg, K.N. 2015. Gene therapy in dentistry: Tool of genetic engineering. Revisited. Arch. Oral Biol. 60(3): 439–446.

Han, P., Ivanovski, S., Crawford, R. and Xiao, Y. 2015. Activation of the canonical Wnt signaling pathway induces cementum regeneration. J. Bone Miner. Res. 30(7): 1160–1174.

Huber, P.E., Debus, J., Latz, D., Zierhut, D., Bischof, M., Wannenmacher, M. et al. 2001. Radiotherapy for advanced adenoid cystic carcinoma: neutrons, photons or mixed beam? Radiother. Oncol. 59(2): 161–167.

Huri, P.Y., Temple, J.P., Hung, B.P., Cook, C.A. and Grayson, W.L. 2015. Bioreactor technology for oral and craniofacial tissue engineering. pp. 117–130. *In*: Stem Cell Biology and Tissue Engineering in Dental Sciences. Academic Press.

Intini, G. 2010. Future approaches in periodontal regeneration: gene therapy, stem cells, and RNA interference. Dent. Clin. North Am. 54: 141–155.

Jin, Q., Anusaksathien, O., Webb, S.A., Printz, M.A. and Giannobile, W.V. 2004. Engineering of tooth-supporting structures by delivery of PDGF gene therapy vectors. Mol. Ther. 9(4): 519–526.

Jin, Q.M., Anusaksathien, O., Webb, S.A., Rutherford, R.B. and Giannobile, W.V. 2003. Gene therapy of bone morphogenetic protein for periodontal tissue engineering. J. Periodontol. 74(2): 202–213.

Jooss, K. and Chirmule, N. 2003. Immunity to adenovirus and adeno-associated viral vectors: implications for gene therapy. Gene Ther. 10(11): 955–63.

Kaigler, D., Cirelli, J.A. and Giannobile, W.V. 2006. Growth factor delivery for oral and periodontal tissue engineering. Expert. Opin. Drug Deliv. 3(5): 647–662.

Khurshid, Z., Naseem, M., Sheikh, Z., Najeeb, S., Shahab, S. and Zafar, M.S. 2016. Oral antimicrobial peptides: Types and role in the oral cavity. Saudi Pharm. J. 24(5): 515–524.

Kichenbrand, C., Velot, E., Menu, P. and Moby, V. 2019. Dental Pulp Stem Cell-Derived Conditioned Medium: An Attractive Alternative for Regenerative Therapy. Tissue Eng. Part B Rev. 25(1): 78–88.

Kim, T.K. and Eberwine, J.H. 2010. Mammalian cell transfection: The present and the future. Anal. Bioanal. Chem. 397(8): 3173–8.

Kwon, D.H., Bennett, W., Herberg, S., Bastone, P., Pippig, S., Rodriguez, N.A. et al. 2010. Evaluation of an injectable rhGDF-5/PLGA construct for minimally invasive periodontal regenerative procedures: a histological study in the dog. J. Clin. Periodontol. 37(4): 390–397.

Lee, J.H., Um, S., Jang, J.H. and Seo, B.M. 2012. Effects of VEGF and FGF-2 on proliferation and differentiation of human periodontal ligament stem cells. Cell Tissue Res. 348(3): 475–484.

Li, F. and Niyibizi, C. 2012. Cells derived from murine induced pluripotent stem cells (iPSC) by treatment with members of TGF-beta family give rise to osteoblasts differentiation and form bone *in vivo*. BMC Cell Biol. 13(1): 35.

Li, Z., Jiang, C.M., An, S., Cheng, Q., Huang, Y.F., Wang, Y.T. et al. 2014. Immunomodulatory properties of dental tissue-derived mesenchymal stem cells. Oral Dis. 20(1): 25–34.

Madry, H., Kaul, G., Cucchiarini, M., Stein, U., Zurakowski, D., Remberger, K. et al. 2005. Enhanced repair of articular cartilage defects *in vivo* by transplanted chondrocytes overexpressing insulin-like growth factor I (IGF-I). Gene Ther. 12(15): 1171.

Maeda, H., Wada, N., Tomokiyo, A., Monnouchi, S. and Akamine, A. 2013. Prospective potency of TGF-β1 on maintenance and regeneration of periodontal tissue. pp. 283–367. *In*: International Review of Cell and Molecular Biology (Vol. 304). Academic Press.

Mele, L., Vitiello, P.P., Tirino, V., Paino, F., De Rosa, A., Liccardo, D. et al. 2016. Changing paradigms in craniofacial regeneration: current and new strategies for the activation of endogenous stem cells. Front. Physiol. 7: 62.

Nakashima, M., Iohara, K., Ishikawa, M., Ito, M., Tomokiyo, A., Tanaka, T. et al. 2004. Stimulation of reparative dentin formation by *ex vivo* gene therapy using dental pulp stem cells electrotransfected with growth/differentiation factor 11 (Gdf11). Hum. Gene Ther. 15(11): 1045–1053.

Nanci, A. 2012. Development of the tooth and its supporting tissues. pp. 79–107. *In*: Nanci, A. (ed.). Ten Cate's Oral Histology: Development, Structure, and Function. Seventh Ed. Mosby Press, St. Louis, Mo, USA.

Ohba, S., Hojo, H. and Chung, U.I. 2012. Bioactive factors for tissue regeneration: state of the art. Muscles Ligaments Tendons J. 2(3): 193–203.

Okawa, R., Iijima, O., Kishino, M., Okawa, H., Toyosawa, S., Sugano-Tajima, H. et al. 2017. Gene therapy improves dental manifestations in hypophosphatasia model mice. J. Periodontal. Res. 52(3): 471–478.

Padial-Molina, M. and Rios, H.F. 2014. Stem cells, scaffolds and gene therapy for periodontal engineering. Curr. Oral Health Rep. 1(1): 16–25.

Park, J.B., Matsuura, M., Han, K.Y., Norderyd, O., Lin, W.L., Genco, R.J. et al. 1995. Periodontal regeneration in class III furcation defects of beagle dogs using guided tissue regenerative therapy with platelet-derived growth factor. J. Periodontol. 66(6): 462–477.

Park, S.Y., Kim, K.H., Gwak, E.H., Rhee, S.H., Lee, J.C., Shin, S.Y. et al. 2015. *Ex vivo* bone morphogenetic protein 2 gene delivery using periodontal ligament stem cells for enhanced re-osseointegration in the regenerative treatment of peri-implantitis. J. Biomed. Mater. Res. A. 103(1): 38–47.

Passineau, M. 2017. Salivary gland gene therapy in experimental and clinical trials. pp. 217–228. *In*: Salivary Gland Development and Regeneration. Springer, Cham.

Rao, S.M., Ugale, G.M. and Warad, S.B. 2013. Bone morphogenetic proteins: Periodontal regeneration. N. Am. J. Med. Sci. 5(3): 161.

Scheller, E.L., Villa-Diaz, L.G. and Krebsbach, P.H. 2012. Gene therapy: implications for craniofacial regeneration. J. Craniofac. Surg. 23(1): 333.

Seow, Y. and Wood, M.J. 2009. Biological gene delivery vehicles: Beyond viral vectors. Mol. Ther. 17(5): 767–777.

Shi, S., Mercer, S. and Trippel, S.B. 2010. Effect of transfection strategy on growth factor overexpression by articular chondrocytes. J. Orthop. Res. 28(1): 103–109.

Siddique, N., Raza, H., Ahmed, S., Khurshid, Z. and Zafar, M.S. 2016. Gene therapy: A paradigm shift in dentistry. Genes (Basel) 7(11): 98.

Sood, S., Gupta, S. and Mahendra, A. 2012. Gene therapy with growth factors for periodontal tissue engineering—A review. Med. Oral Patol. Oral Cir. Bucal. 17(2): e301.

Taba Jr, M., Jin, Q., Sugai, J.V. and Giannobile, W.V. 2005. Current concepts in periodontal bioengineering. Ortho. Craniofac. Res. 8(4): 292–302.

Van den Bergh, J.P.A., Ten Bruggenkate, C.M., Groeneveld, H.H.J., Burger, E.H. and Tuinzing, D.B. 2000. Recombinant human bone morphogenetic protein-7 in maxillary sinus floor elevation surgery in 3 patients compared to autogenous bone grafts: A clinical pilot study. J. Clin. Periodontol. 27(9): 627–636.

Van Tendeloo, V.F., Van Broeckhoven, C. and Berneman, Z.N. 2001. Gene therapy: principles and applications to hematopoietic cells. Leukemia 15(4): 523–44.

Venezia, E., Goldstein, M., Boyan, B.D. and Schwartz, Z. 2004. The use of enamel matrix derivative in the treatment of periodontal defects: a literature review and meta-analysis. Crit. Rev. Oral Biol. Med. 15(6): 382–402.

Warnock, J.N., Daigre, C. and Al-Rubeai, M. 2011. Introduction to viral vectors. pp. 1–25. *In*: Viral Vectors for Gene Therapy. Humana Press.

Waehler, R., Russell, S.J. and Curiel, D.T. 2007. Engineering targeted viral vectors for gene therapy. Nat. Rev. Genet. 8(8): 573–87.

Wikesjö, U.M., Sorensen, R.G., Kinoshita, A., Jian Li, X. and Wozney, J.M. 2004. Periodontal repair in dogs: Effect of recombinant human bone morphogenetic protein-12 (rhBMP-12) on regeneration of alveolar bone and periodontal attachment: A pilot study. J. Clin. Periodontol. 31(8): 662–670.

Williams, R.C. 1990. Periodontal disease. New Eng. J. Med. 322(6): 373–382.

Yin, C., Dang, H.N., Gazor, F. and Huang, G.T.J. 2006. Mouse salivary glands and human β-defensin-2 as a study model for antimicrobial gene therapy: technical considerations. Int. J. Antimicrob. Agents 28(4): 352–360.

Zhu, L., Dissanayaka, W.L. and Zhang, C. 2019. Dental pulp stem cells overexpressing stromal-derived factor-1α and vascular endothelial growth factor in dental pulpregeneration. Clin. Oral Investig. May; 23(5): 2497–2509. doi: 10.1007/s00784-018-2699-0. Epub 2018 Oct 12. PubMed PMID: 30315421.

13

Injectable Scaffolds for Oral Tissue Regeneration

Suárez-Franco J.L.* and *Cerda-Cristerna B.I.*

Introduction

Regenerative medicine is a field of medicine that has seen a great increase in recent years, this due to the potential it has in repairing or replacing damaged tissues and organs whether caused by trauma, age or illness, as well as congenital defects. The bio-engineering of tissues is a discipline that has been developed during the last 25 years, this discipline, is based on the synergy of three fundamental fields of science; such as cellular engineering, where the cells with the most promising potential due to their capacity for self-replication, their immuno-modulation and, most importantly, their capacity to become any tissues, are stem cells, which have been isolated and characterized by almost any organ; the synthesis of new materials that act as scaffolding or template simulating the physical, chemical and mechanical properties of the native extracellular matrix, these scaffolds are made of three different materials such as: metallic, ceramic and polymeric, being the second factor within bio-technical engineering and finally the addition or functionalization with molecules that induce the biological response of the cells towards the phenotype of the tissue or organ that it is trying to repair or replenish (Chatterjee et al. 2011; Bhattarai et al. 2018).

This chapter focuses on the synthesis of polymeric scaffolds by injection techniques, which simulate the architectural properties of the native extracellular matrix. The techniques described in the literature, such as electrospinning, will be discussed in depth, spinning by propulsion of air to the design of scaffolds by means of 3D and 4D printing. The advantages, disadvantages and challenges of each of these synthesis techniques will be addressed (Ashammakhi et al. 2018; Vazquez-Vazquez et al. 2019).

Electro-spinning

The technique of electro-spinning, is a technique developed in the year 1900, which is an economic technique and more common to develop fibrillary scaffolds in micrometric and nanometric scales. This technique uses electrostatic forces to deposit polymer solutions, thus forming the said scaffolds.

School of Dentistry, Universidad Veracruzana, Abasolo S/N, Río Blanco, Veracruz, México.
* Corresponding author: jsuarez@uv.mx

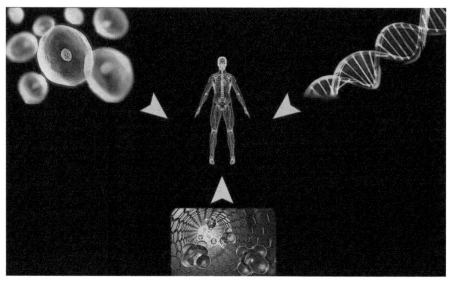

Figure 13.1. Regenerative medicine, which are confluencing biomolecules, scaffolds and cell engineering.

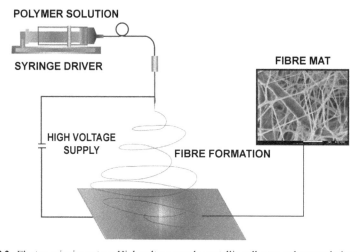

Figure 13.2. Electro-spinning set up. High voltage supply, metallic collector, polymer solution and pump.

The electro spinning technique consists of a device composed of four key elements: a peristaltic pump containing a syringe with polymer solution, a capillary (needle) to direct the polymer solution into an electro magnetic field, a high-voltage source to eliminate the surface tension of the polymeric solution and a metal collector where the fibers will be deposited, and in this way the nano/micro fibrillary polymeric scaffold is formed (Matthews et al. 2002; Yin et al. 2013).

During the electro-spinning process, the polymer solution stops through the needle, applying an electric potential to eliminate the surface tension and the visco elastic force, resulting in the formation of the Taylor cone. While the solvent evaporates during deposition, the continuous flow of the polymer is collected in a suitable metallic device (Bhattarai et al. 2018). The handling of the parameters of such polymer concentration, the distance between the collector and needle, the speed of deposition and electric field might affect the physico-chemical results in the handling of the properties of the synthetic scaffold. Furthermore, if a rotor coupled to the metal collector is used, it results in obtaining scaffolds with aligned fibers. It is because of these characteristics that the technique of electro spinning, is a simple, flexible and low cost option for the formation of scaffolds that can be applied to different areas of research (Alvarez-Perez et al. 2010; Levorson et al. 2013; Yin et al. 2013). The characterization of the scaffolds are widely reported in literature, such as SEM, TEM, RAMAN and FTIR are the mostly frequently reported.

Air Jet Spinning

Recently, a new technique has been reported, while not completely a method based on a flow of liquid driven by a plunger, it could be considered injectable, since it uses a polymer solution propelled by a flow of air or gas. This technique has been refiled as spinning by air propulsion, in this process with which three-dimensional polymeric scaffolds have been obtained, consists of a device, which is integrated by materials sold in hardware stores; this device has as it's main feature the use of an air brush connected to a source of air or gas, such as an air compressor or gas cylinder, mainly containing nitrogen or argon and finally a collector that can be of any material lined with aluminum or waxed paper. Like electro-spinning, the properties of the resulting fibrillar scaffolds by means of air pressure spinning vary according to certain parameters such as: polymer concentration, the different types of solvents used for the preparation of the solution, the distance from the tip of the air brush to the collector and the pressure of the air flow used (Medeiros et al. 2009; Abdal-hay et al. 2013).

One of the advantages of the synthesis by means of spinning by propulsion of gas, lies in the simplicity of the device, in how easy it is to obtain the assembly of the said device and how safe it is to operate it in comparison with other techniques that use high voltage for their operation. Among the disadvantages is that in spite of obtaining scaffolds in nanometric and micrometric scales, it is not possible to regulate the fiber diameter size with accuracy, in comparison with the electrospinning technique, in which this is one of its main advantages besides that in the technique of spinning by propulsion of air, only scaffolds with fibers without a defined orientation are obtained, on the other hand with electro-spinning scaffolds can even be obtained with fibers oriented definidamente, which resemble certain characteristics of tissues of the body (Tutak et al. 2013).

Within the biological characterization, it has been reported that mesenchymal stem cells isolated from bone marrow, when placed on these scaffolds and compared with the biological response on scaffolds synthesized by electro-spinning, are similar. It has been reported that in defects in animal models, when implanting these scaffolds, the specimens present inflammation which is possibly due to the concentration of the polymer and also to the non-evaporated solvents, however, after this response the scaffolds show dimensional stability, which is a requirement for tissue regeneration (Granados-Hernández et al. 2018; Suarez-Franco et al. 2018).

In recent times, a machine has been designed with the same principle of spinning by gas propulsion, this technique has been referred to as a blow spinning solution, in which a semi-automated device, that in addition to integrating the original components of spinning by gas propulsion, has incorporated a steel/aluminum nozzle, especially designed for this purpose, which has the advantage of easy disassembly and cleaning, since in the technique of spinning by air propulsion, it is a disadvantage that increases the concentration of polymer and the time of deposit, by giving rise to variations in the physical properties of the scaffolding, in addition to the nozzle, a mobile base was also incorporated, which allows automated movements in both horizontal and vertical direction and finally a pump injection to control the constant flow of the polymer solution; the other components are the same as the gas propulsion spinning technique. In this way the physical properties such as: fiber diameter size, orientation and no droplet formation are controlled in comparison with gas propulsion spinning; while the chemical properties of the scaffolding are not affected, something that occurs in the same way with electro-spinning and gas-propulsion spinning (Hell et al. 2018).

In dentistry, scaffolds synthesized by gas pressure spinning, the biological response of isolated mesenchymal cells of periodontal ligament has been characterized, where it was found that the response of biological activity such as adhesion and proliferation is related to the diameter of the

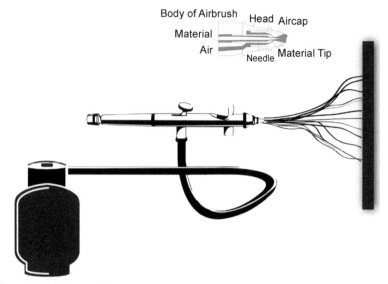

Figure 13.3. Air jet spinning set up. Air/gas supply, collector and commercial available air brush.

fibers that in turn depends on the modification of the concentration of the polymer solution, reporting that the lower the polymer concentration, the thinner the fibers which influence the expression of molecules related to the cell adhesion process previously reported on scaffolds synthesized by the electro-spinning technique (Abdal-hay et al. 2013; Suarez-Franco et al. 2018).

3D Printing and Bioprinting

Recent advances in tissue engineering owe their success to the development of novel biomaterials-based strategies mimicking native tissues, organ shapes and physiology. These biomaterials are capable of harnessing the innate abilities of cells to sense their local environment through cell–cell and cell–extracellular matrix (ECM) interactions and self assemble into complex networks. Many studies have focused on tuning the bulk properties (i.e., biomolecule concentrations, mechanical properties, cell to cell interactions) of these materials, which assumes that the native cellular environment is homogeneous across multiple length scales. Although modern biomaterials permit the investigation of complex cellular behaviors such as Mesenchymal Stem Cell (MSCs) differentiation and Epithelial-Mesenchymal Transition (EMT) to accurately replicate the heterogeneous nature of native cellular environments. Much of this is due to the inherent difficulty of precisely replicating three dimensional (3D) environments. Although many methods have been developed to generate complex two-dimensional (2D) patterns, biochemical and mechanical cues, 2D culture conditions may not be friendly for cell types. Furthermore, 2D fabrication techniques often cannot be readily translated into 3D culture systems. Therefore, there is a need to adapt these 2D methods and/or create entirely new methods to mimic the complex 3D cellular environment (Liu et al. 2018; Morissette Martin et al. 2019).

Recently, several methods have been developed to spatially encode local properties to 3D materials-based culture systems, and these methods are generically referred to as 3D biofabrication techniques. Such procedures are capable of either constructing or patterning materials, with a high degree of control, by finely tuning and defining material geometries, localization of biomolecular cues, and/or mechanical properties. In doing so, they have created complex material geometries to resemble endogenous tissues. Similarly, biofabrication-based patterning techniques have immobilized controlled concentrations of adhesive ligands, growth factors, or other signaling molecules to mimic cellular architectures *in vivo*. By enabling this precise control over local and bulk material properties,

Figure 13.4. 3D bioprint machine for polymeric intelligent biomaterials.

these methods can create new biomaterials that better replicate the complex and heterogeneous nature of endogenous tissues and organs. This will help to elucidate the gap between the state of tissue engineering and the unrealized hope of true artificial organs and tissues. In addition, this level of control could allow for a more complete model of cancerous or diseased states in cells and tissues to assess current or develop new, therapeutic strategies (Bajaj et al. 2014).

4D Printing and Bioprinting

Four dimensional (4D) printing was first introduced in 2013 and immediately spurred great attention in various research areas including, but not limited to, smart materials and biomedical research (Yang et al. 2019). Most 4D structures are developed by incorporating shape transformation within a material/structural design, which contributes to the 4D definition: 3D printing of objects which can, immediately after printing, self-transform in function or form when exposed to a predetermined stimulus, including osmotic pressure, exposure to heat, current, ultraviolet light or other energy sources. Shape memory materials have the inherent capacity to fix a temporary shape and recover their permanent structure under suitable stimuli, which is extremely similar in principle with the 4D dynamic process. 3D printing of shape memory materials is reported as novel 4D printing in recent articles; this is becoming a unique and rapidly expanding, research area in 4D printing. One argument in this field is whether controlled degradation of 3D printed constructs can be classified as a 4D effect, a claim noted in some articles. 3D printing has demonstrated great potential in biomedical fields. If biodegradation is included as a tunable mechanism of incorporating a time-dependent effect, 3D printing of timed-release therapeutics and other biodegradable structures fall under 4D printing. A point of critique in this argument is that the 3D printed structures noted above completely disappear in the dynamic processes. On the contrary, in shape or functional transformation through the 4D process, most of the 3D printed structures remain intact; that is, the 3D fabricated structures are the carrier of the shape or functionality shift as a product of the material properties of the print medium. As such, degradation of 3D printed constructs will not be treated as a 4D effect in this context. Although the surrounding environments may change significantly during the 4D process, the configuration or function before and after the stimulation should be structurally or functionally stable. For example, if a 3D fabricated construct has a primary conformation 'A' in air and changes to conformation 'B' in water, both conformations should be structurally stable, without external forces to maintain its structure (Ashammakhi et al. 2018; Yang et al. 2019).

Based on the concept of 4D printing, the fabricated structure performs a self-assembly process immediately after 3D printing; this self-assembly is a process in which a pre-existing form dynamically changes to another structure as a consequence of external stimuli. Actually, 3D printing of self-assembly structures existed prior to the definition of 4D printing. For example, shape-changing architectures were inkjet printed by patterning light-absorbing ink onto a prestrained polystyrene substrate; the ink acted like hinges to induce an autonomous shape change while they transferred the absorbed heat to the underlying substrate. Additionally, 3D aqueous droplets, with a volume of ~ 65 pL, were ejected into a lipid containing oil bath and programmed into pre-designed, bilayered networks which swelled or shrank in response to water flow, resulting from osmotic-pressure differences; a variety of shapes were fabricated through the use of droplet networks. 4D printing, however, is leading to greater attention, in various research fields, after the definition was coined (Miao et al. 2017; Morissette Martin et al. 2019).

The repair and regeneration of craniofacial tissues continue to be a challenge for clinicians and biomedical engineers. Reconstruction of pathologically damaged craniofacial tissues is often required because of tumors, traumas, or congenital malformations. There are constructive procedures for craniofacial tissue regeneration which are usually complex because the craniofacial region is a complex construct, consisting of bone, cartilage, soft tissue and neurovascular bundles. For instance, to reconstruct damaged craniofacial bones, surgical procedures are available. Autologous bone grafts have been considered the reference standard for bone regenerative therapies. Together with allogenic bone grafts, this type of bone graft material comprises more than 90% of grafts performed. However, these grafting procedures have numerous disadvantages, including hematomas, donor site morbidity, inflammation, infection and high cost (Morissette Martin et al. 2019).

However, the biomechanical properties of the tissues regenerated through these treatment options are mediocre compared with those of native articular cartilage. Furthermore, the repair and regeneration of muscle tissue (for example, tongue muscle) after traumatic injuries frequently exhibit a challenging clinical situation in the craniofacial region. Substantial esthetic and functional issues will arise if a significant amount of tissue is lost, because of the inability of the native muscle tissue to regrow and fill the defect site. To find an alternative treatment option for the reconstruction of craniofacial tissue, clinicians and scientists have been analyzing new approaches in craniofacial tissue regeneration to maximize patient benefit and minimize related complications. Craniofacial tissue regeneration using Mesenchymal Stem Cells (MSCs) presents an advantageous alternative therapeutic option. MSCs are multipotent cells that are capable of multiple lineage differentiation based on the presence of inductive signals from the microenvironment. MSCs reside in a wide spectrum of postnatal tissue types and have been successfully isolated from several orofacial tissues. Studies have confirmed the self-renewal and multilineage differentiation capacities of orofacial-derived MSCs and have shown that they have better growth properties than Bone Marrow Mesenchymal Stem Cells (BMMSCs). Therefore, dental MSCs are attractive for craniofacial applications as they may be better at differentiating into craniofacial tissues.

Biomaterials are widely used to engineer the physiochemical properties of the extracellular cell microenvironment to tailor niche characteristics and direct cell phenotype and differentiation. Such interactions between stem cells and biomaterials have largely been studied by introducing the cells into 2- or 3-dimensional scaffolds, or by encapsulating cells within hydrogel bio-materials. Alginate hydrogel has been used extensively as a vehicle for stem cell delivery in tissue regeneration. The ability to control the spatial presentation of alginate enables the examination of the effects of alginate hydrogel on stem cell differentiation in a systematic way. In the current chapter, the application of dental-derived MSCs and alginate hydrogel for potential applications in craniofacial tissue regeneration is emphasized (Yang et al. 2019).

Figure 13.5. Differences among 3D and 4D resulting printing.

Injectable Scaffolds and Regenerative Endodontics

Regeneration of dental pulp is the major challenge in endodontics. Contemporary research on endodontics focuses its attention on regenerative therapy based on the biological ability of the pulp tissue to regenerate itself with the help of stimuli. The regenerative biological activity occurs because of human stem cells in the pulp are able to differentiate into odontoblasts. Four cell types associated to dental pulp have been identified as a source of autologous cells to differentiate into odontoblasts: Dental Pulp Stem Cells (DPSCs), Stem Cells from Exfoliated Deciduous Teeth (SHED), Stem Cells from Apical Papilla (SCAP), and Dental Follicle Stem Cells (DFST). Strategies are investigated intensively to use those cells with tissue engineering to regenerate a dental pulp after disinfection of the root canal or even to regenerate a necrotic pulp. The trio scaffold and encapsulated cells plus encapsulated active molecules is the most common model investigated to promote pulp regeneration. As a first step, pulp regeneration requires physical-chemical disinfection of the root canal; irrigation with NaOCl or application of antimicrobial paste are clinical procedures to disinfect a root canal without removing the pulp tissue. Disinfection might be promoted also by a scaffold releasing an antimicrobial molecule. After the disinfection, the second step is inducing in the pulp the activation of key factors of pulp regeneration: (1) promotion of angiogenesis and its consequent vascularization of a matrix, (2) promotion of cell differentiation into odontoblasts, and (3) promotion of release of active molecules to trigger cell differentiation.

Injectable scaffolds are ideal systems to help activating the key factors for pulp regeneration, they are biomaterials with potential application in regenerative endodontics. Dental pulp is located in the root canal with an intricated morphology, with a micrometric diameter and with irregular walls; an injectable scaffold results ideal to be introduced into the complex root canal to carry inductors for pulp regeneration. Injectable scaffolds are a soft viscous matrix made of a natural or a synthetic polymer; thus, injectable scaffolds are hydrogels, in this chapter injectable scaffold or hydrogel refers to the same type of biomaterial. The injectable scaffolds for purposes of regenerative endodontics are formulated with different polymers and are loaded with different cells and different active molecules. The variety of formulations has been tested with both *in vitro* and *in vivo* models, the evidence from tests have shown very interesting results. It has been seen that injectable scaffolds induce biological responses in several cell lines and also in animal models. The ability to promote cell differentiation or tissue regeneration is associated with the composition and to the load of the hydrogel.

Injectable Scaffolds and In Vitro Observations

Injectable scaffolds can be designed to induce angiogenesis and a consequent vascularization of pulp tissue. In this regard, the use of a hydrogel encapsulating and releasing angiogenic growth factors is a good strategy for pulp regeneration. An example for this is an injectable hyaluronic acid hydrogels (HACPL) reinforced with cellulose nanocrystals and enriched with Platelet Lysate (PL) (Silva et al. 2018). The PL releases two angiogenic factors: Platelet-Derived Growth Factor (PDGF) and the Vascular Endothelial Growth Factor (VEGF). The HACPL has been used to encapsulate Human

Dental Pulp Cells (HDPC) and Human Umbilical Vein Endothelial Cells (HUVEC), after encapsulation the two different cell lines kept its viability and its metabolic activity, and were also able to develop a cell-to-cell network and a cell-matrix contact. Very interestingly, an elegant *in vivo* experiment in a chick chorioallantoic membrane assay showed that the HACPL induced blood vessel formation when the HDPC were encapsulated in the scaffold; the PL caused the formation of blood vessels (Silva et al. 2018). Thus injectable scaffolds that release angiogenic factors are attractive systems to create an environment suitable for blood vessel formation in a pulp tissue.

Another requirement for pulp regenerative therapy is cell seeding into the root canal by the injectable scaffold. The Human Dental Pulp Derived Cells (HDPDC) should maintain its metabolic activity and its ability to grow and to spread. Injectable scaffolds have shown that they are systems allowing these biological actions. A commercial injectable scaffold (PuraMatrix®) based on a polymer of the amino acid sequence (arginine-alanine-aspartic acid-alanine) showed its potential suitability to encapsulate dental pulp cells. The hydrogel with co-cultures of DPSCs and Human Umbilical Vein Endothelial Cells (HUVECs) was injected into root segments in the subcutaneous space of the dorsum of immuno-deficient mice. After the implantation, a pulp-like tissue grew in the middle and in apical third of the root segments; moreover odontoblast-like cells were observed and were identified as an odontoblastic lineage (Dissanayaka et al. 2015). The hydrogel was also used to entrap SHED and to inject the cell-hydrogel mix into root canals of human premolars. The engineered tissue appeared, and showed odontoblasts-like cells with a proliferative activity, these cells were able to form dentin. Interestingly, SHED in the hydrogel formed a connective-like tissue with blood vessels near to the predentin of the root canals (Rosa et al. 2013). Odontoblast-like cells and engineered tissue are pieces on the building of the pulp tissue, the success of its production induced by an injectable scaffold is definitively a valuable evidence.

Hyaluronic Acid (HA)-based injectable hydrogels have been explored for promoting cell differentiation including cell division and cell migration, angiogenesis and wound healing (Chircov et al. 2018). The HA is a common compound of the pulp tissue, and induces *in vitro* alkaline phosphate (ALP) activity in dental pulp cells (Chen et al. 2016). The HA has properties for acting like a good matrix for injectable hydrogels and HA-based injectable scaffolds have showed an appropriate capability to encapsulate precursor cells of odontoblasts. For instance, a HA-based Injectable Scaffold (HAISr) promoted *in vitro* differentiation of Stem Cell of the Apical Papilla (SCAP). The injectable scaffold encapsulating SCAP in osteogenic medium induced ALP activity significantly in cells that were identified as odontoblast-like. When SCAP contacted the injectable scaffold, its cell viability increased from 6 hours to 72 hours, however, the scaffold induced a lower cell viability compared with a control group (SCAP alone) (Chrepa et al. 2017). The HAISr is an injectable scaffold that the FDA approved for use in human subjects on cosmetic medicine (Restylane®), thus HA commercial scaffolds are an alternative to explore regenerative endodontics. Another commercial HA-based injectable hydrogel (HyStem-C®, not for human use) has been evaluated on its effect on hDPSCs. Hydrogel includes polyethylene glycol diacrylate and gelatin in its composition. When the hDPSCs were encapsulated in the injectable scaffold, the proliferation increased significantly from day one to day four, and the increase was higher when human fibronectin was added in the scaffold composition. The cell viability increased significantly from day one to day four (Jones et al. 2016). The potential for the injectable scaffolds on human dental pulp derived cells is large. Blood vessel formation, cell growing and cell differentiation and release of active molecules are some of the biological responses achieved.

Injectable Scaffolds and In Vivo Observations

Treatment of periapical lesions is another application investigated for injectable scaffolds in regenerative endodontics. A periapical lesion occurs because a root canal treatment fails and bacterial infection spreads into the periapical tissues, a granuloma or as a cyst might appear with a periapical bone defect. Hence a surgical treatment is needed to remove the periapical lesion and the surgical niche need tissue regeneration. Injectable scaffolds have shown *in vivo*—dog animal model—that have been able to help to heal the periapical tissue. An injectable scaffold based on DL-poly-lactide-co-glycolide with ester terminal plus collagen and lornonoxicam (anti-inflammatory drug) and risodronate (bisphosphonate) (PLGA-CLR) induced healing of periapical tissues. The PLGA-CLR was injected into an induced periapical lesion in a dog, and after 2 months, the periapical tissues showed signs of repairing in comparison with a control (with no treatment). The inclusion of collagen in the matrix favored the formation of porosity in the scaffold, the porosity helps in cell adhesion and also to release the lornoxicam and the risodronate. The addition of drugs to the injectable scaffold is interesting because the LNX is a COX-2 inhibitor, hence it reduces the inflammatory process while the risodronate inhibits the osteoclastic activity to promote a healing of the tissues (Shamma et al. 2017).

The clinical application of injectable scaffolds is reported in the scientific literature. For clinical use, selected cases are chosen to apply the strategy. First, the DPSCs cells are culture from immature root end, or from the apical papilla. Then the cells were encapsulated in polyethylene glycol/PLGA injectable scaffolds. The scaffolds were used to treat a large radiolucency in a lower led second premolar of a 20 year-old woman. In this case, the scaffold was injected into the root canal with a chemically disinfected pulp (5.25% NaOCl) to induce apical closure and thickening of root walls. The strategy was also used to treat a large periapical lesion, thus the scaffold was injected into the bony defect at the periacal zone of a left lateral maxillary incisor treated previously with a root canal treatment. After 6 months, the bone defect totally disappeared with complete healing (Shiehzadeh et al. 2014). While the results are of interest, it should be considered that showing a clinical success is evidence that should be clinically contextualized. The case reports involved young and healthy age subjects, that are expected to have a positive biological response because of a healthy immunological and healing activity. Moreover, the time reported for tissue repairing are similar to those reported in a treatment with no tissue engineering strategies. Controlled clinical trials are necessary to investigate the truly impact of injectable scaffolds on endodontic treatments for patients.

Injectable Scaffolds and Regenerative Periodontics

Periodontal diseases are complex pathological entities involving factors such as bacteria, biofilm and inflammation. These factors cause loss of periodontal tissues that might result in loss of teeth. Hence, regenerative periodontal therapy aims to induce formation of gum, bone, cementum and periodontal ligament, it is indeed a challenging therapy. Bone regeneration is the major activity investigated for use of injectable scaffolds on regenerative periodontics; hydrogels based on natural or synthetic polymers are investigated *in vitro* and *in vivo* and are presented beneficial properties to induce bone formation. Injectable scaffolds are important for bone tissue regeneration because they can be injected or molded to adapt its form in a bone defect. The injectable scaffolds fill the bone defect to induce bone repairing, and its advantages are: (1) reducing the operation time, (2) reducing the damaging of large muscle retraction, (3) reducing postoperative pain and scar size, (4) promoting a rapid recovery, and (5) reducing cost (Zhao et al. 2010).

Injectable Scaffolds and In Vitro Observations

Osteoconductive materials or osteoinductive materials are entrapped into the scaffold to promote cell growth, cell differentiation and cell attachment to the scaffold. In addition to its biocompatibility properties, the mechanical properties are also important for bone repair, because the scaffold for periodontal bone repair might receive loading forces, in this regard injectable scaffolds are designed to have suitable mechanical properties and suitable regenerative properties.

Chitosan-Based Injectable Scaffolds (ChBIS) have been explored because of its properties for inducing bone repairing. A ChBIS loaded with calcium phosphate cement tetracalcium phosphate and dicalcium phosphate anhydrous has shown to be useful for bone regeneration (Moreau and Xu 2009). The chitosan gave important flexural strength and elastic modulus to the injectable scaffold. Rat bone-marrow-derived Mesenchymal Stem Cells (MSCs) adhered on the scaffold and showed a normal polygonal morphology, the attaching of the cells was noticed for its cytoplasmatic extension and cell-cell junctions were also observed. After 14 days of contact, the MSCs showed a 99% of viability. The ALP activity increased and the MSCs were differentiated into an osteogenic lineage (Moreau and Xu 2009). The ChBIS have also been formulated with alginate beads (AlB) loaded with human Umbilical Cord Mesenchymal Stem Cells (hUCMSCs). The alginate beads (73–465 µm; mean diameter 207 µm) gave important flexural strength to the scaffold as well as elastic modulus and work-of-fracture; the ChBIS-AlB was also easily injectable from a 10-gauge needle. The hUCMSCs were viable for 14 days (longest experimental time) and the same number of days was observed for an important ALP activity. Very interestingly, the hUCMSCs caused showed mineral synthesis with an increase from day 1 to day 14 (Zhao et al. 2010). The HydroMatrix® hydrogel has been investigated as a scaffold for human Periodontal Ligament Stem Cells (hPDLSCs) (Nagy et al. 2018). The cells proliferated in the hydrogel, and also showed adhesion on the surface. Cell viability with a fibroblast-like morphology was observed at 24 hours, 48 hours and 72 hours. The ALP activity increased from day 7 to day 21, and the osteogenic differentiation of PDLSCs was identified because of gene expression of ALP, Runx2 and osterix. Very interesting, mineralized nodules were detected on the scaffold after 14 days (Nagy et al. 2018). Hence, commercial and home-made injectable scaffolds are able to promote *in vitro* regeneration of bone.

Injectable Scaffolds and In Vivo Observations

Injectable scaffolds have been already tested on *in vivo* models. Regenerative periodontal therapy has been studied for many decades, thus animal models for periodontal regeneration have been usually used for years. The hydrogels have been applied in rats or dogs. Some examples are described next. A ChBIS prepared with β-glycerophosphate (GC) (Ch/GC 9:1) was used to grow human periodontal ligament cells (HPDLCs) in an *in vivo* model in a canine model of periodontal regeneration to evaluate furcal healing. The scaffold was injected in the furcation area, and after 12 weeks of treatment, an 80% increase in bone was observed, with osteoblasts, cell-like newly formed cementum and newly regenerated periodontal ligament were observed at histological examination (Zang et al. 2014). A ChBIS loaded with bone morphogenetic protein 6 (BMP 6) also showed bone formation. The injectable scaffold was thermosensitive, it was a fluid at 4°C and was in a semi-solid phase once it was injected in an 37°C environment. The injectable scaffold was tested in a periodontal defect model in rats for 6 weeks. Micro-CT analysis showed that the treated defects developed bone regeneration with a significant bone volume and bone bridging and trabecular bone. Histological analysis demonstrated formation of a new bone, of cementum, connective tissue and of periodontal ligament (Chien et al. 2018). Gelatin-based Injectable Scaffolds (GbIS) have shown induction of bone repairing. A hydrogel based on gelatin with PLGA microspheres loaded with simvastatin promoted bone formation in rats (Li et al. 2019). The simvastatin promoted bone regeneration associated with the ability of statins to induce bone morphogenetic protein 2 gene expression in osteoblast and marrow cells, its release from the microspheres in the GbIS promotes bone formation (Li et al. 2019). The GbIS injected in sockets in mandible of rats after tooth extraction increased bone mineral density that grew significantly from 1 week to 8 weeks, both radiographical and histological examinations demonstrated the bone formation (Li et al. 2019).

Conclusion

Although it is true that in recent years various injectable techniques have been developed for the production of scaffolds that simulate the properties of the native extracellular matrix in addition to the experimentation with mesenchymal stem cells and cells with certain lineages, regenerative medicine still has challenges to face, because to date it has not been possible to replicate an organ that possesses the physiological properties of an organ in its native state. So for this purpose the harmonic and synergic interaction that only through its multiple signaling can grant. However, the advances are great and encouraging, since they have been synthesizing ever better materials which are biointelligent that can observe the interactions of cells, molecules and scaffolds, which reproduce the natural characteristics of the organ in question. These advances give hope to millions of patients who are waiting for a tissue or organ to restore their health and thus improve the expectation and quality of life for them.

References

Abdal-hay, A., Sheikh, F.A. and Lim, J.K. 2013. Air jet spinning of hydroxyapatite/poly(lactic acid) hybrid nanocomposite membrane mats for bone tissue engineering. Coll. Surf. B: Bioint. 102: 635–643.

Alvarez-Perez, M.A., Guarino, V., Cirillo, V. and Ambrosio, L. 2010. Influence of gelatin cues in PCL electrospun membranes on nerve outgrowth. Biomacromolecules 11(9): 2238–2246.

Ashammakhi, N., Ahadian, S., Zengjie, F., Suthiwanich, K., Lorestani, F., Orive, G. et al. 2018. Advances and future perspectives in 4D bioprinting. Biotech. J. 13(12): 1800148.

Bajaj, P., Schweller, R.M., Khademhosseini, A., West, J.L. and Bashir, R. 2014. 3D biofabrication strategies for tissue engineering and regenerative medicine. Ann. Rev. Biomed. Eng. 16: 247–276.

Bhattarai, R.S., Bachu, R.D., Boddu, S.H.S. and Bhaduri, S. 2018. Biomedical applications of electrospun nanofibers: drug and nanoparticle delivery. Pharmaceutics 11(1): 5.

Chatterjee, K., Young, M.F. and Simon, Jr. C.G. 2011. Fabricating gradient hydrogel scaffolds for 3D cell culture. Combinat. Chem. High Through. Screen. 14(4): 227–236.

Chen, K.L., Yeh, Y.Y., Lung, J., Yang, Y.C. and Yuan, K. 2016. Mineralization effect of hyaluronan on dental pulp cells via CD44. J. Endod. 42(5): 711–716.

Chien, K.H., Chang, Y.L., Wang, M.L., Chuang, J.H., Yang, Y.C., Tai, M.C. et al. 2018. Promoting induced pluripotent stem cell-driven biomineralization and periodontal regeneration in rats with maxillary-molar defects using injectable BMP-6 hydrogel. Sci. Rep. 8(1): 114.

Chircov, C., Grumezescu, A.M. and Bejenaru, L.E. 2018. Hyaluronic acid-based scaffolds for tissue engineering. Rom. J. Morphol. Embryol. 59(1): 71–76.

Chrepa, V., Austah, O. and Diogenes, A. 2017. Evaluation of a commercially available hyaluronic acid hydrogel (Restylane) as injectable scaffold for dental pulp regeneration: An *in vitro* evaluation. J. Endod. 43(2): 257–262.

Dissanayaka, W.L., Hargreaves, K.M., Jin, L., Samaranayake, L.P. and Zhang, C. 2015. The interplay of dental pulp stem cells and endothelial cells in an injectable peptide hydrogel on angiogenesis and pulp regeneration *in vivo*. Tissue Eng. Part A 21(3-4): 550–563.

Granados-Hernández, M.V., Serrano-Bello, J., Montesinos, J.J., Alvarez-Gayosso, C., Medina-Velázquez, L.A., Alvarez-Fregoso, O. et al. 2018. *In vitro* and *in vivo* biological characterization of poly(lactic acid) fiber scaffolds synthesized by air jet spinning. J. Biomed. Res. Mat. Res. B: Appl. Biomater. 106(6): 2435–2446.

Hell, A.F., Simbara, M.M.O., Rodrigues, P., Kakazu, D.A. and Malmonge, S.M. 2018. Production of fibrous polymer scaffolds for tissue engineering using an automated solution blow spinning system. Res. Biomed. Eng. 34(3): 273–278.

Jones, T.D., Kefi, A., Sun, S., Cho, M. and Alapati, S.B. 2016. An optimized injectable hydrogel scaffold supports human dental pulp stem cell viability and spreading. Adv. Med. (7363579).

Levorson, E.J., Raman Sreerekha, P., Chennazhi, K.P., Kasper, F.K., Nair, S.V. and Mikos, A.G. 2013. Fabrication and characterization of multiscale electrospun scaffolds for cartilage regeneration. Biomed. Mater. (Bristol, England) 8(1): 014103–014103.

Li, X., Liu, X., Ni, S., Liu, Y., Sun, H. and Lin, Q. 2019. Enhanced osteogenic healing process of rat tooth sockets using a novel simvastatin-loaded injectable microsphere-hydrogel system. J. Craniomaxillofac. Surg.

Liu, F., Chen, Q., Liu, C., Ao, Q., Tian, X., Fan, J. et al. 2018. Natural polymers for organ 3D bioprinting. Polymers 10(11): 1278.

Matthews, J.A., Wnek, G.E., Simpson, D.G. and Bowlin, G.L. 2002. Electrospinning of collagen nanofibers. Biomacromolecules 3(2): 232–238.

Medeiros, E.S., Glenn, G.M., Klamczynski, A.P., Orts, W.J. and Mattoso, L.H.C. 2009. Solution blow spinning: A new method to produce micro- and nanofibers from polymer solutions. J. Appl. Polym. Sci. 113(4): 2322–2330.

Miao, S., Castro, N., Nowicki, M., Xia, L., Cui, H., Zhou, X. et al. 2017. 4D printing of polymeric materials for tissue and organ regeneration. Materials today (Kidlington, England) 20(10): 577–591.

Moreau, J.L. and Xu, H.H. 2009. Mesenchymal stem cell proliferation and differentiation on an injectable calcium phosphate-chitosan composite scaffold. Biomaterials 30(14): 2675–2682.

Morissette Martin, P., Grant, A., Hamilton, D.W. and Flynn, L.E. 2019. Matrix composition in 3-D collagenous bioscaffolds modulates the survival and angiogenic phenotype of human chronic wound dermal fibroblasts. Acta Bio. 83: 199–210.

Nagy, K., Lang, O., Lang, J., Perczel-Kovach, K., Gyulai-Gaal, S., Kadar, K. et al. 2018. A novel hydrogel scaffold for periodontal ligament stem cells. Interv. Med. Appl. Sci. 10(3): 162–170.

Rosa, V., Zhang, Z., Grande, R.H. and Nor, J.E. 2013. Dental pulp tissue engineering in full-length human root canals. J. Dent. Res. 92(11): 970–975.

Shamma, R.N., Elkasabgy, N.A., Mahmoud, A.A., Gawdat, S.I., Kataia, M.M. and Abdel Hamid, M.A. 2017. Design of novel injectable *in-situ* forming scaffolds for non-surgical treatment of periapical lesions: *In-vitro* and *in-vivo* evaluation. Int. J. Pharm. 521(1-2): 306–317.

Shiehzadeh, V., Aghmasheh, F., Shiehzadeh, F., Joulae, M., Kosarieh, E. and Shiehzadeh, F. 2014. Healing of large periapical lesions following delivery of dental stem cells with an injectable scaffold: new method and three case reports. Indian J. Dent. Res. 25(2): 248–253.

Silva, C.R., Babo, P.S., Gulino, M., Costa, L., Oliveira, J.M., Silva-Correia, J. et al. 2018. Injectable and tunable hyaluronic acid hydrogels releasing chemotactic and angiogenic growth factors for endodontic regeneration. Acta Biomater. 77: 155–171.

Suarez-Franco, J.L., Vázquez-Vázquez, F.C., Pozos-Guillen, A., Montesinos, J.J., Alvarez-Fregoso, O. and Alvarez-Perez, M.A. 2018. Influence of diameter of fiber membrane scaffolds on the biocompatibility of hPDL mesenchymal stromal cells. Dental Mater. J. 37(3): 465–473.

Tutak, W., Sarkar, S., Lin-Gibson, S., Farooque, T.M., Jyotsnendu, G., Wang, D. et al. 2013. The support of bone marrow stromal cell differentiation by airbrushed nanofiber scaffolds. Biomaterials 34(10): 2389–2398.

Vazquez-Vazquez, Febe Carolina, Osmar Alejandro Chanes-Cuevas, David Masuoka, Jesús Arenas Alatorre, Daniel Chavarria-Bolaños, José Roberto Vega-Baudrit, Janeth Serrano-Bello and Marco Antonio Alvarez-Perez. 2019. Biocompatibility of developing 3D-printed tubular scaffold coated with nanofibers for bone applications. J. Nanomater. 2019: 13.

Yang, G.H., Yeo, M., Koo, Y.W. and Kim, G.H. 2019. 4D Bioprinting: Technological advances in biofabrication. Macromol. Biosci. 19(5): 1800441.

Yin, A., Zhang, K., McClure, M.J., Huang, C., Wu, J., Fang, J. et al. 2013. Electrospinning collagen/chitosan/poly(L-lactic acid-co-ϵ-caprolactone) to form a vascular graft: Mechanical and biological characterization. J. Biomed. Mat. Res. A 101A(5): 1292–1301.

Zang, S., Dong, G., Peng, B., Xu, J., Ma, Z., Wang, X. et al. 2014. A comparison of physicochemical properties of sterilized chitosan hydrogel and its applicability in a canine model of periodontal regeneration. Carbohydr. Polym. 113: 240–248.

Zhao, L., Weir, M.D. and Xu, H.H. 2010. An injectable calcium phosphate-alginate hydrogel-umbilical cord mesenchymal stem cell paste for bone tissue engineering. Biomaterials 31(25): 6502–6510.

14

Clinical Progresses in Regenerative Dentistry and Dental Tissue Engineering

*Rohan Shah** and *Manasi Shimpi*

Introduction

Tissue engineering can be defined as an interdisciplinary field that applies the principles of engineering and life sciences towards the development of biological substitutes that restore, maintain or improve tissue function (Langer and Vacanti 1993).

Endodontic therapy accounts for a high rate of success in retention of teeth. However, many teeth are not restorable because of apical resorption, fracture, incompletely formed roots or carious destruction of coronal structures. To tackle this issue, a novel approach to restore tooth structure based on biology, i.e., regenerative endodontic procedures by the application of tissue engineering can be carried out. Presently, two concepts exist in regeneration endodontics to treat non-vital infected teeth. One is the active pursuit of pulp-dentin regeneration to implant or regrow pulp (tissue engineering technology) and the other, in which new living tissue is expected to form from the tissue present in the teeth itself, allowing continued root development (revascularization)(Banshal and Banshal 2011).

History of Tissue Engineering

The regenerative capacity of a real living creature was recorded as early as 330 BC when Aristotle observed that a lizard could grow back the lost tip of its tail. In the late 1700s, the scientist Spallanzani reported that a newt could regenerate an entire limb. Since then, the study of regeneration in lower life forms has laid the grounds for understanding the regenerative capabilities and potential of humans. Bacteria and the single-celled protozoans regenerate complete organisms with each cell division. Many multicellular invertebrates also exhibit extensive regenerative abilities. When cut in half, the Atworm planaria can grow a new head from one piece and a new tail from the other. However, progression up the evolutionary ladder is generally accompanied by a reduction in regenerative capacity.

Bharati Vidyapeeth Dental College and Hospital, Pune, Maharashtra, India.
  Email: manasi.shimpi92@gmail.com
* Corresponding author: drrohanshah@gmail.com

With the introduction of a new scientific method came new understanding of the natural world. The systematic unravelling of the secrets of biology was coupled with the scientific knowledge of disease and trauma. Artificial or prosthetic materials for replacing limbs, teeth and other tissues resulted in the partial restoration of lost function. Also, the concept of using one tissue as a replacement for another was developed. This concept was termed tissue engineering and was loosely applied to the use of prosthetic devices and the surgical manipulation of tissues. In essence, new and functional living tissue is fabricated using living cells, which are usually associated, in one way or another, with a matrix or scaffolding to guide tissue development (Vacanti 2006). New sources of cells, including many types of stem cells, have been identified in the past several years, igniting a new interest in the field. The emergence of stem cell biology has led to a new term, regenerative medicine and dentistry.

Components of Tissue Engineering

The three key components (Fig. 14.1) for tissue engineering are (Langer and Vacanti 1993).

    I. Stem cells—to respond to growth factors
    II. Scaffolds to represent extracellular matrix (ECM)
    III. Growth factors—signals for morphogenesis

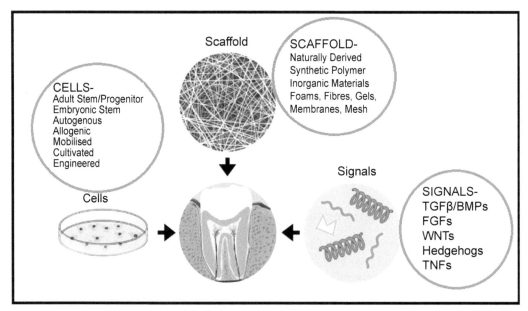

Figure 14.1. The key components for tissue engineering.

Stem Cells

Stem cell biology has become an essential field for the understanding of tissue regeneration and implementation of regeneration. For many years the promise of stem cells has been talked about but, like the child who doesn't get a penny, we have been left disappointed with heightened expectation and nothing to show. In today's era, a fascinating area of medical science is stem cell research. The stem cell is the origin of life. As stated first by the great pathologist Rudolph Virchow, "All cells come from pre-existing cells."

    Stem cells can be defined as clonogenic cells capable of both self-renewal and multiple-lineage differentiation by Grothos et al. 2002.

The plasticity of the stem cell defines its ability to produce cells of different tissues (Rendon and Watt 2003). In short, stem cells are unprogrammed cells in the human body that can be described as 'shape-shifters'. These cells can change into other types of cells.

Types of Stem Cells

- Embryonic stem cells—located within the inner cell mass of the blastocyte stage of development.
- Postnatal cells—that have been isolated from various tissues including bone marrow, neural tissues, dental pulp and periodontal ligament.

Embryonic/fetal Stem Cells

This type of cell is derived from embryos that are typically 4–5 days old called blastocytes. These cells are pluripotent, i.e., they have capacity to form all tissues (MacArthur and Oreffo 2005). Thriving culture of stem cells from human embryos was reported for the first time in 1988 (Thompson et al. 1998).

Pluripotent embryonic stem cells could be directly implanted into the patient's tissues, where they would then differentiate into specific cell types after encountering the appropriate niche (MacArthur and Oreffo 2005).

The greater plasticity of embryonic stem cells makes these cells more valuable among researchers for developing new therapies (Gardner 2002).

The main problem with embryonic stem cells is ethical and legal issues, so now the hot topic among researchers/scientists is stem cell therapy using postnatal stem cells donated either by patients themselves or their close relatives.

Adult Stem Cells/Post-natal Stem Cells

- Autologous stem cells—are obtained from the same individual to whom they will be implanted.

The sources include bone marrow, peripheral blood, fat removed by liposuction, periodontal ligament, oral mucosa and dental pulp.

Various sources for postnatal dental stem cells (Simon et al. 2008):

- Permanent teeth-: Dental Pulp Stem Cells (DPSC): derived from the third molar.
- Deciduous teeth-: Stem cells from human-exfoliated deciduous teeth (SHED), i.e., stem cell present within the pulp tissue of deciduous teeth.
- Periodontal ligament-: periodontal ligament stem cells (PDLSCs).
- Stem Cells from Apical Papilla (SCAP).
- Stem cells from a supernumerary tooth—mesiodens or stem cells from teeth extracted for orthodontic purposes.
- Dental Follicle Progenitor Cells (DFPCs) (Vacanti 2006).
- Stem cells from human Natal Dental Pulp (hNDP).

Unique characteristics of adult stem cells (Langer and Vacanti 1993):

1) They exist as undifferentiated cells and maintain this phenotype by the environment and/or the adjacent cell populations until they are exposed to and respond to appropriate signals.
2) They have an ability to self-replicate for prolonged periods.
3) They maintain their multiple differentiation potential throughout the life of the organism.

Newly discovered 'chameleon' cells are modified by the environment when the cells cross the border between two tissues (Grove 1921).

Dental Pulp Stem Cells (DPSCS)

The first type of dental pulp stem cell was isolated from the human pulp tissue and termed as 'post-natal' Dental Pulp Stem Cells (DPSCs) by Gronthos et al. 2000.

The stem cell population in the pulp is tiny: approximately 1% of the total cells (Smith et al. 1995).

Isolation of heterogeneous populations of DPSCs—

One crucial feature of pulp cells is their odontoblastic differentiation potential which is why they are called odontoblastoid cells, as these cells appear to synthesize and secrete dentin matrix like the odontoblasts cells they replace. Human pulp cells can be induced *in vitro* to differentiate into cells of odontoblastic phenotype, characterized by polarized cell bodies and accumulation of mineralized nodules (Couble et al. 2000). DPSCs isolated with enzyme treatment of pulp tissues form CFU Fs with various characteristics (Gronthos et al. 2000; Huang et al. 2006). If seeded onto dentin, some DPSCs are capable of generating new stem cells or multilineage differentiation into odontoblasts, adipocytes and neural-like cells. This stem cell behavior occurs following cryopreservation, signifying the potential use of frozen tissues for stem cell isolation (Zhang et al. 2006). Pulp cells can proliferate and differentiate into odontoblast-like cells processes, extending into dentinal tubules when in contact with chemo-mechanically treated dentine surface in an *in vitro* situation. This is a requirement for the secretion of new dentine (Huang et al. 2006).

Transplanted *ex vivo* expanded DPSCs mixed with hydroxyapatite/tricalcium phosphate (HA/TCP) form ectopic pulp-dentin like tissue complexes in immunocompromised mice (Gronthos et al. 2000; Batouli et al. 2003).

These pools of heterogeneous DPSCs form vascularized pulp-like tissue and are surrounded by a layer of odontoblast-like cells expressing dentin sialophosphoprotein (DSPP), which produces dentin containing dentinal tubules similar to those in natural dentin. Over time, the amount of dentin gets thicker. When DPSCs are seeded onto human dentin surfaces and implanted into immunocompromised mice, a reparative dentin-like structure is deposited on the dentin surface (Batouli et al. 2003).

Carinci and his colleagues identified a subpopulation of stem cells from human dental pulp with osteogenic potential forming bonelike tissue *in vivo*. They termed these cells 'Osteoblasts Derived from Human Pulpar Stem Cells' (ODHPSCs) and used microarrays to compare the genetic profiles of these cells with those of normal osteoblasts (Carinci et al. 2008).

Human Dental Pulp Stem Cells (hDPSCs) have been isolated from adult tooth pulp of the third molars. Takeda et al. in 2008 successfully isolated hDPSCs from developing third molars, extracted before their eruption. They observed that the hDPSCs separated from the crown—completed stage showed higher proliferation potential compared with those from later stages. Combined with previous data (Batouli et al. 2003), these facts suggest that the dental pulp tissues contained stem cells capable of generating dentin-like structures, even when isolated at the early stage of tooth development.

Dental pulp cells can be reprogrammed into induced-pluripotent stem cells (iPS) at a higher rate compared with other cell types of human origin tried so far (Yan et al. 2010).

Stem Cells From Exfoliated Deciduous Teeth (SHED)

Types of stem cells from exfoliated deciduous teeth include:

- **Adipocytes:** they can be used to treat cardiovascular disease, spine and orthopedic conditions, congestive heart failure, Crohn's disease, and to be used in plastic surgery (Perry et al. 2008).
- **Chondrocytes and Osteoblasts:** have successfully been used to grow bone and cartilage suitable for transplant (Miura et al. 2003).
- **Mesenchymal:** have the potential to treat degenerative neuronal disorders such as Alzheimer's and Parkinson's diseases, cerebral palsy (Perry et al. 2008).

Potential Clinical Applications of Stem Cell Therapy with SHED

This comprehensive list of diseases and conditions currently being treated using stem cells include Stem Cell Disorders, Acute and chronic Leukaemias, Myeloproliferative Disorders, Myelodysplastic Syndromes, Lymphoproliferative Disorders, Inherited Erythrocyte Abnormalities, Liposomal Storage Diseases, Histiocytic Disorders, Phagocyte Disorders, Congenital Immune System Disorders, Inherited Platelet Abnormalities, Plasma Cell Disorders and malignancies (Mao et al. 2006).

Advantages of banking SHED cells

- Provides a guaranteed matching donor (autologous transplant) for life. There are many advantages of autologous transplantation including no immune reaction and tissue rejection of the cells, no immunosuppressive therapy needed, and significantly reduced risk of infectious diseases.
- Saves cells before natural damage occurs.
- Painless and straightforward for both child and parent.
- Less than one-third of the cost of cord blood storage.
- SHED are adult stem cells and are not the subject of the same ethical concerns as embryonic stem cells (Jay 2008).
- SHED cells are complementary to stem cells from cord blood. While cord blood stem cells have proven valuable in the regeneration of blood cell types, SHED are able to regenerate solid tissue types that cord blood cannot—such as potentially repairing connective tissues, dental tissues, neuronal tissue and bone (Mao et al. 2006).
- SHED may also be useful for close relatives of the donor such as grandparents, parents, uncles and siblings (Jay 2008).

Collection, Isolation, and Preservation of SHED

The technique is non-invasive and straightforward involving collection, isolation and storage of SHED.

Step 1: Tooth Collection (Casagrande et al. 2010)

Since, SHED banking is a proactive decision made by the parents, the first step is to inform them to keep the tooth (fulfilling the above mentioned criteria) in sterile saline solution and inform the tooth bank or attending dentist of the bank. The tooth exfoliated should have a pulp red in color, indicating that the pulp received blood flow up until the time of removal, which is indicative of cell viability. If the pulp is gray, the blood flow to the pulp has likely been compromised, and thus the stem cells are susceptibly necrotic and are no longer viable for recovery. Teeth that have become very mobile, either through trauma or disease (e.g., Classes III or IV mobility), often have a severed blood supply, and are not candidates for stem cell recovery. This is why recovery of stem cells from primary teeth is preferred after extraction than the tooth that is 'hanging on by a thread' with mobility. Pulpal stem cells should not be harvested from teeth with apical abscesses, tumors or cysts.

In the event of a scheduled procedure, the dentist visually inspects the freshly-extracted tooth to confirm the presence of a healthy pulpal tissue and the tooth or teeth is transferred into the vial containing a hypotonic phosphate-buffered saline solution, which provides nutrients and helps to prevent the tissue from drying out during transport (up to four teeth in the one vial). Placing a tooth into this vial at room temperature induces hypothermia.

The vial is then carefully sealed and placed into the thermite, a temperature phase change carrier, after which the carrier is then placed into an insulated metal transport vessel. The thermette, along with the insulated transport vessel maintains the sample in a hypothermic state during transportation. This procedure is described as sustentation.

Store-A-Tooth, a company involved in tooth banking, uses the Save-A-Tooth device same as that used for transportation of avulsed teeth for transporting stem cells from the dental office to the laboratory.

The viability of the stem cells is both time and temperature-sensitive, and careful attention is required to ensure that the sample will remain viable. The time from harvesting to arrival at the processing storage facility should not exceed 40 hours.

The same steps are performed by the attending assistant of the tooth bank, if it is not a scheduled extraction for the collection of specimens.

Step 2: Stem Cell Isolation (Freshney et al. 2007):

When the tooth bank receives the vial, the following protocol is followed.

A) The tooth surface is cleaned by washing three times with Dulbecco's Phosphate Buffered Saline without Ca++ and Mg++ (PBSA).

B) Disinfection is done with disinfection reagent such as povidone iodine and again washed with PBSA.

C) The pulp tissue is isolated from the pulp chamber with small sterile forceps or dental excavator. Stem cell-rich pulp can also be flushed out with salt-water from the center of the tooth.

D) Contaminated pulp tissue is placed in a sterile petri dish which is washed at least thrice with PBSA.

E) The tissue is then digested with collagenase Type I and Dispase for 1 hour at 37°C. Trypsin-EDTA can also be used.

F) Isolated cells are passed through a 70 μm filter to obtain single-cell suspensions.

G) Then the cells are cultured in a Mesenchymal Stem Cell Medium (MSC) medium which consists of alpha modified minimal essential medium with 2 mM glutamine and supplemented with 15% fetal bovine serum (FBS), 0.1 Mm L-ascorbic acid phosphate, 100 U/ml penicillin and 100 ug/ml streptomycin at 37°C and 5% CO_2 in air. Usually, isolated colonies are visible after 24 hours.

H) Different cell lines can be obtained such as odontogenic, adipogenic and neural by making changes in the MSC medium.

I) If cultures are obtained with unselected preparation, colonies of cells with a morphology resembling epithelial cells or endothelial cells can be established. Usually cells disappear during the course of successive cell passages. If contamination is extensive, three procedures can be performed:

   1) Retrypsinizing culture for a short time so that only stromal cells are detached because epithelial or endothelial-like cells are more strongly attached to the culture flask or dish.

   2) Changing the medium 4–6 hours after subculture because stromal cells attach to the culture surface earlier than contaminating cells.

   3) Separate stem cells using Fluorescence Activated Cell Sorting (FACS), in which STRO-1 OR CD 146 can be used. This is considered the most reliable.

Confirmation of the current health and viability of these cells are given to the donor's parents.

Step 3: Stem Cell Storage

In the light of present research, either of the following two approaches is used for stem cell storage.

a) Cryopreservation

b) Magnetic freezing

Cryopreservation

It is the process of preserving cells or whole tissues by cooling them to sub-zero temperatures. At these freezing temperatures, biological activity is stopped, as are any cellular processes that lead to cell death (Oh et al. 2005). SHED can be successfully stored long-term with cryopreservation and remain viable for use. These cells can be cryopreserved for an extended period of time, and when needed, carefully thawed to maintain their viability. Cells harvested near the end of log-phase growth (approximately 80–90% confluent) are best for cryopreservation. The sample is divided into four cryo-tubes, and each part is stored in a separate location in a cryo-genic system so that even in the unlikely event of a problem with one of storage units, there will be another sample available for use. The cells are preserved in liquid nitrogen vapor at a temperature of less than –150°C. This protects the cells and maintains their latency and potency. In a vial, $1-2 \times 10^6$ cells in 1.5 ml of freezing medium is optimum. Too low or high cell number may decrease recovery rate.

Suchánek et al. (2007) established a protocol of Dental Pulp Stem Cells (DPSCs) isolation and to cultivate DPSCs either from adult and exfoliated tooth and compared these cells with Mesenchymal Progenitor Cell (MPCs) cultures. The results proved that the DPSCs and MPCs were highly proliferative, clonogenic cells that can be expanded beyond Hayflick's limit and remain cytogenetically stable.

Magnetic Freezing

Hiroshima University uses magnetic freezing rather than cryogenic freezing. This technology is called CAS and exploits the little-known phenomenon that applying even a weak magnetic field to water or cell tissue will lower the freezing point of that body by up to 6–7 degrees celsius. The idea of CAS is to completely chill an object below the freezing point without freezing occurring, thus ensuring, distributed low temperature without the cell wall damage caused by ice expansion and nutrient drainage due to capillary action, as generally caused by conventional freezing methods. Then, once the object is uniformly chilled, the magnetic field is turned off, and the object snap-freezes (TT-450—Stem Cells and Teeth Banks—http://www.japaninc.com/tt450).

The Hiroshima University company was the first to formulate this new technology. Using CAS, Hiroshima University claims that it can increase the cell survival rate in teeth to a high of 83%. This compares to 63% for liquid nitrogen (–196 degrees C), 45% for ultra-cold freezing (–80 degrees C), and just 21.5% for a household freezer (–20 degrees C). Maintaining a CAS system is a lot cheaper than cryogenics and more reliable as well.

The best candidates for SHED are moderately resorbed canine and incisors with the presence of healthy pulp. In children, other sources of easily accessible stem cells are supernumerary teeth, mesiodens, over-retained deciduous teeth associated with congenitally missing permanent teeth and prophylactically removed deciduous molars for orthodontic indications (Thomson et al. 1998).

According to Miura et al. 2003; Mao et al. 2006; Seo et al. 2008, deciduous tooth stem cells are an easily accessible stem cell source and capable of robust *ex vivo* expansion for several potential clinical applications.

In vivo characterization of SHED—Production of dentin pulp-like structures but without a complex formation. *Ex vivo* expanded SHED transplanted into immunocompromised mice yield human-specific odontoblast-like cells directly associated with a dentin-like structure. The regenerated dentin expresses dentin-specific DSPP. However, unlike DPSCs, SHED is unable to regenerate a complete dentin-pulp like complex *in vivo* (Miura et al. 2003).

Osteo-inductive capacity—One striking feature of SHED is that they are capable of inducing recipient murine cells to differentiate into bone-forming cells, which is not a property attributed to DPSCs following transplantation *in vivo*. When single-colony-derived SHED clones were transplanted into immunocompromised mice, only one-fourth of the clones had the potential to generate ectopic dentin-like tissue equivalent to that produced by multicolony-derived SHED (Miura et al. 2003).

However, all single-colony derived SHED clones tested are capable of inducing bone formation in immunocompromised mice. While SHED could not differentiate directly into osteoblasts, they appeared to induce new bone formation by forming an osteo-inductive template to recruit murine host osteogenic cells (Miura et al. 2003). With the osteo-inductive potential, SHED can repair critical-sized calvarial defects in mice with substantial bone formation (Seo et al. 2008). These findings imply that deciduous teeth may not only provide guidance for the eruption of permanent teeth, as generally assumed, but may also be involved in inducing bone formation during the eruption of permanent teeth.

Zheng et al. 2009 provided the first evidence that stem cells derived from miniature pig deciduous teeth (SPD) are capable of regenerative critical-size defects in the oro-facial bone in large animal models, specifically swine, and may potentially serve as an alternative stem-cell-based approach in the reconstruction of alveolar and orofacial bone defects.

SHED appear to represent a population of multipotent stem cells that are perhaps more immature than other post-natal stromal stem cell populations. SHED express neuronal and glial cell markers, which may be related to the neural-crest-cell origin of the dental pulp (Chai et al. 2000). Scientists believe that these stem cells behave differently than post-natal (adult) stem cells. SHED are capable of extensive proliferation and multipotent differentiation, which makes them an essential resource of stem cells for the regeneration and repair of craniofacial defects, tooth loss and bone regeneration (Chai et al. 2000).

Periodontal Ligament Stem Cells (PDLSCs)

This was first identified by Seo et al. 2004. Earlier evidence has shown that PDL contains cell populations that can differentiate into either cementum-forming cells (cementoblasts) or bone-forming cells (osteoblasts) (Isaka et al. 2001).

The presence of multiple cell types within PDL suggests that this tissue contains progenitor cells that maintain tissue homeostasis and regeneration of periodontal tissue. Enzyme digestion treatment of PDL releases a population of clonogenic cells with characteristics of postnatal stem cells (Seo et al. 2004).

The successful isolation and characterization of PDLSCs have led to the identification of tendon MSCs by the same approaches (Bi et al. 2007).

Instead of forming the entire tooth, even a bio-root with periodontal ligament tissues has been generated by utilizing SCAP along with the PDLSCs. This bio-root is encircled with periodontal ligament tissue and has a natural relationship with the surrounding bone.

Stem Cells from Apical Papilla (SCAP)

This was first identified by Sonoyama et al. 2008. Dental stem cells can also be extracted from the apical papilla of exfoliated primary teeth (SCAP) (Bluteau et al. 2008).

Apical papilla refers to the soft tissue at the apices of developing permanent teeth (Sonoyama et al. 2006). Apical papilla is more apical to the epithelial diaphragm, and there is an apical cell-rich zone lying between the apical papilla and the pulp (Rubio et al. 2005).

Similar to DPSCs and SHED, *ex vivo* expanded SCAP can undergo odontogenic differentiation *in vitro*. SCAP express lower levels of DSP, in comparison with DPSCs. SCAP represent early progenitor cells.

Conservation of these stem cells when treating immature teeth may allow the continuous formation of the root to its completion.

The tissue is loosely attached to the apex of the developing root and can be easily detached with a pair of tweezers. Importantly these are stem/progenitor cells located in both dental pulp and the apical papilla, but have somewhat different characteristics. Because of the apical location of the apical papilla, this tissue may be beneficial by its collateral circulation, which enables it to survive during the process of pulp necrosis.

The Potential Role of SCAP in Continued Root Formation: The role of the apical papilla in root formation may be observed in clinical cases. An immature human incisor was injured and the crown fractured with pulp exposure. During treatment, the apical papilla was retained while the pulp was extirpated. The continued root-tip formation was observed after root canal treatment. But still further investigation is needed to verify the radiographic evidence of continued apical papilla or if it is merely cementum formation. Although the finding suggests that apical root papilla is likely to play a pivotal role in root formation (Seo et al. 2007).

The Potential Role of SCAP in Pulp Healing and Regeneration: The open apex provided an excellent communication from the pulp space to the periapical tissues; therefore, it may be possible for periapical disease to occur while the pulp is only partially necrotic and infected. Along the same line of reasoning, stem cells in pulp tissue and in apical papilla may also have survived the infection and allowed regeneration of pulp and root maturation to occur. The infection could have spread through survived pulp tissue reaching the periapex. It should be noted that prolonged infection may eventually lead to total necrosis of the pulp and apical papilla; under these conditions, apexogenesis or maturogenesis, a term that encompasses not just the completion of root-tip formation but also the dentin of the root, would then be unlikely.

The Potential Role of SCAP in Replantation and Transplantation: The fate of human pulp space after dental trauma has been observed in clinical radiographs. Andreasen et al. 1995 and Kling et al. 1986 showed excellent radiographic images of the ingrowth of bone and periodontal ligament (PDL) (next to the inner dentinal wall) into the canal space with arrested root formation after the replantation of avulsed maxillary incisors, suggesting a complete loss of the viability of pulp, apical papilla and/or HERS. Some cases showed the partial formation of the root accompanied with ingrowth of bone and PDL into the canal space, and in some cases the teeth continued to develop roots to their completion, suggesting that there was partial or total pulp survival after the reimplantation (Seo et al. 2008).

SCAP can be used for bioroot engineering, in one of the animal studies, it showed negative results, whereas another animal study showed 33.3% success rate (Seo et al. 2008).

Dental Follicle Precursor Cells (DFPCS)

Dental follicle is a loose vascular connective tissue composed of a heterogenous layer of an ectomesenchymal cells surrounding the enamel organ and the dental papilla of the developing tooth germ in early stages of tooth development prior to eruption (Lesot et al. 1993). This tissue contains progenitor cells that form the periodontium, i.e., cementum, PDL and alveolar bone. Precursor cells have been isolated from human dental follicles of impacted third molars. Recently, human dental follicle progenitor cells showed hard tissue-forming potential in immunocompromised rats (Yagyuu et al. 2010). Dental follicle stem cells may provide a cell source for tissue engineering. Similar to other dental stem cells, these cells form low numbers of adherent clonogenic colonies when released from the tissue following enzymatic digestion (Morsczeck et al. 2005).

Stem cells are also isolated from aging teeth, but it is observed that the number of cells and their proliferation rate decreases with age and it is maximum when only the crown is formed (germ stage).

Scaffolds

Physical scaffolds are a crucial component of tissue engineering. Scaffold provides the framework for cell growth and differentiation at a local site. The scaffold might be implanted alone or in combination with cells and growth factors. When implanted, a scaffold allows for cell migration and organization (Stewart Sells 2004).

The tooth slice/scaffold model constitutes a powerful and nimble approach for mechanistic studies designed to understand the processes of dental pulp stem cell differentiation. It allows for the evaluation of the impact of biological processes that regulate stem cell differentiation on the formation of tubular dentin and the vascularization and innervation of engineered pulp tissues.

Molecular Signaling and Growth Factors

Morphogenic signals leading to the differentiation and function of odontoblasts and ameloblasts are known to be mediated by specialized molecules.

Molecular signals flow between odontogenic cells and guide them to position themselves along mineralization fronts, differentiate and start secreting new molecules (proteins). These proteins constitute extra-cellular matrices, which will eventually be mineralized into dentin and enamel. Examples of molecular signals are Bone Morphogenetic Proteins (BMPs) and Amelogenin.

Odontoblasts are induced by a process that involves signals from the epithelium to synthesize extracellular matrix proteins required for dentin formation. The cells from the oral epithelium layer (oral ectoderm) secrete BMPs during the early stages of odontogenesis (dental lamina). These first BMP molecules signal a reciprocal response from the cells of the odontogenic mesenchyme and establish epithelial mesenchymal crosstalk that is an absolute requirement for tooth morphogenesis (Kassai et al. 2005). This crosstalk is mediated by diffusible molecules (BMPs) that find specific receptors (BMPRs) present in cells that are prepared to respond to the stimulus. Once these receptors are engaged, intracellular signaling occurs, and the odontoblasts begin secreting the extracellular matrix proteins that will eventually be mineralized into dentin.

Growth Factors

Growth factors are proteins that bind to receptors on the cell and induce cellular proliferation and/or differentiation (Wingard and Demetric 1999). Growth factors are essential in the cellular signaling for odontoblasts differentiation and stimulation of dentin matrix secretion. The odontoblasts secrete these growth factors which are deposited within the dentin matrix, where they remain in active form through interaction with other components of the dentin matrix (Wingard and Demetric 1999). Many growth factors are quite versatile, stimulating cellular division in numerous cell types, while others are more cell-specific.

The morphogenetic signaling networks include the five major classes of evolutionarily conserved genes:

1) Bone Morphogenetic Proteins (BMPs),
2) Fibroblast Growth Factor (FGFs),
3) Wingless- and int-related proteins (Wnts),
4) Hedgehog proteins (Hhs),
5) Tumor Necrotic Factor (TNF).

Growth factors released by platelet-rich plasma upon activation are summarized as below (Table 14.1)

Enamel Matrix Protein (Emdogain)

Enamel Matrix Derivative (EMD) in the form of Emdogain (commercially available enamel matrix protein) incite natural regenerative processes in mesenchymal tissues. The EMD induced processes mimic normal odontogenesis and participate in reciprocal ectodermal–mesenchymal signaling that controls these processes (Hammartson 1997).

Emdogain is a purified acidic extract of developing embryonal enamel derived from six-month-old piglets. It acts as a tissue-healing modulator that mimics the events which occur during root development and to help stimulate regeneration (Heijl et al. 1997).

The table below (Table 14.2) summarizes the biomaterials that may be used for dental tissue engineering.

Table 14.1. Various growth factors released by platelet rich plasma.

▪ **PDGF (Platelet derived growth factor)** ▪ **Function**	Produced by the alpha granules of the platelet PDGF is a potent regulatory growth factor and a sentinel growth factor that begins nearly all wound healing. ▪ To stimulate cell replication (mitogenesis) of healing capable stems and pre-mitotic partially differentiated cells which are part of the pulpal tissue ▪ PDGF also causes replication of endothelial cells, causing budding of new capillaries (angiogenesis) ▪ Activates TGF-beta ▪ Stimulates neutrophils and macrophages ▪ Stimulates chemotaxis ▪ Stimulates collagen synthesis and collagenase activity
▪ **TGF-a & b (Transforming growth factor-alpha & beta)**	▪ TGF regulates proliferation and differentiation of multiple cell types. TGF found in platelets is subdivided into TGFβ1 and TGFβ2, which are the more generic connective tissue growing factors involved with dentine formation influencing odontoblasts to lay down dentine through the process of dentinogenesis
▪ **EGF (Epithelial growth factor)**	▪ 53-amino acid polypeptide chain ▪ Released during platelet degranulation ▪ Stimulate re-epithelialization, angiogenesis and collagenase activity
▪ **FGF (Fibroblast growth factor)**	▪ Stimulate angiogenesis and dentinogenesis ▪ Stimulate endothelial cell proliferation ▪ Stimulate collagen synthesis ▪ Stimulate wound contraction ▪ Stimulate dentine synthesis ▪ Stimulate epithelialization ▪ Produces keratinocyte growth factor
▪ **IGF (Insulin growth factor**	▪ Key regulator of cell metabolism and growth. Stimulates proliferation and differentiation functions in osteoblasts
▪ **PDEGF**	▪ Platelet-derived epidermal growth factor
▪ **Interleukin 1**	▪ Stimulates lymphocyte proliferation ▪ Influences collagenase activity
▪ **High concentration of leukocytes**	(neutrophils, eosinophils) for microbicidal events
▪ **High concentration of wound macrophages and other phagocytic cells, for biological debridement**	▪ Histamines, Serotonin, ADP, Thromboxane A2 and other vasoactive and chemotactic agents ▪ High platelet concentration and native fibrinogen concentration for improved hemostasis

Gene Therapy

Gene therapy is a term that has appeared with increasing frequency in both the popular and the scientific literature. It is commonly used in reference to any clinical application of the transfer of a foreign gene. Human gene therapy is defined as the treatment of disorder or disease through the transfer of engineered genetic material into human cells, often by viral transduction (Scheller and Krebsbach 2009).

Methods of Gene Transfer

There are two general methods for transferring genes into cells:

Viral Techniques

Many viruses could be used for gene transfer, yet only a few retroviruses, adenoviruses, adeno-associated viruses and herpes viruses-have been widely employed.

Table 14.2. Biomaterials for dental pulp tissue engineering (Rao 2004).

Materials	Engineering approach	Results	References
PGA, collagen I, alginate	Pulp fibroblasts seeded onto different materials, cell culture *in vitro*	Pulp-like tissue after 45 to 60 days on PGA	Mooney et al. 1996 Bohl et al. 1998
HA/TCP	Stem cells from dental pulp (SHED, DPSC) mixed with HA/TCP powder transplanted into nude mice	Generation of dentin or bone (SHED) and dentin-pulp-like complexes (DPSC)	Gronthos et al. 2000 Miura et al. 2003
Collagens I and III, chitosan, gelatin	Human dental pulp cells seeded into different materials for comparison *in vitro*	Adhesion and proliferation: Col I > Col III > Gelatin >> Chitosan ALP Activity: Col I > Col III > Gelatin >> Chitosan Mineralization: Col I > Col III > Gelatin Oc, Dspp, and Dmp-1 expression on collagen	Kim et al. 2009
Collagen I with Dmp-1	Collagen scaffolds laden with Dmp-1 and dental pulp stem cells were placed in dentin disks with a simulated furcal perforation and transplanted subcutaneously into nude mice	Formation and organization of new pulp tissue	Prescott et al. 2008
PLA	SHED seeded onto PLA scaffolds into tooth slices, subcutaneous transplantation into nude mice	Formation of vascularized soft connective, pulp-like tissue and new tubular dentin	Cordeiro et al. 2008 Sakai et al. 2010
PLGA	SCAP and DPSC seeded onto PLGA into root canals sealed with MTA on one side, subcutaneous implantation into nude mice for 3–4 months	Formation of a pulp-like tissue, deposition of dentin along the root canal wall	Huang et al. 2010

Non-viral Methods

The non-viral or physical methods are considered to be highly safe but can be much less efficient mechanisms for gene transfer. Bruce and Fitzgerald 1995 mentioned the two most promising methods of physical gene transfer: liposomes (mostly bags of lipids containing the DNA) and macromolecular conjugates (the negatively charged DNA mixed with a large positively charged molecule that is linked to a specific cell ligand). These methods are capable of transferring relatively large genes, but expression is quite transient. Non-viral technique involves either electroporation or ultrasound method for gene delivery. Ultrasound mediated gene delivery is found to be successful both *in vivo* and *in vitro*, but electroporation method is found to be successful only *in vitro*.

Clinical Aspects of Gene Therapy

There are three main strategies for gene delivery: *in vivo*, *in vitro* and *ex vivo*. Though the most direct method is *in vivo* injection, this approach lacks the improved patient safety of *in vitro* and *ex vivo* methods. Alternatively, matrix-based delivery allows for tissue-specific gene delivery, higher localized loading of DNA or virus, and increased control over the structural microenvironment (Dang and Leong 2006). Thus far, human *in vivo* clinical trials have introduced adenovirus, AAV, retrovirus, and herpes simplex virus by intravenous (IV) injection, intra-tissue injection or lung aerosol (Kemeny et al. 2006).

Gene therapy is especially suited for long term delivery of a transgene to persons with a single genetic deficiency that is not amenable to protein or pharmacokinetic treatment.

Gene therapy for craniofacial regeneration: In contrast to traditional replacement gene therapy, craniofacial regeneration via gene therapy seeks to use genetic vectors as supplemental building blocks for tissue growth and repair. The synergistic combination of viral gene therapy with craniofacial tissue engineering, will significantly enhance our ability to repair and regenerate tissues *in vivo*.

Gene therapy for Head and Neck Squamous Cell Carcinoma (HNSCC): Though the treatment of HNSCC does not directly fall in the category of craniofacial regeneration, it is the best-developed use of gene therapy in the craniofacial region. There are three main strategies to target any solid tumor with gene therapy.

First, immunomodulatory therapy seeks to increase the visibility of the tumor cells to the immune system to enhance targeting of the tumor via the introduction of specific gene expression.

Second, oncolytic viruses have been developed that can selectively target, multiply in, and destroy cancer cells (Dambach et al. 2006).

Third, suicide genes such as herpes simplex thymidine kinase can be introduced to cancer cells to increase their susceptibility to anti-viral drugs such as acyclovir (Niculescu-Duvaz and Springer 2005).

Gene therapy for regeneration of mineralized tissues: Animal-model-based gene therapy and engineering of individual craniofacial structures such as bone and cartilage have firmly established a productive relationship, and novel approaches to regeneration of complex mineralized tissues such as tooth (Nakashima et al. 2006) and TMJ (Rabie et al. 2007) are just beginning to emerge. Clinical protein delivery of PDGF-B or Bone Morphogenetic Proteins (BMPs) at periodontal defect sites are well-known to enhance repair and healing of bone and gingival (Kaigler et al. 2006).

Successful engineering of teeth and the TMJ is challenging and requires the generation of functional interfaces. The introduction of BMPs *in vivo* to exposed pulp tissue has been proposed as a novel strategy for odontoblast transduction to enhance dentin regeneration and repair (Nakashima et al. 2006).

Gene therapy for salivary glands: There are many reasons for the reduction of salivary flow. In addition to direct repair non-functional glandular tissue, researchers are working to develop an engineered salivary gland substitute that could be implanted in place of the parotid gland (Aframian and Palmon 2008).

Indeed, attempts to restore salivary flow by *in vivo* transduction of adenovirus encoding AQP1 into remaining glandular tissue of persons treated with radiation for head and neck cancer is the first human craniofacial repair gene therapy clinical trial and is currently ongoing (Baum et al. 2006; NIH 2008).

Gene therapy for wound healing-mucosa: Engineering of skin and mucosal equivalents is essential for the aesthetic reconstruction of individuals disfigured by trauma, several surgery or severe burns.

Clinically, a product known as Gene-Activated Matrix (GAM) has been developed as an enhanced skin graft substitute. GAM for wound specific delivery of adenovirus vector encoding PDGF-B to improve healing of diabetic ulcers is currently in Phase II clinical trials (NIH 2008). It is reasonably expected that these developments could be expanded to enhance wound healing and tissue repair in the craniofacial region (Jin et al. 2004).

Gene therapy to vaccinate against caries: The eradication of dental caries or periodontal disease may be successful if gene transfer therapy can mediate humoral and cellular immune responses to the pathogenic bacteria involved in these disease processes. This type of gene transfer therapy is called DNA vaccination because DNA-containing antigens which can mediate an immune response are delivered in a plasmid to the target bacteria. Caries vaccine strategies may use the mucosal immune system in newborn infants, which is functional before the appearance of their first teeth, as an effective way to induce immunity against the colonization of teeth by mutans streptococci and protection against subsequent dental caries.

Revascularization

As mentioned earlier, two concepts exist in regenerative endodontics. Revascularization being the second concept. It simply means the regeneration of tissues from cells present within the teeth.

Revascularization studies have established the following prerequisites:

- Revascularization occurs most predictably in teeth with open apices and necrotic pulp secondary to trauma.
- Apex is open more than 1.5 mm.
- Bacteria should be removed from the canal by any of the following methods:
  - 3 mix—MP triple antibiotic paste consisting of ciprofloxacin, metronidazole and minocycline.
  - Calcium hydroxide, formocresol.
- Effective control seal.
- Matrix into which new tissue can grow.
- Patient should be young.
- Use of anesthetic without a vasoconstrictor when trying to induce bleeding.
- No instrumentation of canals.
- Sodium hypochlorite is used as an irrigant.
- Formation of blood clot probably serves as a protein scaffold permitting 3-dimensional in growth of tissue.

Microbial Challenges to Pulp Regeneration

Pulp regeneration is considered in cases where the dental pulp has been destroyed because of microbial irritation. The dental pulp frequently succumbs to irreversible disease caused by caries, trauma, congenital abnormalities or consequences of previous dental procedures. In the initial phases of pulpal inflammation, when the immunological responses are still intact, and the bacterial invasion of the tissues is minimal, several vital pulp procedures in young immature teeth have been shown to be effective, thus maintaining the radicular pulp's ability to continue tooth formation (Cvek 1992). However, following pulp necrosis, an infection spreads throughout the tissue and may extend into the periapical region, creating an environment that is not conducive to tissue regeneration without substantial disinfection procedures.

Recent case reports have proposed the use of topical antibiotics for root canal disinfection to take advantage of their efficacy and biocompatibility (Banchs and Trope 2004; Petrino et al. 2010).

Applications of Dental Tissue Engineering

- **Stem cells for pulp tissue engineering and regeneration:** With the emergence of tissue-engineering sciences, dental pulp tissue regeneration has been explored with the use of various biomaterials, where pulp cells grown on polyglycolic acid (PGA) formed pulp-like tissue in both *in vitro* and *in vivo* models (Mooney et al. 1996; Buurma et al. 1999). One concern is that implanting stem cells/scaffolds into root canals that have a blood supply only from the apical end may compromise vascularization to support the vitality of the implanted cells in the scaffolds. The ability of dental pulp stem cells, apical papilla stem cells and periodontal ligament stem cells to produce pulp-dentin like complexes *in vivo* suggests potential applications involving stem cells, growth factors and scaffolds for apexification and apexogenesis.

- **Regenerative dentistry: Making dentin**—Dental clinicians know that no material is available today to mimic all the physical, mechanical and esthetic properties of enamel and dentin. Demineralized dentin powder also has an intrinsic capability to induce mineralization. When applied directly to areas of pulp exposure, demineralized dentin induces the local formation of mineralized tissues. An understanding of which components of dentin powder had the inductive capability began in the early 1990s when it was discovered that specific fraction of dentin, which presumably contained Bone Morphogenic Proteins (BMP) activity, induced reparative dentin formation (Trembley A 1744).

- **Regenerative dentistry: Making enamel**—Cells that specialize in the making of enamel (ameloblasts) are no longer present in teeth with complete crown development. Therefore, an *in situ* cell-based strategy to regenerate enamel is not feasible. However, researchers' creativity and ingenuity have recently allowed for the development of synthetic enamel that is fundamentally based on the use of the principles of tissue engineering and nanotechnology. Amelogenin allowed for the synthesis of elongated crystals. The combination of amelogenin and fluoride allowed for the formation of rod-like apatite crystals with dimensions that resemble the ones observed in natural enamel (Trembley A 1744).

- **Regenerative dentistry: Making dental pulp**—The maintenance of dental pulp vitality is an underlying goal of most restorative procedures. A tissue engineering-based approach that results in new pulp tissue could potentially allow for the completion of vertical and lateral root development and perhaps, prevent the premature loss of these teeth (Trembley A 1744).

- **Regeneration of functional tooth by tissue engineering technique:** A tissue-engineering approach that uses cultured cells and biodegradable polymer scaffolds. Tissue-specific cells are isolated from a biopsy specimen, expanded in culture and combined with a porous biodegradable polymer scaffold. The cells adhere to the scaffold, proliferate and, over time, form a new tissue that can be returned to the tissue donor or another patient. Regenerative dental medicine uses an integrated sciences approach, involving developmental and molecular/cellular biology, molecular genetics and chemical engineering. More recently, scientists have reported the successful bioengineering of whole tooth crowns composed of accurately formed enamel, dentin and pulp tissues (Young et al. 2005).

- **Regeneration of functional tooth structure:** Loss of teeth has many causes including physical trauma, gum disease, tooth decay and genetic defects. Missing teeth can result in movement of the remaining teeth, difficulty in chewing and a lack of self-confidence. Currently there are several approaches to replacing teeth including the use of dentures, bridges and implants, all of which are based on non-biological techniques and none of which is without problems.

- **Tissue engineering a tooth:** The principles of early odontogenesis are being used to devise methods to generate teeth that can be used for replacement in humans.

  Sharpe's approach is to replace embryonic cells that make teeth in the embryo with cultured cells that can be isolated from a patient.

- **Implantation of tooth rudiments into the diastema:** If tooth germs or teeth are grown or cultured prior to implantation, it will be crucial to establish a method for implanting these into the mouth.

- **Functional bioengineered tooth root formation:** The current bioengineered tooth model exhibits crowns that are much more developed than the roots. To improve bioengineered tooth root formation, hybrid tooth/bone constructs were prepared to test whether the co-coordinated alveolar bone formation could improve bioengineered tooth root development (Lynch et al. 2006).

Future, Prospects and Hurdles

Dental applications of engineered tissues will be seen in the following few years. However, reconstruction of complex tissue defects made up of multiple cell types has not yet been attempted in the craniofacial complex, even in preclinical trials. Such a goal (for example, engineering a complete and functional salivary gland) will likely take about 10 to 15 years.

- Understanding the interactions between the stem cell and immune system. Immunosuppressive allogeneic MSCs may present an abundant cell source for clinical applications. Further research is needed to determine whether allogenic dental MSCs may suppress recipient host short- and long-term immune-rejection.

- More recently, engineering of dental pulp and dentin with pulp-derived stem cells has made considerable progress (Cordeiro et al. 2008; Huang et al. 2010; Sakai et al. 2010). With a wide range of biomaterials choices, the question is how far we can optimize our strategies for dentin-pulp complex engineering with the help of novel and smart biomaterials, which are tunable and tailor-made for this specific approach.

- Banking teeth as an autologous cell source and the potential use of allogenic stem cells both require further research to determine the ultimate benefits to our patients.

- Skin tissue and cartilage are becoming available for specific medical applications, and strategies to engineer bony tissues are close to receiving FDA approval. Dental applications of engineered tissues will be seen in a few years. However, reconstruction of complex tissue defects made up of multiple cell types has not yet been attempted in the craniofacial complex, even in preclinical trials.

  - **Mineralized tissue defects:** Many problems managed by general dentists or specialists are prime candidates for tissue-engineering solutions, including fractures of bones and teeth, craniofacial skeletal defects, destruction of the pulp-dentin complex and periodontal disease. BMPs and other growth-factor–rich preparations are being applied with a variety of natural and synthetic scaffolds (Nakashima and Akamine 2005).

  - **Engineering salivary gland function:** For many patients, in particular, those whose salivary epithelial cells have been replaced with fibrotic tissue, there is no adequate treatment available. Researchers believe that there is a realistic opportunity to develop a first-generation artificial salivary gland suitable for initial clinical testing relatively soon (within about 10 years).

- **Design of the artificial salivary gland envisioned:** The device is shaped like a blind-end tube and composed of three essential elements: a biodegradable substratum; a coating of an extracellular matrix protein on the luminal (internal) surface of the substratum; and a polarized epithelial cell layer consisting of autologous graft cells.

- **Nerve and vascular regeneration in dentine–pulp complex**

  - Nerve Regeneration:
    Dental pulp is richly innervated. The main nerve supply enters the pulp through the apical foramen along with vascular elements. The innervation of the pulp has a critical role in hemostasis of dental pulp. Invasion of immune and inflammatory cells into sites of injury in the pulp is stimulated by sensory nerves.

  - Vascular Regeneration:
    The vascular system in dental pulp plays a vital role in nutrition and oxygen supply and as a conduit for the removal of metabolic waste. The cellular elements of the blood vessels such as endothelial cells, pericytes and associated cells contribute to pulpal homeostasis along with the nerves. Thus, the vascular contribution to the regeneration of dentin–pulp complex is immense.

Summary

Despite the conclusive findings and wealth of data provided by *in vitro* and *in vivo* approaches in the field of dental regeneration, further research studies are required before pulp regeneration and even tooth restoration can be applied in dentistry. However, all data also confirm realistic feasibility of dental tissue repair in the near future. It is obvious that our knowledge in dental tissue engineering will expand rapidly. In this context, it has been demonstrated that present dental pulp stem/progenitor cells have the ability to differentiate *in vitro* as well as *in vivo* into odontoblast-like cells. Furthermore, the application of bioactive glasses incorporated into a biodegradable polymer matrix also seems to be a suitable material as a regenerating dental substitute. In particular, the material would provide stability and a stimulation effect on hard tissue formation. The next step has to be the design of a 'smart' and appropriate growth factors release system for diffusion through a residues dentin matrix after cavity preparation. Future experiments should be focused on the design of a highly sophisticated biological based scaffold system, which would significantly improve tooth viability and health maintenance in dentistry. Therefore, the development of a 3D-material as pulp capping agents meeting prerequisites mentioned above, and aiming at the dentin–pulp complex regeneration, are currently a high priority in research investigations.

The future of dental tissue engineering is summarized in the table (Table 14.3) given below:

Table 14.3. Techniques used in dental tissue engineering with their advantages and disadvantages.

Sr. no.	Technique	Advantages	Disadvantages
1	**Root–canal revascularization:** Open up teeth apex to 1 mm to allow bleeding into root canals	Lowest risk of immune rejection Lowest risk of pathogen transmission	Minimal of reports published to date Potential risk of necrosis if tissue becomes reinfected
2	**Stem cell therapy:** Autologous or allogenic stem or cells are delivered to teeth via injectable matrix	Quick Easy delivery Least painful Cell are easy to harvest	Low cell survival Cell do not produce new functioning pulp High risk of complications
3	**Pulp implant:** Pulp tissue is grown in the laboratory in sheet and implanted surgically	Sheets of cells are easy to grow More stable than an injection of dissociated cells	Sheets lack vascularity so only small constructs are possible Must be engineered to fit root canal precisely
4	**Scaffold implant:** Pulp cells are seeded onto a 3-D scaffold made of polymers and surgically implanted	Structure supports cell organization Some materials may promote vascularization	Low cell survival after implantation Must be engineered to fit root canal precisely
5	**3-D cell printing:** Ink-jet-like device dispenses layers of cells in a hydrogel which is surgically implanted	Multiple cell types can be precisely positioned	Must be engineered to fit root canal precisely Early-stage research has yet to prove functional *in vivo*
6	**Injectable scaffolds:** Polymerizable hydrogels, alone or containing cell suspension are delivered by injection	Easy delivery May promote regeneration by providing substitute for extracellular matrix	Limited control over tissue formation Low cell survival Early-stage research has yet to prove functional *in vivo*
7	**Gene therapy:** Mineralizing genes are transfected into the vital pulp cells of necrotic and symptomatic	May avoid cleaning and shaping root canal May avoid the need to implant stem cells	Most cells in a necrotic tooth are already dead Difficult to control Risk of health hazards Not approved by the FDA

Conclusion

Dentistry is entering an exciting era in which many of the advances in biotechnology/tissue engineering offer opportunities for exploitation in a novel and more effective therapy. Tissue engineering using the triad of dental pulp stem cells, morphogens and scaffolds may provide an innovative and biologically based approach for generation of clinical materials and treatment of dental diseases. The challenges of introducing endodontic tissue engineered therapies are substantial; the potential benefits to patients and the profession are ground breaking. Better understanding of cell interactions and growth along with further research can make dental tissue engineering a reality in the near future.

References

Aframian, D.J. and Palmon, A. 2008. Current status of the development of an artificial salivary gland. Tissue Eng. Part. B Rev. 14: 187–198.

Andreasen, J.O., Borum, M.K., Jacobsen, III and Andreasen, F.M. 1995. Replantation of 400 avulsed permanent incisors.2 factors related to pulp healing. Endo. Dent. Traumatol. 11: 59–68.

Banchs, F. and Trope, M. 2004. Revascularization of immature permanent tooth with apical periodontitis: New treatment protocol? J. Endod. 30: 196–200.

Banshal, R. and Banshal, R. 2011. Regenerative endodontics: A state of the art. Indian J. Dent. Res. Jan-Feb; 22(1): 122–131.

Batouli, S., Miura, M., Brahim, J., Tsutsui, T.W., Fisher, L.W., Gronthos S. et al. 2003. Comparison of stem-cell-mediated osteogenesis and dentinogenesis. J. Dent. Res. 82: 976–981.

Baum, B.J., Zheng, C., Cotrim, A.P., Goldsmith, C.M., Atkinson, J.C., Brahim, J.S. et al. 2006. Transfer of the AQP1 cDNA for the correction of radiation induced salivary hypofunction. Biochim. Biophys. Acta 1758: 1071–1077.

Bi, Y., Ehirchiou, D., Kilts, T.M., Inkson, C.A., Embree, M.C., Sonoyama, W. et al. 2007. Identification of tendon stem/progenitor cells and the role of the extracellular matrix in their niche. Nat. Med. 13: 1219–1227.

Bluteau, G., Luder, H.U., De Bari, C. and Mitsiadis, T.A. 2008. Stem cells for tooth engineering. Eur. Cell Mater. 16: 1–9.

Bohl, K.S., Shon, J., Rutherford, B. and Mooney, D.J. 1998. Role of synthetic extracellular matrix in development of engineered dental pulp. J. Biomater. Sci. Polym. Ed. 9: 749–764.

Bruce, R. and Fitzgerald, M. 1995. A new biological approach to vital pulp therapy. Crit. Rev. Oral Biol. Med. 6(3): 218–229.

Buurma, B., Gu, K. and Rutherford, R.B. 1999. Transplantation of human pulpal and gingival fibroblasts attached to synthetic scaffolds. Eur. J. Oral. Sci. 107: 282–289.

Carinci, F., Papaccio, G., Laino, G., Palmieri, A., Brunelli, G., D'Aquino, R. et al. 2008. Comparison between genetic portraits of osteoblasts derived from primary cultures and osteoblasts obtained from human pulpar stem cells. J. Craniofac. Surg. 19: 616–625.

Casagrande, L., Demarco, F.F., Zhang, Z., Araujo, F.B., Shi, S. and Nör, J.E. 2010. Dentin-derived BMP-2 and odontoblast differentiation. J. Dent. Res. 89: 603–608.

Chai, Y., Jiang, X., Ito, Y., Bringas P. Jr, Han, J., Rowitch, D.H. et al. 2000. Fate of the mammalian cranial neural crest during tooth and mandibular morphogenesis. Development 127: 1671–1679.

Charles A. Vacanti. 2006. History of tissue engineering and a glimpse into its future: Tissue Engineering 12(5): 1137–1143.

Cordeiro, M.M., Dong, Z., Kaneko, T., Zhang, Z., Miyazawa, M., Shi, S. et al. 2008. Dental pulp tissue engineering with stem cells from exfoliated deciduous teeth. J. Endod. 34: 962–969.

Couble, M.L., Farges, J.C., Bleicher, F., Perrat-Mabillon, B., Boudeulle, M. and Magloire, H. 2000. Odontoblast differentiation of human dental pulp cells in explant cultures. Calcif. Tissue Int. 66: 129–138.

Cvek, M. 1992. Prognosis of luxated non-vital maxillary incisors treated with calcium hydroxide and filled with gutta-percha. A retrospective clinical study. Endod. Dent. Traumatol. 8: 45–55.

Dambach, M.J., Trecki, J., Martin, N. and Markovitz, N.S. 2006. Oncolytic viruses derived from the gamma34.5-deleted herpes simplex virus recombinant R3616 encode a truncated UL3 protein. Mol. Ther. 13: 891–898.

Dang, J.M. and Leong, K.W. 2006. Natural polymers for gene delivery and tissue engineering. Adv. Drug. Deliv. Rev. 58: 487–499.

Freshney, I.R., Stacey, G.N. and Auerbach, J.M. 2007. Culture of human stem cells. Chapter 8: 187–207.

Gardner, R.L. 2002. Stem cells: potency, plasticity and public perception. J. Anat. 200(Pt 3): 277–82.

Gronthos, S., Mankani, M., Brahim, J., Robey, P.G. and Shi, S. 2000. Postnatal human dental pulp stem cells (DPSCs) *in vitro* and *in vivo*. Proc. Natl. Acad. Sci. USA 97: 13625–13630.

Gronthos, S., Brahim, J., Li, W., Fisher, L.W., Cherman, N., Boyde, A. et al. 2002. Stem cell properties of human dental pulp stem cells. J. Dent. Res. 81(8): 531–535.

Grove, C.J. 1921. Nature's method of making perfect root fillings following pulp removal, with a brief consideration of the development of secondary dentin. Dent. Cosmos. 63: 968–982.

Hammarstrom 2 years. 1997. Enamel matrix and cementum development, repair and re-methods. Journal of Clinical Periodonto. 658–668.

Heijl, L., Heden, G., Svärdström, G. and Östgren, A. 1997. Enamel matrix derivative (EMDOGAINA) in the treatment of intrabony periodontal defects. J. Clin. Periodontol. 24: 705–714.

Huang, A.H., Chen, Y.K., Lin, L.M., Shieth, T.Y. and Chan, A.W. 2008. Isolation and characterization of dental pulp stem cells from a supernumerary tooth. J. Oral. Pathol. Med. 39: 571–574.

Huang, G., Sonoyama, W., Chen, J. and Park, S. 2006. *In vitro* characterization of human dental pulp cells: various isolation methods and culturing environments. Cell Tissue Res. 324: 225–236.

Huang, G.T.-J., Sonoyama Wataru, Liu, Yi, Liu, H., Wang, S. and Shi, S. 2008. The hidden treasure in apical papilla: The potential role in pulp/dentin regeneration and bioroot engineering. J. Endod. 34: 645–651.

Huang, G.T., Yamaza, T., Shea, L.D., Djouad, F., Kuhn, N.Z., Tuan, R.S. et al. 2010. Stem/progenitor cell-mediated *de novo* regeneration of dental pulp with newly deposited continuous layer of dentin in an *in vivo* model. Tissue Eng. Part A 16: 605–615.

Isaka, J., Ohazama, A., Kobayashi, M., Nagashima, C., Takiguchi, T., Kawasaki, H. et al. 2001. Participation of periodontal ligament cells with regeneration of alveolar bone. J. Periodontol. 72: 314–323.

Jay, B. Reznick. 2008. Continuing Education: Stem Cells: Emerging Medical and Dental Therapies for the Dental Professional. Dentaltown Magazine. Oct: 42–53.

Jin, Q., Anusaksathien, O., Webb, S.A., Printz, M.A. and Giannobile, W.V. 2004. Engineering of tooth-supporting structures by delivery of PDGF gene therapy vectors. Mol. Ther. 9: 519–526.

Kaigler, D., Cirelli, J.A. and Giannobile, W.V. 2006. Growth factor delivery for oral and periodontal tissue engineering. Expert. Opin. Drug. Deliv. 3: 647–662.

Kassai, Y., Munne, P., Hotta, Y., Penttila, E., Kavanagh, K., Ohbayashi, N. et al. 2005. Regulation of mammalian tooth cusp patterning by ectodin. Science 309(5743): 2067–2070.

Kemeny, N., Brown, K., Covey, A., Kim, T., Bhargava, A., Brody, L. et al. 2006. Phase I, open-label, dose-escalating study of a genetically engineered herpes simplex virus, NV1020, in subjects with metastatic colourectal carcinoma to the liver. Hum. Gene Ther. 17: 1214–1224.

Kim, N.R., Lee, D.H., Chung, P.H. and Yang, H.C. 2009. Distinct differentiation properties of human dental pulp cells on collagen, gelatin, and chitosan scaffolds. Oral Surgery, Oral Medicine, Oral Pathology, Oral Radiology, and Endodontology 108(5): e94–100.

Kling, M., Cvek, M. and Mejare, I. 1986. Rate and predictability of pulp revascularization in therapeutically reimplanted permanent incisors. Endod. Dent. Traumatol. 2: 83–9.

Langer, R. and Vacanti, J.P. 1993. Tissue Engineering. Science 260: 920–926.

Lesot, H., Bègue-Kirn, C., Kübler, M.D., Meyer, J.-M., Smith, A.J., Cassidy, N. et al. 1993. Experimental induction of odontoblast differentiation and stimulation during reparative processes. Cells Mater. 3: 201–217.

Lynch, S.E., Marx, R.E., Nevins, M. and Lynch, L.A. 2006. Tissue Engineering: Applications in Oral and Maxillofacial Surgery and Periodontics. 2nd ed. Chicago: Quintessence Publishing, 11–15.

Mao, J.J., Giannobile, W.V., Helms, J.A., Hollister, S.J., Krebsbach, P.H., Longaker, M.T. et al. 2006. Craniofacial tissue engineering by stem cells. J. Dent. Res. 85(11): 966–979.

Martin-Rendon, E. and Watt, S.M. 2003. Exploitation of stem cell plasticity. Transfus Med. 13: 325–349.

MacArthur, B.D. and Oreffo, R.O.C. 2005. Bridging the gap. Nature 433: 19.

Miura, M., Gronthos, S., Zhao, M., Fisher, L.W., Robey, P.G. and Shi, S. 2003. SHED: stem cells from human exfoliated deciduous teeth. Proc. Natl. Acad. Sci. USA 100: 5807–5812.

Mooney, D.J., Powell, C., Piana, J. and Rutherford, B. 1996. Engineering dental pulp-like tissue *in vitro*. Biotechnol. Prog. 12: 865–868.

Morsczeck, C., Gotz, W., Schierholz, J., Zeilhofer, F., Kuhn, U., Mohl, C. et al. 2005. Isolation of precursor cells (PCs) from human dental follicle of wisdom teeth. Matrix Biol. 24: 155–165.

Nakashima, M. and Akamine, A. 2005. The application of tissue engineering to regeneration of pulp and dentin in endodontics. J. Endod. 31: 711–718.

Nakashima, M., Iohara, K. and Zheng, L. 2006. Gene therapy for dentin regeneration with bone morphogenetic proteins. Curr. Gene Ther. 6: 551–560.

Niculescu-Duvaz, I. and Springer, C.J. 2005. Introduction to the background, principles, and state of the art in suicide gene therapy. Mol. Biotechnol. 30: 71–88.

NIH. ClinicalTrials.gov: US National Institutes of Health. 2008: http:// www.clinicaltrials.gov/.

Oh, Y.H., Che, Z.M., Hong, J.C., Lee, E.J., Lee, S.J. and Kim, J. 2005. Cryopreservation of human teeth for future organization of a tooth bank—a preliminary study. Cryobiology 51(3): 322–9, Epub.

Perry, B.C., Zhou, D., Wu, X., Yang, F.C., Byers, M.A., Chu, T.M. et al. 2008. Collection, cryopreservation, and characterization of human dental pulp-derived mesenchymal stem cells for banking and clinical use. Tissue Eng. Part C Methods 14(2): 149–156.

Petrino, J.A., Boda, K.K., Shambarger, S., Bowles, W.R. and Mc Clanahan, S.B. 2010. Challenges in regenerative endodontics: a case series. J. Endod. 36: 536–541.

Prescott, R.S., Alsanea, R., Fayad, M.I. et al. 2008. *In vivo* generation of dental pulp-like tissue by using dental pulp stem cells, a collagen scaffold, and dentin matrix protein 1 after subcutaneous transplantation in mice. J. Endod. 34: 421–426.

Rabie, A.B., Dai, J. and Xu, R. 2007. Recombinant AAV-mediated VEGF gene therapy induces mandibular condylar growth. Gene Ther. 14: 972–980.

Rao, M.S. 2004. Stem sense: a proposal for the classification of stem cells. Stem Cells Dev. 17: 435–462.

Rendon, M.E. and Watt, S.M. 2003. Exploitation of stem cell plasticity. Transfus. Med. 13: 325–349.

Rubio, D., Garcia-Castro, J., Martin, M.C., de la Fuente, R., Cigudosa, J.C., Lloyd, A.C. et al. 2005. Spontaneous human adult stem cell transformation. Cancer Res. 65: 3035–3039.

Ruch, J.V. 1987. Determinism of odontogenesis. Cell Biol. Rev. 14: 1–112.

Rutherford, B. and Mark Fitzgerald. 1995. A new biological approach to vital pulp therapy: Crit. Rev. Oral. Biol. Med. 6(3): 218–229.

Sakai, V.T., Zhang, Z., Dong, Z., Neiva, K.G., Machado, M.A., Shi, S. et al. 2010. SHED differentiate into functional odontoblasts and endothelium. J. Dent. Res. 89: 791–796.

Scheller, E.L. and Krebsbach, P.H. 2009. Gene therapy: Design and prospects for craniofacial regeneration. J. Dent. Res. 88(7): 585–596.

Seo, B.M., Miura, M., Gronthos, S., Bartold, P.M., Batouli, S., Brahim, J. et al. 2004. Investigation of multipotent postnatal stem cells from human periodontal ligament. Lancet 364: 149–155.

Seo, B.M., Miura, M., Sonoyama, W., Coppe, C., Stanyon, R. and Shi, S. 2005. Recovery of stem cells from cryopreserved periodontal ligament. J. Dent. Res. 84: 907–12.

Seo, B.M., Sonayama, W., Yamaza, T., Coppe, C., Kikuiri, T., Akiyama, K. et al. 2008. SHED repair critical-size calvarial defects in mice. Oral. Dis. 14: 428–434.

Simon, S., Cooper, P., Smith, A., Picard, B., Ifi, C.N. and Berdal, A. 2008. Evaluation of a new laboratory model for pulp healing: preliminary study. Int. Endod. J. 41: 781–790.

Smith, A.J., Cassidy, N., Perry, H., Bègue-Kirn, C., Ruch, J.V. and Lesot, H. 1995. Reactionary dentinogenesis. Int. J. Dev. Biol. 39: 273–280.

Sonoyama, W., Liu, Y., Yamaza, T. Tuan, R.S., Wang, S., Shi, S. et al. 2008. Characterization of apical papilla and its residing stem cells from human immature permanent teeth: A pilot study. J. Endod. 34: 166–71.

Stewart Sells. 2004. Stem Cells Handbook: Humana Press.

Suchánek, J., Soukup, T., Ivancaková, R., Karbanová, J., Hubková, V., Pytlík, R. et al. 2007. Human dental pulp stem cells—isolation and long term cultivation. Acta Medica (Hradec Kralove) 50(3): 195–201.

Takeda, K., Oida, S., Ichijo, H. et al. 1994. Molecular cloning of rat bone morphogenetic protein (BMP) type IA receptor and its expression during ectopic bone formation induced by BMP. Biochem. Biophys. Res. Commun. 204: 203–209.

Takeda, T., Tezuka, Y. and Horiuchi, M. 2008. Characterization of dental pulp stem cells of human tooth germs. J. Dent. Res. 87(7): 676–681.

Thomson, J.A., Itskovitz-Eldor, J., Shapiro, S.S., waknitz, M.A., Swiergie, J.J., Marshall, V.S. et al. 1998. Embryonic stem cell lines derived from human blastocytes. Science 282: 1145–1147.

Trembley, A. Mémoires pour servir à l'histoire d'un groupe de polypes d'eau douce à bras en forme de cornes. Leiden: J & H Verbeek:1744.

TT-450—Stem Cells and Teeth Banks, ebiz news from Japan. https://www.japaninc.com/tt450.

Vacanti, C.A. 2006. History of tissue engineering and a glimpse into its future. Tissue Engineering 12(5): 1137–1143.

Wingard, J.R. and Demetric, G.D. 1999. Clinical Applications of Cytokines and Growth Factors. New York : Springer.

Yagyuu, T., Ikeda, E., Ohgushi, H., Tadokoro, M., Hirose, M., Maeda, M. et al. 2010. Hard tissue-forming potential of stem/progenitor cells in human dental follicle and dental papilla. Arch. Oral. Biol. 55: 68–76.

Yan, X., Qin, H., Qu, C., Tuan, R.S., Shi, S. and Huang, G.T. 2010. iPS cells reprogrammed from human mesenchymal-like stem/progenitor cells of dental tissue origin. Stem Cells Dev. 19: 469–480.

Young, C.S., Abukawa, H., Asrican, R., Ravens, M.S., Troulis, M.J., Kaban, L.B. et al. 2005. Tissue engineered hybrid tooth and bone. Tissue Eng. 11: 1599–1610.

Zhang, W., Walboomers, X.F., Shi, S., Fan, M. and Jansen, J.A. 2006. Multilineage differentiation potential of stem cells derived from human dental pulp after cryopreservation. Tissue Eng. 12: 2813–2823.

Zheng, Y., Liu, Y., Zhang, C.M., Zhang, H.Y., Li, W.H., Shi, S. et al. 2009. Stem cells from deciduous tooth repair mandibular defect in swine. J. Dent. Res. Mar. 88(3): 249–254.

Index

T - #0517 - 071024 - C236 - 254/178/11 - PB - 9780367626945 - Gloss Lamination